Bionic Limb Reconstruction

Oskar C. Aszmann • Dario Farina
Editors

Bionic Limb Reconstruction

Springer

Editors
Oskar C. Aszmann
Division of Plastic and Reconstructive Surgery
Medical University of Vienna
Vienna
Austria

Dario Farina
Department of Bioengineering
Imperial College London
London
UK

ISBN 978-3-030-60745-6 ISBN 978-3-030-60746-3 (eBook)
https://doi.org/10.1007/978-3-030-60746-3

This Springer imprint is published by the registered company Springer Nature Switzerland AG
The registered company address is: Gewerbestrasse 11, 6330 Cham, Switzerland

Contents

Introduction

Even though the hand comprises only 1% of our body weight, the biggest part of our sensory-motor cortex is related to its control. The loss of a hand thus presents not only a loss of the most important tool allowing us to creatively interact with our environment but also leaves a dramatic sensory-motor deficit that challenges our central nervous system (CNS). Reconstruction of hand function is therefore not only an essential part of restoring body integrity and functional wholeness but also closes the loop of our neural circuits diminishing phantom sensation and neural pain. If biology fails to restore meaningful function, today we can resort to complex mechatronic replacements that have functional capabilities that in some respects even outperform biological alternatives, such as conservative reconstructive measures or hand transplantation. However, as with replantation and transplantations, the challenge of bionic replacement is solid skeletal attachment and connecting the prosthesis to our neural circuits to achieve natural, intuitive control and also provide basic sensory feedback. In recent years, we have developed a number of strategies to improve neural interfacing, signal extraction, interpretation and stable mechanical attachment that are important parts of our current research. This book presents a compilation of our own expertise throughout the last 10 years of research and the vast clinical experience that we have gained in endeavouring the best possible solutions for patients that have suffered hand loss or grave injuries leading to essentially useless extremities. This book gives an overview of recent advances in bionic reconstruction, surgical refinements over technological interfacing, skeletal fixation, and modern rehabilitation tools that allow quick integration of prosthetic replacement. Furthermore, it seeks to provide the reader with the information needed to decide as to whether a patient may be a good candidate for a bionic reconstruction.

The development of any medical innovation is subject to controversies and must be evaluated against the traditional concepts that have historically been employed to meet these needs. For the upper extremity, body-powered prostheses and EMG-driven devices were developed and provided simple functional tools. Biological replacement via hand transplantations has been attempted as early as 1963, but until today immunosuppressive regimes have critical side effects and the re-amputation rate is as high as 30%. Both fields of research have made substantial progress and both reconstructive methods must be considered depending on the specific situation, risk profile, and need of the patient. Obviously, a hand transplantation meets the supreme reconstructive principle to reconstruct "like with like" best, as it offers a hand of flesh and blood with immediate, intuitive control, sensory feedback, and a sense of ownership that at this time cannot be achieved by any prosthetic device. Aside from myoelectric prostheses, controlled by at least two electromyography (EMG) signals from remnant stump muscles, also passive, body-powered devices are currently used in classic prosthetic reconstruction. Passive prostheses range from stable or adjustable cosmetic hands, with silicone cover and natural appearance, to prosthetic tools, which are mainly hooks or grasping devices. So-called body-powered prostheses can perform simple grasping tasks by external cables attached to the prosthetic arm, driven by body movement. This serves as an assistant hand to the dominant hand, but it is obviously not capable of performing different grasps or hand movements.

Obviously the aforementioned replacements are far away from the goal of "replacing like with like" and come dreadfully short in reconstructing the functional capacity of a real hand and restoring body integrity.

In this book we present our efforts and experience in overcoming these deficits and present methods and tools that help us to understand the principles of neural and skeletal integration of modern mechatronics. New treatment options also bring up new ethical considerations. Obviously, sound indications and a clear risk-benefit analysis are the basis of any ethical decision making. When facing an amputee both indication and risk-benefit analysis present a very clear scenario; however, bionic replacement of existing body parts brings on a broad new spectrum of ethical questions that need to be considered. Today there is a vast array of technology-based body enhancements available to us: exoskeletal support systems for the work force in the automobile industry or in military service, sensory enhancement via night vision, augmented reality via visual feedback, and immense support systems of both information and data management available via voice or touchpad technology. The human body has limitations and at the same time immense capabilities to adapt and interact with high-end technological devices. In extremity reconstruction, our very first bionic patient was so frustrated with his own biological hand (that was devastated in an electrocution accident) that we sought for mechatronic alternatives. We remember him crying for joy when he realized that this mechatronic replacement actually worked and he again could move fingers, even though they were not his own.

Loss of extremity function in the western world today is dramatically more frequent due to the consequences of opulence with a neurologically and metabolically challenged society. Here patients have lost the ability to move their arm and hand after a brain stroke, but through cognitive nerve transfers patients are enabled to control machines that act like arms. Patients in a wheelchair can command various devices with their own biosignals via bluetooth and may ambulate with activated exoskeletons. Even today we have virtual-based rehabilitation environments that help our central nervous systems to handle the cognitive challenges that go along with these new technologies.

How far can we go then? We will likely not be able to alter the nature of our existence as being fragile and limited, but to a certain extent technology can help overcome disabilities by exploiting the ability of our CNS to adapt to new bodily realities, which in part can be replaced by mechatronic devices. This book explores the limits and possibilities of modern reconstructive schemes and seeks to provide hope to all those who have irreversibly lost parts of their functional capacities that today can be restored by technological advances.

Vienna, Austria — Oskar C. Aszmann
London, UK — Dario Farina

Part I

Severe Injuries of the Upper Extremity

General Considerations on Upper Limb Amputation and Its Levels

1

Clemens Gstoettner, Stefan Salminger, and Oskar C. Aszmann

Amputation describes the loss of a body part due to injury or disease, most commonly referring to the loss of a limb. Millions of people worldwide are currently living with amputation of the upper extremity [25]. Since trauma is the leading cause, especially young people are affected, often suffering from severe impairments of their working status and independence in daily life [11, 19]. Prosthetic replacement aims to restore the various functions of the lost hand—a task which becomes increasingly challenging with ascending level of limb loss.

Epidemiology and Socioeconomic Implications

It is estimated that approximately 1.6 million people with limb amputation were living in the USA in 2005, a number expected to more than double by 2050 [25]. Upper limb amputation, making up for a third of that number, can be further divided into minor (92%) and major (8%) amputations (see below). Prevalence of major upper limb amputation ranges from 11.6 to 13.9 per 100,000 in studies conducted in Norway and the USA, respectively [15, 25]. Trauma is by far the most common cause, followed by vascular disease and cancer. In general, the typical patient receiving major upper limb amputation is healthy and young, compared to the morbid and elderly population of lower limb amputees [20]. Males are affected by major traumatic amputations much more frequently than women, and injuries are most often the result of motor vehicle trauma or machinery-related accidents. The most common levels for amputation are transhumeral and transradial, while through-joint amputations are less frequent [7]. The incidence of traumatic amputation increases when a country is actively involved in war. Mortality after war-related injuries has been decreasing in recent years because of advances in field care, resulting in more survivors that have sustained devastating injuries. This is believed to be the reason for the growing number of homecoming soldiers with multiple amputations, as was seen in recent conflicts in Iraq and Afghanistan [9]. Figure 1.1a–c gives an overview of the epidemiology of amputation levels and injury mechanisms in the USA.

As most acquired amputees lose their arm at a young age, often in their twenties, reintegration

C. Gstoettner (✉)
Clinical Laboratory for Bionic Extremity Reconstruction, Department of Surgery, Medical University of Vienna, Vienna, Austria
e-mail: clemens.gstoettner@meduniwien.ac.at

S. Salminger · O. C. Aszmann
Clinical Laboratory for Bionic Extremity Reconstruction, Department of Surgery, Medical University of Vienna, Vienna, Austria

Division of Plastic and Reconstructive Surgery, Department of Surgery, Medical University of Vienna, Vienna, Austria

O. C. Aszmann, D. Farina (eds.), *Bionic Limb Reconstruction*,
https://doi.org/10.1007/978-3-030-60746-3_1

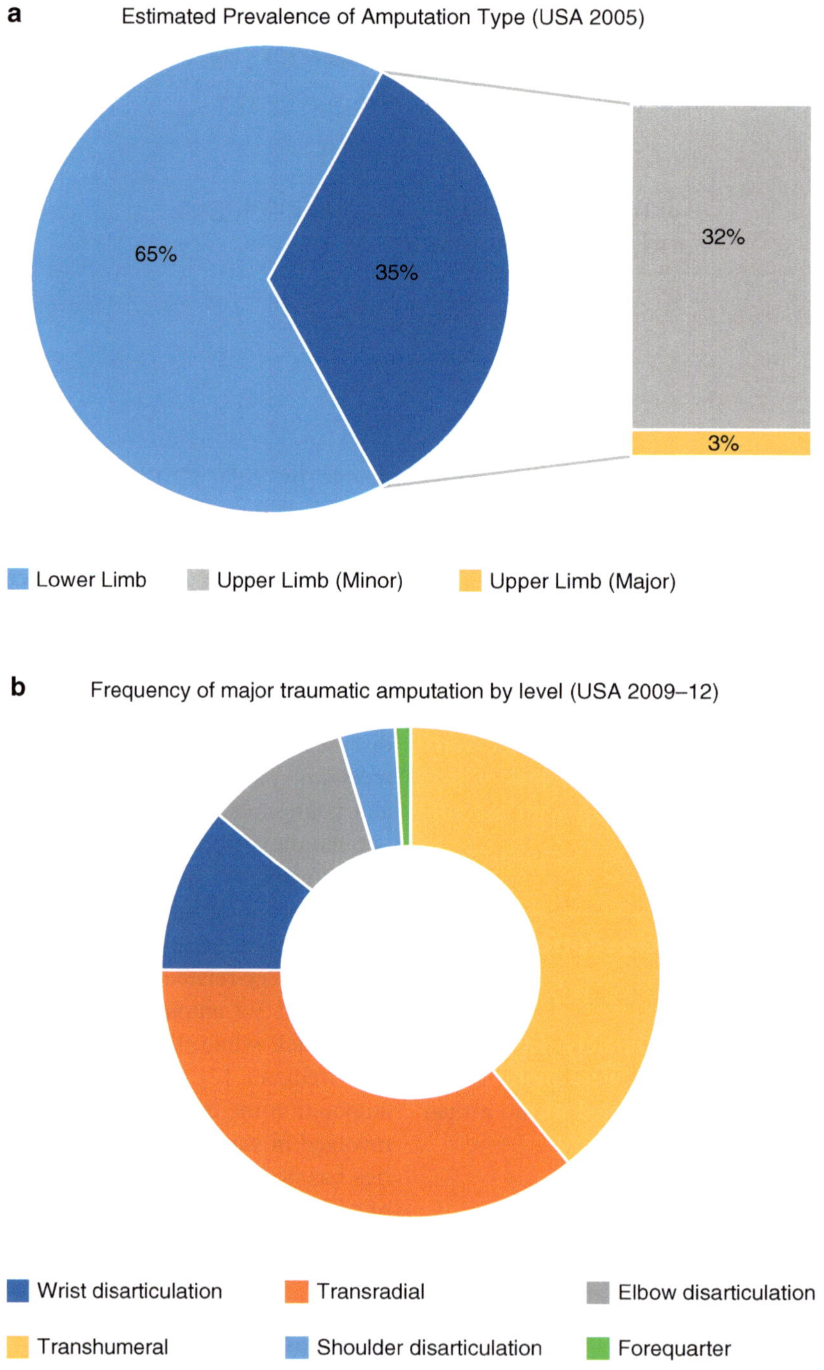

Fig. 1.1 (**a**) Distribution of amputation types among patients living with limb loss in the USA [25]. (**b**) Frequency of upper limb amputation levels among traumatic amputees [7]. (**c**) Distribution of trauma mechanisms among traumatic amputations. *MVT* motor vehicle trauma [7]

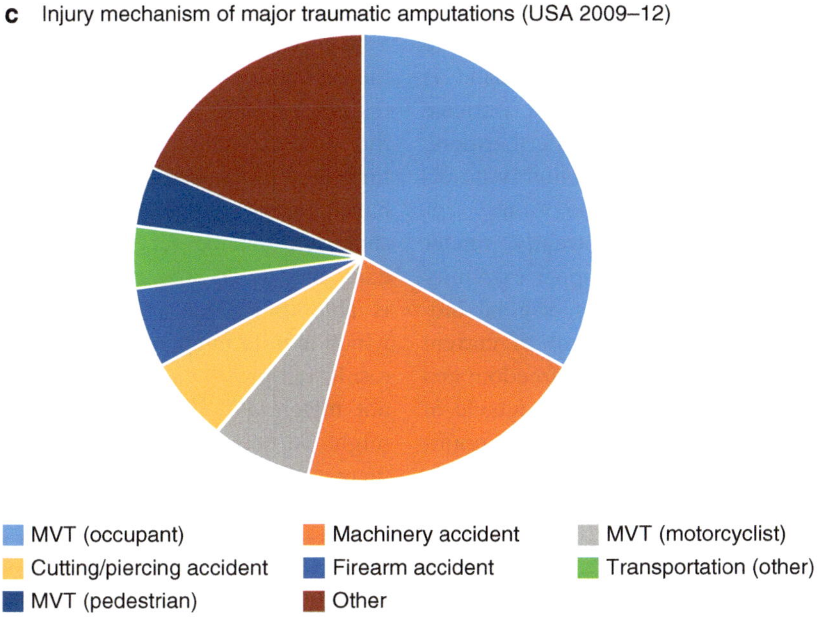

Fig. 1.1 (continued)

into working life is an important factor with major implications at a socioeconomic level. In general, quality of life is scored lower by upper limb amputees compared to the general population, an effect which is partly mediated by the reduced ability to work [14]. Chapter 2 will go into more detail regarding the sequelae of amputation at a personal level, highlighting in particular the psychological factors. Somewhere between 50% and 75% of people return to work after upper limb amputation [18]. Among others, good general health and the regular use of a prosthetic device was identified as a positive predictive factor for work participation in this population. Furthermore, around 95% of upper limb amputees suffer from some kind of amputation-related pain, mostly phantom pain, residual limb pain, or back pain [6]. Especially phantom limb pain has been shown to be positively influenced by prosthetic usage [21]. There are other detrimental long-term consequences of amputation which may be limited or prevented through regular prosthesis use, including scoliosis, atrophy of the remaining muscles, joint stiffness and arthrosis. However, depending on the level of amputation, prosthetic usage rates from 39% to 81% can be found in the present literature [15, 23]. This wide variance may be due to the heterogeneous cohorts regarding levels of amputation and terminal devices included [4].

General Considerations on Upper Limb Amputation

When dealing with upper limb amputation, a general distinction can be made between minor amputations, defined here as affecting one or more fingers or parts of the hand, and major amputations, which range from transcarpal to forequarter. Ascending the different amputation levels from distal to proximal, the higher we go the more severe the functional impairment. With each joint patients lose, they are less able to control the position of the remaining limb in three-dimensional space, and the ability to interact with surroundings is increasingly diminished. The complexity and versatility of motor and sensory function in the upper limb and, in particular, the hand has major implications on the concepts of amputation surgery and prosthetic reconstruction. Compared to the leg, which is mainly used for weight-bearing and walking, the hand is continuously performing much more intricate and diverse movement patterns. Furthermore, sensibility of the hand is far more highly developed, occupying a major part of the primary sensory cortex. These factors currently render prosthetic reconstruction of the upper extremity inadequate when compared to its original biologic counterpart, as motor control of current devices is limited to few degrees of freedom and sensory feedback is not yet possible outside of research settings. While there have been significant developments in prosthetic research in recent years, ranging from improvements of the biotechnological interface to the mechanics of the devices, functionality for the patients has not improved greatly. Also, many of these advances have so far remained solely research-related, delivering limited benefit in standard patient care.

For the above reasons, limb salvage remains the first goal when caring for patients with extensive injuries of the arm. Reconstructive efforts should aim at preserving as much function as possible. Even if at a later stage it becomes clear that biological reconstruction cannot yield the desired functional outcomes, amputation and prosthetic limb replacement will always remain an option, giving the patient the chance to actively let go of the useless extremity instead of perceiving it being taken away [2]. In limb salvage surgery, there is an increasing trend toward the use of free tissue transfers, including functional and composite free flaps [13]. Such modern concepts in microsurgery have made it possible to cover bigger wounds and reconstruct devastated functional units of the limb. In major traumatic amputations, replantation of the limb may be tried whenever feasible, as outcomes regarding patient satisfaction are generally better than with prosthetics [17]. While distal replantations are favourable, even in above-elbow amputations, reasonable results can be expected [12]. Apart from traumatic injuries, similar considerations apply to reconstruction in upper limb tumour surgery. In certain cases of advanced sarcoma, where the more distal parts of the limb remain unharmed, a very useful but often underrecognised option is the replantation of the hand after wide tumour resection (see Fig. 1.2) [22]. While the resulting limb will inevitably be shorter, this is not a major limitation in the upper extremity. As long as the hand shows decent motor and sensory function, a short arm can be very useful in most tasks of daily living. The resection-replantation concept is able to deliver a biological, sensate hand, which is fully integrated into the body image—a result that prosthetic replacement currently cannot match. While similar functional outcomes might be achieved through allotransplantation, there is a general agreement that these procedures are only indicated for bilateral amputees, because of the many risks associated with foreign tissue rejection and especially long-term immunosuppression (see Chap. 8).

If amputation is unavoidable, the surgeon must be aware that the decisions taken during amputation surgery will pave the way for prosthetic reconstruction and have major implications on the overall rehabilitation process. The concepts of limb salvage surgery should also be employed to retain as much amputation length as is necessary or even to preserve an additional joint. In certain cases, tissue from the amputated extremity parts may be used to gain length or provide coverage. This concept of using free or pedicled fillet flaps was termed spare-part surgery [16]. Where this is not possible, traditional free or local flaps may also be used to retain amputation length and improve residual limb function [3]. Shape and size of the stump must be formed in a way that will enable stable prosthetic attachment. Soft tissue coverage needs to be sufficient to prevent painful pressure points but not excessive, which would lead to less stable prosthetic attachment and poor myoelectric pickup. Distal muscle ends are generally either fixed at the bone (myodesis) or sutured to their antagonists (myoplasty). When dealing with the remaining nerves of the stump, at the very least, traction resection neurotomy should be employed to prevent painful neuroma formation at an area of loading. However, targeted muscle reinnervation offers an improved solution for this issue, not only pre-

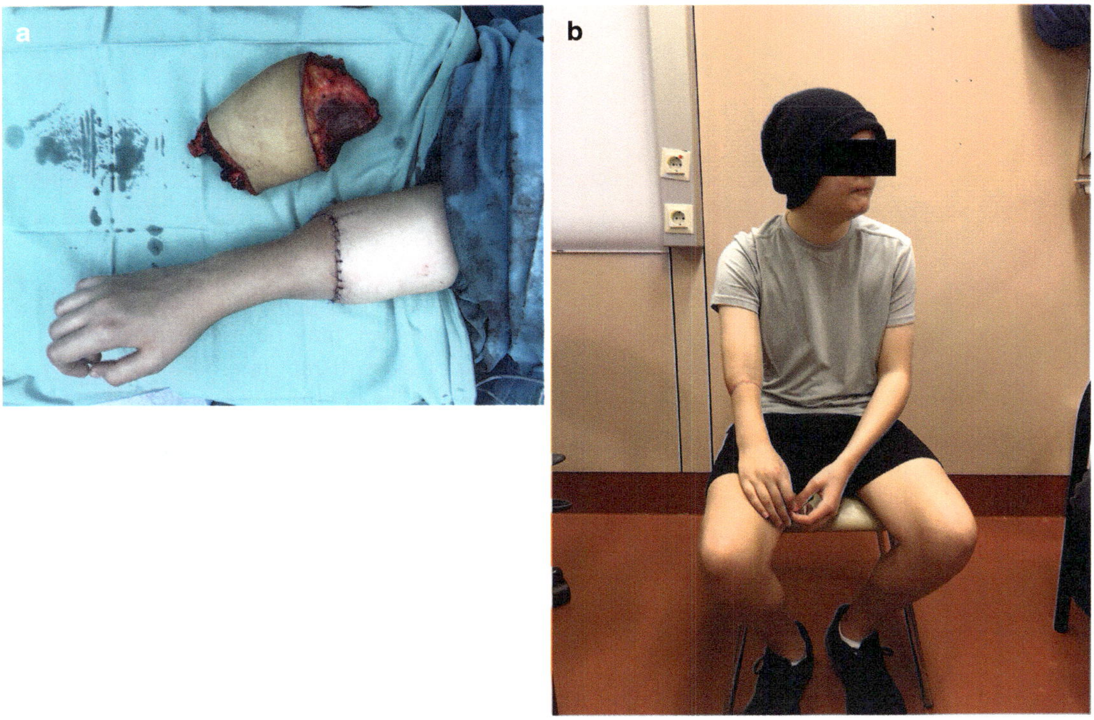

Fig. 1.2 This young patient suffered from a malign tumour of his right elbow. Since the hand was unharmed, it could be reattached to the humerus after tumour resection (**a**). This represents a valuable alternative to amputation in selected cases, preserving the biological, sensate hand. In contrast to the lower limb, a shorter arm does not greatly impair function and may even be hard to spot on a first glance (**b**)

venting neuromas but also greatly improving the information transfer between amputee and prosthesis. Part III of this book will go into more detail regarding the surgical creation of the functional interface between man and prosthesis, and Chap. 15 will give an overview on the concept of osseointegration, which has greatly expanded the possibilities of prosthetic attachment.

Different Levels of Upper Limb Amputation

Minor

Amputations involving one or more fingers are the most common in the upper limb. Wherever feasible, replantation can have very good functional and aesthetic results, with success of the procedure in about two thirds of the cases [8]. However, the decision whether single-digit replantation should be performed depends on the affected finger and level, thumb and index being the primary indications [24]. Secondary biological reconstruction in single-digit amputations is mainly performed for the thumb, given its indispensable role for prehensile function. Pollicisation is commonly achieved via local index/middle finger or free toe transfer. Regarding prosthetic options, silicone finger prostheses may be fitted if the stump is long enough, providing aesthetically pleasing results. Even if the stump is very short, prosthetic fingers may be attached via osseointegration (see Fig. 1.3). If more than one finger is affected, individualised biological and/or prosthetic concepts are necessary, with the main goal to achieve stable grasping function. Once all fingers are lost, biological reconstruction is generally not an option. As through-hand amputations are in most cases difficult to fit with functional prostheses, a shortening to a transcarpal or even transradial level may be considered. This should be discussed together by

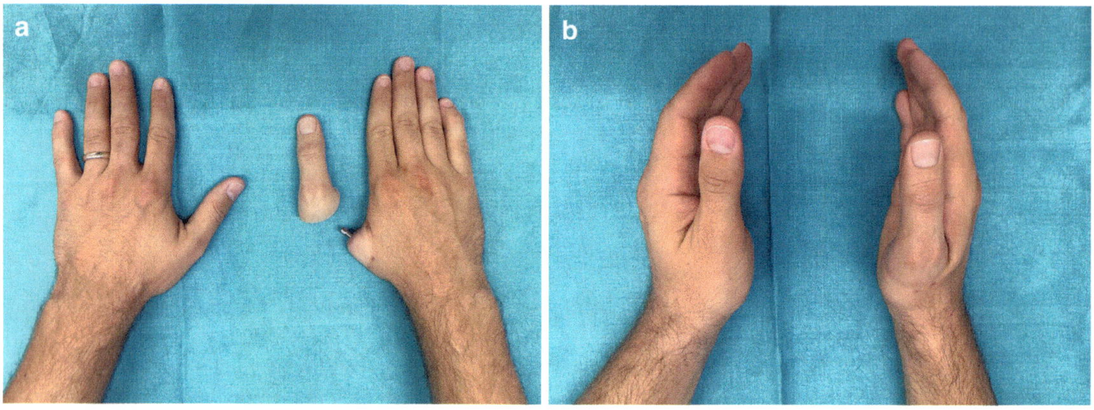

Fig. 1.3 After resection of a sarcoma of his right thumb, this patient wished for a prosthetic reconstruction. Because of the very short metacarpal stump, osseointegration was the only option to facilitate attachment of a silicone finger (**a**). Modern prostheses can be designed to closely replicate the contralateral thumb (**b**)

surgeon, prosthetist and patient, considering demands of the patient and the different devices available for each level of amputation.

Carpal Disarticulation

Prosthetic reconstruction after carpal disarticulation is a challenge to the prosthetist due to length restriction of the prosthetic replacement and frequently the presence of terminal neuromas. For functional myoprostheses, it is difficult to fit all the necessary prosthetic components while maintaining an acceptable length of the hand, which does not markedly exceed the sound limb. An advantage to the transradial level is that the ability for full pro- and supination may be improved. Wherever possible, the thick palmar skin should be used for terminal stump coverage, as bony landmarks of the distal radius and ulna might lead to painful pressure spots during socket use. Depending on patient preference and quality of the stump, shortening to a transradial level might be considered, especially if prosthetic fitting is unlikely to achieve the desired results.

Transradial

This is one of the most frequent amputation levels in the upper limb. There is much experience for this level, from the surgical as well as prosthetic viewpoint. For prosthetic reconstruction, ideal stump length is for most patients somewhere between 16 and 18 cm as measured from the lateral epicondyle, which will also retain the main muscles for pro- and supination and provides a muscular coverage over the terminal bone stumps. There is a multitude of prosthetic options available, and functional outcomes with transradial devices are generally favourable, even though patient satisfaction is still higher after successful replantation [17]. Using direct myoelectric control, two independent signals can be employed for addressing two degrees of freedom, usually hand open/close and wrist rotation. As flexion of the fingers will automatically also trigger activation of wrist extensors to stabilise the wrist, the most precise signals can be extracted by activating wrist flexors or extensors. Here the agonist will automatically quiet all antagonistic activity and thus provide excellent and easy prosthetic control. Co-contraction may be an option for more degrees of freedom (DOF). Recent advances in signal interpretation strategies have led to a large number of transradial amputees in the USA controlling their device through pattern recognition, while in Europe direct approaches are still most frequently used. Body-powered devices also remain a commonly favoured option, delivering the benefit of improved proprioception. Furthermore, there are also some biological alternatives for functional reconstruction in transradial amputees, as is depicted in Chap. 8.

Elbow Disarticulation

Disarticulation of the elbow joint and very distal transhumeral amputation are generally uncommon amputation levels. While preserved epicondyles of the humerus are beneficial for prosthetic suspension and rotational control, prosthetic fitting with any form of artificial elbow joint will inevitably lead to a much longer upper arm, resulting in an awkward and inconveniently shaped limb. To achieve appropriate length, shortening osteotomy of the humerus can be performed, which will allow for a standard transhumeral fitting while retaining the advantages described above [10].

Transhumeral

Through the humerus has been reported to be the most common level of major traumatic upper limb amputation [7]. These patients only have their shoulder joint left for moving the arm in three-dimensional space. Therefore, a larger number of myosignals are needed for intuitive control of more degrees of freedom, even though generally only two are available. This functional mismatch can be greatly improved by TMR surgery, which is able to provide up to six distinct myoelectric signals and therefore allow for 3 degrees of freedom, e.g. elbow flexion/extension, wrist pronation/supination and hand open/close. A further challenge for this amputation level is the mechanic attachment of prosthetic devices. Conventional socket prostheses will in most cases require encasement of the shoulder joint, greatly constraining range of motion. Here, osseointegration is able to retain free shoulder movement (see Fig. 1.4) while providing a much more stable anchorage and easier handling and is increasingly becoming standard care for this indication. In selected cases it can also be employed for very short transhumeral stumps (<5 cm), which would for conventional socket

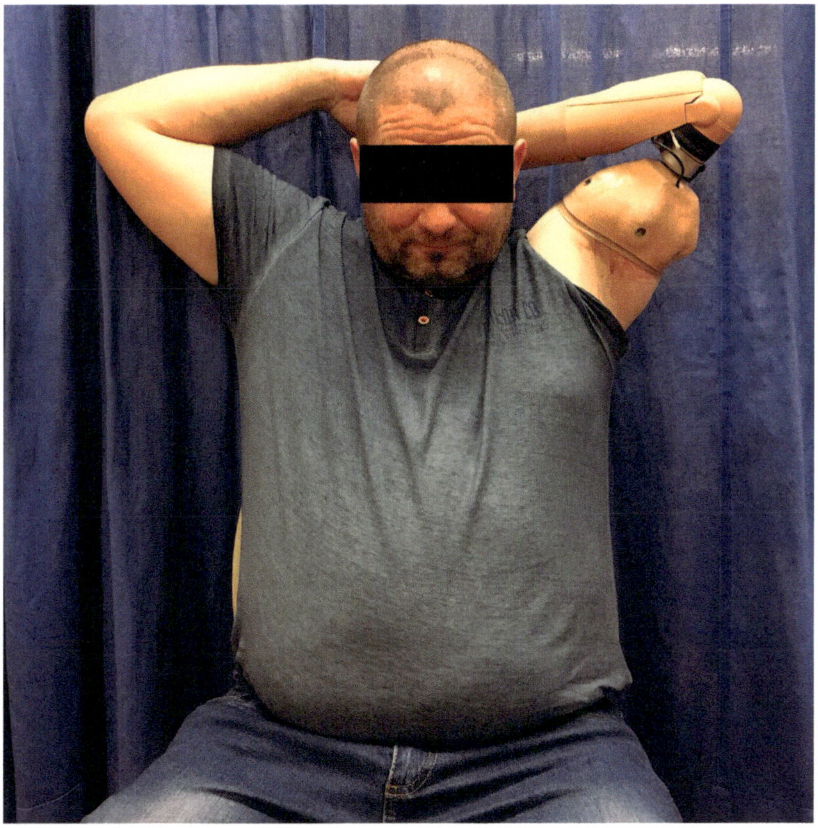

Fig. 1.4 A transhumeral amputee who received osseointegration to enable prosthetic fitting while maintaining full range of motion in the shoulder

fitting be regarded as shoulder disarticulations. In order to preserve shoulder function in short transhumeral stumps, it is essential to retain or reconstruct the insertions of the pectoralis major, latissimus dorsi and deltoid [20].

Shoulder Disarticulation

Shoulder disarticulation, or glenohumeral amputation, is rare but very challenging for treating physicians and prosthetists. As the number of available myoelectric signals is outweighed by the degrees of freedom needed for appropriate prosthetic control, TMR surgery for the functional interface and pain control is often used and should be standardly performed. The anatomy of the pectoralis major and minor innervation offers the possibility for creating four independently addressable myosignals in these muscles alone [1]. Socket fitting is usually very cumbersome and results in heavy, constraining devices which are inconvenient in handling. The acromion is used for suspension, but belts to the other side of the thorax are still necessary, often restricting range of motion in the unaffected limb. However, the use of a prosthesis should be encouraged, as it can prevent painful sequelae of chronic imbalance and is also positively linked to decrease of phantom limb pain, aside from the more obvious aesthetic aspects [21].

Forequarter

Forequarter amputation is fortunately performed very infrequently, nowadays most often for the treatment of malignant tumours [5]. Prosthetic reconstruction will require an individualised approach for each patient. As the entire arm including the clavicle and scapula are missing, the options for device suspension are extremely limited. Any shaft will have to enclose a major part of the thorax and will be severely constraining the remaining upper body. Depending on the muscles available, TMR might be an option if a myoelectric device is considered. In any case, these cases should be referred to a centre that has high expertise in bionic reconstruction and is involved with novel research approaches, as such patients may profit from novel investigational solutions which have not yet entered the market.

References

1. Aszmann OC, Rab M, Kamolz L, Frey M. The anatomy of the pectoral nerves and their significance in brachial plexus reconstruction. J Hand Surg. 2000;25:942–7. https://doi.org/10.1053/jhsu.2000.17818.
2. Aszmann OC, Vujaklija I, Roche AD, Salminger S, Herceg M, Sturma A, Hruby LA, Pittermann A, Hofer C, Amsuess S, Farina D. Elective amputation and bionic substitution restore functional hand use after critical soft tissue injuries. Sci Rep. 2016;6 https://doi.org/10.1038/srep34960.
3. Baccarani A, Follmar KE, De Santis G, Adani R, Pinelli M, Innocenti M, Baumeister S, von Gregory H, Germann G, Erdmann D, Levin LS. Free vascularized tissue transfer to preserve upper extremity amputation levels. Plast Reconstr Surg. 2007;120:971–81. https://doi.org/10.1097/01.prs.0000256479.54755.f6.
4. Biddiss EA, Chau TT. Upper limb prosthesis use and abandonment: a survey of the last 25 years. Prosthetics Orthot Int. 2007;31:236–57. https://doi.org/10.1080/03093640600994581.
5. Dimas V, Kargel J, Bauer J, Chang P. Forequarter amputation for malignant tumours of the upper extremity: case report, techniques and indications. Can J Plast Surg J Can Chir Plast. 2007;15:83–5.
6. Ephraim PL, Wegener ST, MacKenzie EJ, Dillingham TR, Pezzin LE. Phantom pain, residual limb pain, and back pain in amputees: results of a national survey. Arch Phys Med Rehabil. 2005;86:1910–9. https://doi.org/10.1016/j.apmr.2005.03.031.
7. Inkellis E, Low EE, Langhammer C, Morshed S. Incidence and characterization of major upper-extremity amputations in the national trauma data bank. JBJS Open Access. 2018;3:e0038. https://doi.org/10.2106/JBJS.OA.17.00038.
8. Kotsougiani D, Ringwald F, Hundepool CA, Neubrech F, Kremer T, Bickert B, Kneser U, Hirche C. Safety and suitability of finger replantations as a residency training procedure: a retrospective cohort study with analysis of the initial postoperative outcomes. Ann Plast Surg. 2017;78:431–5. https://doi.org/10.1097/SAP.0000000000000945.
9. Krueger CA, Wenke JC, Ficke JR. Ten years at war: comprehensive analysis of amputation trends. J Trauma Acute Care Surg. 2012;73:S438–44. https://doi.org/10.1097/TA.0b013e318275469c.
10. Luccia ND, Marino HLT. Case study: fitting of electronic elbow on an elbow disarticulated

patient by means of a new surgical technique. Prosthetics Orthot Int. 2000;24:247–51. https://doi.org/10.1080/03093640008726556.
11. Luff R, Forrest J, Huntley J. The amputee statistical database for the United Kingdom; 2009.
12. Mattiassich G, Rittenschober F, Dorninger L, Rois J, Mittermayr R, Ortmaier R, Ponschab M, Katzensteiner K, Larcher L. Long-term outcome following upper extremity replantation after major traumatic amputation. BMC Musculoskelet Disord. 2017;18 https://doi.org/10.1186/s12891-017-1442-3.
13. Ninkovic M, Voigt S, Dornseifer U, Lorenz S, Ninkovic M. Microsurgical advances in extremity salvage. Clin Plast Surg. 2012;39:491–505. https://doi.org/10.1016/j.cps.2012.08.003.
14. Østlie K, Magnus P, Skjeldal OH, Garfelt B, Tambs K. Mental health and satisfaction with life among upper limb amputees: a Norwegian population-based survey comparing adult acquired major upper limb amputees with a control group. Disabil Rehabil. 2011;33:1594–607. https://doi.org/10.3109/09638288.2010.540293.
15. Østlie K, Skjeldal OH, Garfelt B, Magnus P. Adult acquired major upper limb amputation in Norway: prevalence, demographic features and amputation specific features. A population-based survey. Disabil Rehabil. 2011;33:1636–49. https://doi.org/10.3109/09638288.2010.541973.
16. Peng Y, Lahiri A. Spare-part surgery. Semin Plast Surg. 2013;27:190–7. https://doi.org/10.1055/s-0033-1360586.
17. Pet MA, Morrison SD, Mack JS, Sears ED, Wright T, Lussiez AD, Means KR, Higgins JP, Ko JH, Cederna PS, Kung TA. Comparison of patient-reported outcomes after traumatic upper extremity amputation: replantation versus prosthetic rehabilitation. Injury. 2016;47:2783–8. https://doi.org/10.1016/j.injury.2016.10.004.
18. Postema SG, Bongers RM, Brouwers MA, Burger H, Norling-Hermansson LM, Reneman MF, Dijkstra PU, van der Sluis CK. Upper limb absence: predictors of work participation and work productivity. Arch Phys Med Rehabil. 2016;97:892–9. https://doi.org/10.1016/j.apmr.2015.12.022.
19. Salminger S, Roche AD, Hruby LA, Sturma A, Riedl O, Bergmeister KD, Aszmann OC. Prosthetic reconstruction to restore function in transcarpal amputees. J Plast Reconstr Aesthetic Surg. 2015; https://doi.org/10.1016/j.bjps.2015.10.029.
20. Tintle LSM, Baechler LMF, Nanos CGP, Forsberg LJA, Potter MBK. Traumatic and trauma-related amputations: Part II: upper extremity and future directions. J Bone Jt Surg Am. 2010;92:2934–45. https://doi.org/10.2106/JBJS.J.00258.
21. Weiss T, Miltner WH, Adler T, Brückner L, Taub E. Decrease in phantom limb pain associated with prosthesis-induced increased use of an amputation stump in humans. Neurosci Lett. 1999;272:131–4.
22. Windhager R, Millesi H, Kotz R. Resection-replantation for primary malignant tumours of the arm. An alternative to fore-quarter amputation. J Bone Joint Surg Br. 1995;77:176–84.
23. Wright TW, Hagen AD, Wood MB. Prosthetic usage in major upper extremity amputations. J Hand Surg. 1995;20:619–22. https://doi.org/10.1016/S0363-5023(05)80278-3.
24. Zhu H, Bao B, Zheng X. A comparison of functional outcomes and therapeutic costs: single digit replantation versus revision amputation. Plast Reconstr Surg. 2017;141(2):244e–9e. https://doi.org/10.1097/PRS.0000000000004024.
25. Ziegler-Graham K, MacKenzie EJ, Ephraim PL, Travison TG, Brookmeyer R. Estimating the prevalence of limb loss in the United States: 2005 to 2050. Arch Phys Med Rehabil. 2008;89:422–9. https://doi.org/10.1016/j.apmr.2007.11.005.

Psychosocial Importance of the Hand and Consequences of Severe Hand Trauma, Amputation and Complete Brachial Plexus Injury

2

Anna Pittermann, Laura A. Hruby, Agnes Sturma, and Oskar C. Aszmann

The immense importance of the human hand is not only based on the extremely complex functional nature of it but also on the very special psychological and social meaning associated with this body part. Loss of hand function secondary to nerve and soft tissue damage and/or the traumatic amputation of a hand due to severe injuries always represents a great impairment for the individual on a functional, social and psychological level. Psychological sequelae result from the trauma experienced but also from physical disabilities, body image disturbances and last but not least changes in social role and social interactions [5].

A. Pittermann (✉)
Department of Clinical Psychology, General Hospital of Vienna, Vienna, Austria

Clinical Laboratory for Bionic Extremity Reconstruction, Department of Surgery, Medical University of Vienna, Vienna, Austria

Division of Plastic and Reconstructive Surgery, Department of Surgery, Medical University of Vienna, Vienna, Austria
e-mail: anna.pittermann@akhwien.at

L. A. Hruby
Clinical Laboratory for Bionic Extremity Reconstruction, Department of Surgery, Medical University of Vienna, Vienna, Austria

A. Sturma
Clinical Laboratory for Bionic Extremity Reconstruction, Department of Surgery, Medical University of Vienna, Vienna, Austria

Department of Bioengineering, Imperial College London, London, UK

O. C. Aszmann
Clinical Laboratory for Bionic Extremity Reconstruction, Department of Surgery, Medical University of Vienna, Vienna, Austria

Division of Plastic and Reconstructive Surgery, Department of Surgery, Medical University of Vienna, Vienna, Austria

Psychosocial Importance of Hand Function

The human hand is of immense functional and social importance. We use our hands to accomplish almost every task in everyday life (daily living activities): we need them to get dressed, to put toothpaste on the toothbrush and to prepare a meal. Additionally, we also use our hands to interact with others in a social context. We hold our partner's hand, we caress our children and we shake hands to agree on something.

The social importance of the hand can also be seen by the many attributions that are made to the look of a hand and to the way a hand is being used. The firmness of a handshake, the appearance of a hand and the gestures made with the hand are all used to judge peoples' appearance and performance and can give a quite immediate impression of their psychological make-up, life circumstances and social status.

O. C. Aszmann, D. Farina (eds.), *Bionic Limb Reconstruction*,
https://doi.org/10.1007/978-3-030-60746-3_2

Just like the face, hands are almost always in our line of vision and therefore are being used as a means to tell us something about the person "attached" to these hands. Also all the numerous idioms referring to hands ("cold hands, warm heart"; "an extra pair of hands"; "have hands tied"; "give one a hand"; "take the law into one's hands"; "be in safe hands"; etc.) show the emotional connotations associated with this part of our body.

From a functional point of view, the hand is an extremely complex tool whose dexterity today cannot be replaced by any technical means.

Severe injuries to the hand itself or the brachial plexus supplying nerval input to it are associated with a loss of function occurring within seconds and always impair the individual functionally, socially and psychologically.

Psychological Sequelae of Amputation and Brachial Plexus Injuries

Individuals who experience a severe injury to their hand and arm or a brachial plexus injury or even an entire loss of parts are confronted with numerous changes in their life. Marked physical disability does not only result from functional impairment but is also related to pain and changes in body image perception. Furthermore, psychological and psychosocial issues may even have a greater impact on one's life and also the lives of the whole family/support system.

From a psychological point of view, patients who are exposed to a trauma have to deal with the general impact the traumatic event has on their psyche first. Traumatic events can lead to flashbacks, intrusion, re-experiencing of the trauma, sleeping disturbances and avoidance symptoms (avoidance of thoughts, feelings or places associated with the trauma) [6, 8]. In the long run, problems like post-traumatic stress disorder, anxiety disorders, adaptation difficulties and substance abuse can occur. All these impairments do not have to be linked to any physical disability in the first line.

Patients with physical disabilities and/or amputations resulting from a trauma additionally have to deal with the psychological sequelae following functional restrictions and alterations in body image (see also [14, 16–18]). Apart from chronic pain (phantom pain, deafferentation pain), these individuals very often face depression and/or anxiety, disturbed body image, negative self-evaluation, sleep disruption, cognitive difficulties and unemployment [6, 9, 15]. In a review of the existing literature, Mckechnie and John [12] found the levels of depression in post-traumatic amputees to be varying between 20.6% and 63%. For anxiety it was 25.45–57%.

Despite occupational retraining a vast number of patients with severe upper extremity trauma are unable to return to their former work or report a worse work situation as a result of the injury, both of which affect their psychological well-being but also their economic situation and independence in everyday life [7, 9].

A visible disfigurement such as an amputation or the physical appearance of a "plexus hand" (atrophic, cold, discolorated) may have severe impact on one's perception of one's body image and lead to self-appearance concerns as well as social appearance concerns [1].

Severe hand injuries also have the tendency to have a psychosocial effect on significant others. Family members as well as friends or co-workers of the affected individual may become overly protective or self-conscious and change their behaviour in presence of the injured. A missing or disfigured hand is a part of the body that can hardly be concealed and is almost always visible to the people surrounding. The affected individual becomes dependent on others even with simple tasks and loses part of his independence and autonomy. These circumstances might lead to a change in social roles and changing interactions within families and other social systems that might make the individual feel a loss of control, helplessness and anxiety. Early psychological intervention should therefore also concentrate on first signs of social isolation, fear and withdrawal.

Psychological Interventions

Individuals exposed to a severe hand trauma are at risk for psychological and behavioural consequences [11, 13, 14, 17]. Grieving is a normal response to amputation or severe hand injury, but the individuals' emotional reactions may be of great variability and do not have to be in direct relation to the objective severity of the injury.

Not all patients who are exposed to trauma develop trauma-related distress [10]—a fact that seems to be especially true in older amputees [3]. The severity and duration of the emotional reaction as well as the psychological strain tell the clinical psychologist about the need for a clinical diagnosis and adequate treatment.

Early identification of severe psychological disturbances may prevent progression of psychological pathology. It is therefore important to assess trauma-related distress in patients as soon as possible to enable appropriate interventions. It is also important to provide individual psychological support to the patient depending on the planned therapy, such as reconstructive surgery or transplantation [2, 17].

Psychological support can help the individual to develop or strengthen effective coping strategies [4], emotional well-being and integration of a new self-concept. Depending on the needs and demands of the patient, training of relaxation techniques, cognitive behavioural therapy (CBT) or EMDR (eye movement desensitisation and reprocessing) might be useful. Psychological counselling should also be offered to close family members if needed, and the patient should be supported in his attempt to return to work or find a new place in society where he/she can feel useful and valuable.

References

1. Aszmann OC, Roche AD, Salminger S, Paternostro-Sluga T, Herceg M, Sturma A, Hofer C, Farina D. Bionic reconstruction to restore hand function after brachial plexus injury: a case series of three patients. Lancet. 2015;385(9983):2183–9.
2. Brown PW. Sacrifice of the unsatisfactory hand. J Hand Surg Am. 1979;4(5):417–23.
3. Frank RG, Kashani JH, Kashani SR, Wonderlich SA, Umlauf RL, Ashkanazi GS. Psychological response to amputation as a function of age and time since amputation. Br J Psychiatry. 1984;144:493–7.
4. Franzblau L, Chung KC. Psychosocial outcomes and coping after complete avulsion traumatic brachial plexus injury. Disabil Rehabil. 2015;37(2):135–43.
5. Grob M, Papadopulos NA, Zimmermann A, Biemer E, Kovacs L. The psychological impact of severe hand injuries. J Hand Surg. 2008;33E(3):358–62.
6. Grunert BK. Hand trauma. In: Sarwer DB, et al., editors. Psychological aspects of reconstructive and cosmetic plastic surgery. Philadelphia, PA: Lippincott Williams & Wilkins; 2006. p. 145–57.
7. Gustafsson M, Ahlström G. Problems experienced during the first year of an acute traumatic hand injury – a prospective study. J Clin Nurs. 2004;13:986–95.
8. Horrowitz MJ. Stress response syndromes. PTSD, grief and adjustment disorders. New York, NY: Jason Aronson; 1997.
9. Kretschmer T, Ihle S, Antoniadis G, Seidel JA, Heinen C, Borm W, et al. Patient satisfaction and disability after brachial plexus surgery. Neurosurgery. 2009;65(41):189–96.
10. MacBride A, Rogers J, Whylie B, Freeman SJ. Psychosocial factors in the rehabilitation of elderly amputees. Psychosomatics. 1980;21(3):258–61.
11. Mayer A, Kudar K, Bretz K, Tihanyi J. Body schema and body awareness of amputees. Prosthetics Orthot Int. 2008;32(3):363–82.
12. Mckechnie PS, John A. Anxiety and depression following traumatic limb amputation: a systematic review. Injury. 2014;45:1859–66.
13. Meyer TM. Psychological aspects of mutilating hand injuries. Hand Clin. 2003;19:41–9.
14. Nichols PJR. Some problems in rehabilitation of the severely disabled. Proc R Soc Med. 1971;64(4):349–53.
15. Rumsey N, Harcourt D. The Oxford handbook of the psychology of appearance. Oxford: Oxford University Press; 2012.
16. Schulz M. Psychologische verarbeitung der amputation. Vasa. 2009;38:72–4.
17. Spyridon P, Galanakos MD, Arjan GJ, Bot MD, Aristides B, Zoubos MD, Panayotis N, Soucacos MD. Psychological and social consequences after reconstruction of upper extremity trauma: methods of detection and management. J Reconstr Microsurg. 2014;30:193–206.
18. Wellington B. Quality of life issues for patients following traumatic brachial plexus injury – part 2 research project. Int J Orthop Trauma Nurs. 2010;14:5–11.

Inner Amputations of the Upper Extremity

3

Laura A. Hruby, Johannes A. Mayer, and Oskar C. Aszmann

Brachial plexus injuries disrupt neural input to the upper extremity. The extent and level of the injury and timing of surgery determine the outcome of either spontaneous healing or the reconstructive efforts that have been undertaken. While upper brachial plexus lesions have an excellent prognosis either way, lower root lesions have a significantly worse outcome. This is owed to lengthy regeneration periods and the complexity of the hand's neuromuscular design. Bionic reconstruction may be of particular help in this unfortunate patient group, since the minute neural input that makes it into the forearm may not be enough for the high demands of a biologic hand, but may be enough to provide biosignals for a prosthetic replacement.

Historical Review of Inner Amputations

The first handwritten records on inner amputations date back more than 2000 years. Homer in his Iliad and Thucydides in his History of the Peloponnesian War provided vivid descriptions of ancient battlefields and war injuries including injuries to the upper extremity resulting in damage of the brachial plexus (BP) [28]. The Roman physician Galen first described the diagnosis and non-surgical treatment of a traumatic BPI [24]. In 1874, the rupture of the superior trunk, which is comprised of the union of the C5 and C6 ventral rami, was first described by Erb as a common presentation of brachial plexus injury (BPI) [27]. At the same time, Klumpke reported of complete BPIs and was the first to associate the Horner syndrome with damage to the T1 root [27], where sympathetic fibers emerge from the spinal cord.

In 1896, Thorburn was the first to report of the surgical treatment of an inner amputation [30]. He treated a 16-year-old girl who was caught by some machinery in a mill and found that due to a retroclavicular lesion "[…] a plastic operation was regarded as possible" [30]. Seven months

L. A. Hruby (✉)
Department of Orthopedics and Trauma Surgery, Medical University of Vienna, Vienna, Austria

Clinical Laboratory for Bionic Extremity Reconstruction, Department of Surgery, Medical University of Vienna, Vienna, Austria
e-mail: laura.hruby@meduniwien.ac.at

J. A. Mayer
Clinical Laboratory for Bionic Extremity Reconstruction, Department of Surgery, Medical University of Vienna, Vienna, Austria

O. C. Aszmann
Clinical Laboratory for Bionic Extremity Reconstruction, Department of Surgery, Medical University of Vienna, Vienna, Austria

Division of Plastic and Reconstructive Surgery, Department of Surgery, Medical University of Vienna, Vienna, Austria

O. C. Aszmann, D. Farina (eds.), *Bionic Limb Reconstruction*,
https://doi.org/10.1007/978-3-030-60746-3_3

after the trauma had occurred, Thorburn performed the first elective brachial plexus repair; 4 years later the girl had recovered useful wrist and elbow flexion; the hand, however, did not recover sensation nor motor function [30].

Before the advent of microsurgical techniques, the common consensus on treating a flail and anesthetic arm was to amputate through the upper arm, stiffen the shoulder joint, and—in rare cases—fit the patient with a passive prosthesis [29, 32]. The increase in civilian brachial plexus lesions due to high-velocity traumata, however, has not only promoted interest in the field but also surgical progress managing these devastating nerve injuries [4]. In the 1970s, Millesi [16] and Narakas [19] were the first to report good functional outcomes following surgical brachial plexus repair using nerve grafts [32]. During the past two decades, the performance of nerve transfers has been expanded and popularized yielding even better results and rendering partial motor recovery possible also in avulsion injuries affecting multiple roots [31].

Etiology and Epidemiology of Inner Amputations

A large number of injuries may cause lesions to the brachial plexus including birth traumata, falls from greater heights, penetrating injuries (gunshot wounds and stabbing injuries), and—most frequently—motor vehicle accidents [26]. Most injuries occur due to stretching and traction force on the brachial plexus [1]. The brachial plexus is barely protected from traction forces due to the loose suspension of the human shoulder girdle [6]. When the head and neck are displaced away from the ipsilateral shoulder with significant force, injuries to the upper trunk or the nerve roots C5 and C6 frequently occur (Fig. 3.1a). The lower trunk or nerve roots C8 and T1 are more likely injured when the arm is forcefully abducted over the head [26] (Fig. 3.1b).

Nerve injury mechanisms may involve stretching, rupture, or avulsion of the nerve root from the spinal cord [22] (Fig. 3.2). In most cases, mixed pattern nerve injuries occur [17]. Avulsion injuries are among the most severe nerve injuries in men, since this preganglionic lesion includes damage to the central nervous system (Fig. 3.2) [6]. In preganglionic (= supraganglionic) nerve root injuries, the nerve is avulsed from the spinal cord separating its motor axons from their respective perikarya located in the anterior or ventral horn [17]. Accordingly, postganglionic (= infraganglionic) injuries, which involve damage to the peripheral nervous system, have a much better prognosis. Global or complete brachial plexopathies affect all five roots, which contribute to the formation of the brachial plexus.

High-speed motor vehicle accidents cause the vast majority of today's brachial plexus injuries (BPIs) [17, 21, 22]. They typically affect young men in the prime of life and otherwise healthy condition [17, 29, 32]. In polytraumatized patients, lesions to the brachial plexus occur in 1.2% [15]. While 0.67% of motor vehicle accidents are the cause for BPIs, they occur in up to 4.2% of motor cycle accidents [15]. Accidents with a motor cycle pose the greatest risk of injury, since severe traction on the brachial plexus may occur with violent arm motion when the motorcycle rider collides with a car or other obstacles [3, 11, 17]. Improved emergency trauma management and intensive care as well as advanced life support techniques have led to a steady increase in civilian brachial plexus lesions due to motorcycle accidents throughout the world [4, 23, 26, 29].

While rarely seen in Europe and Asia, neurapraxia of the brachial plexus may occur during contact sports, especially seen in American football, where a player makes some form of contact with an opposing player utilizing his head, neck, and/or shoulder [8]. The mechanism of injury includes a forceful lateral flexion of the neck with tension on various parts of the brachial plexus. This cervical nerve pinch syndrome is typically known as "stingers" or "burners" by football players and trainers [8].

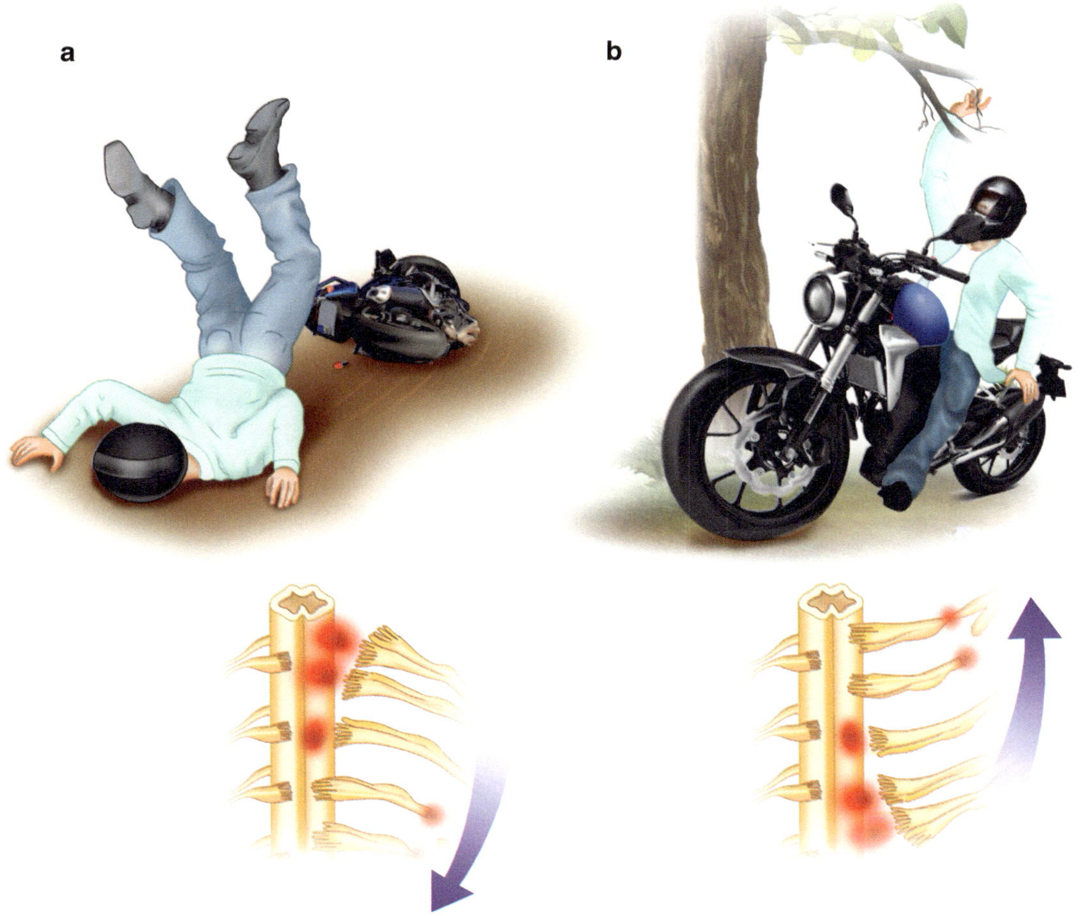

Fig. 3.1 Injury mechanisms resulting in damage to the brachial plexus. (**a**) Schematic illustration of a fall following a motor cycle crash. The head and neck are forcefully displaced away from the ipsilateral shoulder, which causes avulsion of the nerve roots C5 and C6 and simultaneous rupture of the lower roots. (**b**) An upward lift of the arm with significant force will most likely cause avulsion of the lower trunk or roots C8 and T1 and simultaneous stretch injury or rupture of the upper roots

Diagnosis

Clinical Examination Including History of the Accident

Brachial plexus injuries are often overlooked in the emergency trauma setting since patient survival is given the highest priority in the early phase after the accident and the patient is mostly unconscious or sedated. The diagnosis is therefore frequently delayed. The comprehensive clinical examination of a patient combined with profound clinically oriented anatomical knowledge can provide detailed information on the mechanism and level of a brachial plexus injury. A detailed history must be obtained from the patient ranging from the injury mechanism to medical records related to the accident. Any form of recovery since the injury should be inquired, which influences further treatment. Evaluation of the presence of pain is an important component of the physical assessment, since more than 70% of patients with multiple root avulsions suffer from chronic deafferentation pain [20]. Psychosocial sequelae including post-traumatic stress disorder, anxiety and/or depression, social

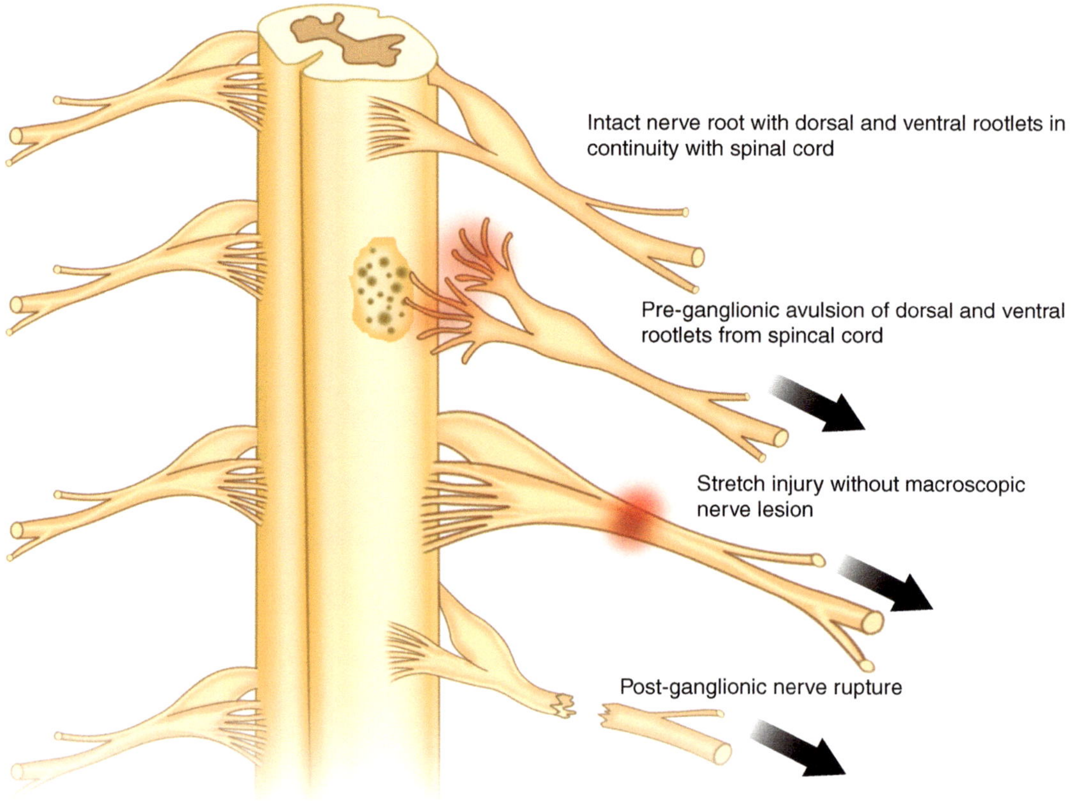

Fig. 3.2 Nerve injury patterns in brachial plexus injuries. Top-down: Intact spinal nerve with dorsal and ventral rootlets attached to the spinal cord; a preganglionic injury showing the avulsion of both ventral and dorsal rootlets proximal to the dorsal root ganglion; stretching of the nerve without macroscopic lesion; a postganglionic rupture of the peripheral nerve. The avulsion of nerve roots proximal to the dorsal root ganglion, i.e., a preganglionic injury, represents the most severe type of a brachial plexus injury

withdrawal, and so forth need to be thoroughly addressed due to the severe socioeconomic hardship that can arise from brachial plexus lesions [7]. The advent of re-traumatization caused by additional surgeries and hospitalizations needs to be considered as well as the patient's coping skills and psychosocial adjustment following the accident [12].

The next step includes a comprehensive clinical examination, which will influence further decision-making and subsequent surgical treatment. Muscle denervation may ascertain a very proximal lesion consistent with a nerve root avulsion injury [32]. These include the serratus anterior muscle innervated by the long thoracic nerve (C5–C7), the levator scapulae and rhomboid muscles innervated by the dorsal scapular nerve (C4–C5), and the hemidiaphragm innervated by the phrenic nerve (C3–C5), which may show ascension in radiologic imaging studies. The suprascapular nerve (C5–C6) branches off the upper trunk and innervates the supra- and infraspinatus muscles. Its denervation leads to a prominent scapular spine, internal rotation of the shoulder, and weakened abduction of the arm.

Unlike older descriptions of the innervation pattern of the pectoral muscles, it has been shown recently that there are three distinct pectoral nerves, which exit the brachial plexus at trunk level [2]. The clavicular portion of the pectoralis major (PM) muscle is innervated by the superior pectoral nerve, which arises shortly distal to the supraclavicular nerve from the superior trunk. Its isolated denervation supports the diagnosis of a superior trunk lesion and/or C5/C6 root avulsion. The middle pectoral nerve arises from the middle trunk and innervates the sternal portion of the PM muscle. The abdominal portion of the PM muscle is innervated by the inferior pectoral nerve, which—when denervated—supports the diagnosis of a lower root avulsion (C8-T1) since it branches off the inferior trunk.

An important clinical sign to localize lesions and guide appropriate investigations is the Claude Bernard-Horner syndrome, which comprises damage to the nerve root T1. Its presence supports the diagnosis of lower root avulsions C8-T1. Preganglionic sympathetic fibers exit the lateral horn of the spinal cord at the T1 level together with the respective spinal nerve. They terminate in the orbit and innervate the superior tarsal muscle, which contributes to eyelid opening, and the dilator pupillae muscle, which causes mydriasis. Damage to the sympathetic fibers at the nerve root level T1 thus results in ptosis, miosis, and (pseudo)-enophthalmos.

The presence of a Tinel-Hoffman's sign (TH sign) in the posterior triangle of the neck radiating to the appropriate dermatomes suggests a rupture lesion distal to the dorsal root ganglion (= postganglionic injury), whereas a patient without a TH sign is more likely to suffer from an avulsion of one or more roots [23]. Furthermore, the early onset of pain is highly associated with nerve root avulsion [5].

The major peripheral nerves originating from the BP (axillary, musculocutaneous, median, ulnar, radial nerves) can be evaluated by classical examination of global upper extremity function. From anesthetic and hypesthetic dermatomes, conclusions can be drawn as to which roots are affected.

Imaging Studies

High-resolution magnetic resonance imaging (MRI) has evolved as the standard diagnostic tool in brachial plexus injuries and has largely replaced CT myelography, which has long been used in the assessment of preganglionic root injuries [13]. MRI allows a structural and microstructural evaluation of the brachial plexus characterizing the location and extent of injury [13]. Its advantages over CT myelography include its noninvasive nature and the lack of need for sedation. Myelography was initially introduced by Murphey in 1947 [18] and allows the radiographic examination of the spinal cord using the injection of a contrast medium to evaluate its location, structure, and pathology. Murphey was the first to describe radiological abnormalities in patients with brachial plexus avulsion injuries [27]. This technique was later combined with computer tomography (CT) to allow highly specific topographical mapping of observed pathologies, such as the formation of scar tissue around the dural sheath, which typically tears along with the avulsed nerve root and is visualized as a pseudomeningocele by CT myelography [32].

Recently, also high-resolution ultrasound has gained relevance in specialist centers in the assessment of brachial plexus injuries [10, 14, 25], making it a valuable tool in defining the nerve lesions in even greater detail [10].

Besides the abovementioned imaging techniques, plain radiographs allow evaluation of bone structures (humerus, clavicle, scapula, ribs) and the height of the diaphragm (phrenic nerve function).

Electrodiagnostic Studies

Electrodiagnostic studies may be used in addition to confirm the diagnosis and to ascertain whether recovery is occurring [9]. Sensory nerve action potentials (SNAPs) may help to distinguish pre- from postganglionic injuries [17]. In preganglionic avulsion injuries, SNAPs appear normal since the perikarya of sensory axons are preserved in the dorsal root ganglion. Contrarily, SNAPs are absent in patients with postganglionic injuries, where nerves rupture distal to the dorsal root ganglion.

Needle electromyography (EMG) can confirm muscle denervation and thereby support the diagnosis of a brachial plexus injury. Denervated muscles show fibrillations, i.e., spontaneous discharges, upon insertion of the needle electrode as compared to healthy muscle, which will not show any electrical activity [32]. Fibrillations occur only 6 weeks post-injury, which should be considered in the planning of such studies. Following nerve reconstruction surgery, the occurrence of active motor units and decreased fibrillation potentials indicate recovery and a good prognosis [9].

References

1. Alnot JY. Traumatic brachial plexus lesions in the adult. Indications and results. Hand Clin. 1995;11(4):623–31.
2. Aszmann OC, Rab M, Kamolz L, Frey M. The anatomy of the pectoral nerves and their significance in brachial plexus reconstruction. J Hand Surg Am. 2000;25(5):942–7. https://doi.org/10.1053/jhsu.2000.17818.
3. Bergerot A, Shortland PJ, Anand P, Hunt SP, Carlstedt T. Co-treatment with riluzole and GDNF is necessary for functional recovery after ventral root avulsion injury. Exp Neurol. 2004;187(2):359–66. https://doi.org/10.1016/j.expneurol.2004.02.003.
4. Bertelli JA, Ghizoni MF. Results and current approach for brachial plexus reconstruction. J Brachial Plexus Peripher Nerve Inj. 2011;6(1):2. https://doi.org/10.1186/1749-7221-6-2.
5. Bruxelle J, Travers V, Thiebaut JB. Occurrence and treatment of pain after brachial plexus injury. Clin Orthop Relat Res. 1988;237:87–95.
6. Carlstedt T. Central nerve plexus injury. London: Imperial College Press; Distributed by World Scientific Publications; 2007.
7. Choi PD, Novak CB, Mackinnon SE, Kline DG. Quality of life and functional outcome following brachial plexus injury. J Hand Surg Am. 1997;22(4):605–12. https://doi.org/10.1016/S0363-5023(97)80116-5.
8. Clancy WG Jr, Brand RL, Bergfield JA. Upper trunk brachial plexus injuries in contact sports. Am J Sports Med. 1977;5(5):209–16. https://doi.org/10.1177/036354657700500508.
9. Giuffre JL, Kakar S, Bishop AT, Spinner RJ, Shin AY. Current concepts of the treatment of adult brachial plexus injuries. J Hand Surg Am. 2010;35(4):678–88. https://doi.org/10.1016/j.jhsa.2010.01.021; quiz 88.
10. Gruber H, Glodny B, Galiano K, Kamelger F, Bodner G, Hussl H, et al. High-resolution ultrasound of the supraclavicular brachial plexus—can it improve therapeutic decisions in patients with plexus trauma? Eur Radiol. 2007;17(6):1611–20. https://doi.org/10.1007/s00330-006-0464-2.
11. Hruby LA, Gstoettner C, Sturma A, Salminger S, Mayer JA, Aszmann OC. Bionic upper limb reconstruction: a valuable alternative in global brachial plexus avulsion injuries—a case series. J Clin Med. 2019;9(1):23. https://doi.org/10.3390/jcm9010023.
12. Hruby LA, Pittermann A, Sturma A, Aszmann OC. The Vienna psychosocial assessment procedure for bionic reconstruction in patients with global brachial plexus injuries. PLoS One. 2018;13(1):e0189592. https://doi.org/10.1371/journal.pone.0189592.
13. Mallouhi A, Marik W, Prayer D, Kainberger F, Bodner G, Kasprian G. 3T MR tomography of the brachial plexus: structural and microstructural evaluation. Eur J Radiol. 2012;81(9):2231–45. https://doi.org/10.1016/j.ejrad.2011.05.021.
14. Mallouhi A, Meirer R, Bodner G. Ultrasonographic features of brachial plexus traumatic rupture. Case illustration. J Neurosurg. 2003;99(2):432. https://doi.org/10.3171/jns.2003.99.2.0432.
15. Midha R. Epidemiology of brachial plexus injuries in a multitrauma population. Neurosurgery. 1997;40(6):1182–8; discussion 8–9.
16. Millesi H. Surgical management of brachial plexus injuries. J Hand Surg Am. 1977;2(5):367–78.
17. Moran SL, Steinmann SP, Shin AY. Adult brachial plexus injuries: mechanism, patterns of injury, and physical diagnosis. Hand Clin. 2005;21(1):13–24. https://doi.org/10.1016/j.hcl.2004.09.004.
18. Murphey F, Hartung W, Kirklin JW. Myelographic demonstration of avulsing injury of the brachial plexus. Am J Roentgenol Radium Ther. 1947;58(1):102–5.
19. Narakas A. Surgical treatment of traction injuries of the brachial plexus. Clin Orthop Relat Res. 1978;133:71–90.
20. Narakas AO. The treatment of brachial plexus injuries. Int Orthop. 1985;9(1):29–36.
21. Nashold BS Jr, Ostdahl RH. Dorsal root entry zone lesions for pain relief. J Neurosurg. 1979;51(1):59–69. https://doi.org/10.3171/jns.1979.51.1.0059.
22. Parry CB. Pain in avulsion lesions of the brachial plexus. Pain. 1980;9(1):41–53.
23. Parry CB. Pain in avulsion of the brachial plexus. Neurosurgery. 1984;15(6):960–5.
24. Robotti E, Longhi P, Verna G, Bocchiotti G. Brachial plexus surgery. An historical perspective. Hand Clin. 1995;11(4):517–33.
25. Shafighi M, Gurunluoglu R, Ninkovic M, Mallouhi A, Bodner G. Ultrasonography for depiction of brachial plexus injury. J Ultrasound Med. 2003;22(6):631–4.
26. Shin AY, Spinner RJ, Steinmann SP, Bishop AT. Adult traumatic brachial plexus injuries. J Am Acad Orthop Surg. 2005;13(6):382–96.

27. Teixeira MJ, da Paz MG, Bina MT, Santos SN, Raicher I, Galhardoni R, et al. Neuropathic pain after brachial plexus avulsion – central and peripheral mechanisms. BMC Neurol. 2015;15:73. https://doi.org/10.1186/s12883-015-0329-x.
28. Terzis JK, Papakonstantinou KC. The surgical treatment of brachial plexus injuries in adults. Plast Reconstr Surg. 2000;106(5):1097–122; quiz 123–4.
29. Terzis JK, Vekris MD, Soucacos PN. Brachial plexus root avulsions. World J Surg. 2001;25(8):1049–61.
30. Thorburn W. A clinical lecture on secondary suture of the brachial plexus. Br Med J. 1900;1(2053):1073–5.
31. Tung TH, Mackinnon SE. Nerve transfers: indications, techniques, and outcomes. J Hand Surg Am. 2010;35(2):332–41. https://doi.org/10.1016/j.jhsa.2009.12.002.
32. Tung THH, Moore AM. Brachial plexus injuries. In: Mackinnon SE, editor. Nerve surgery. New York, NY: Thieme Medical Publishers, Inc.; 2015.

Part II

Functional Restoration in Upper Limb Amputees

4 Body-Powered Prosthetic Systems

Jack Uellendahl

Body power refers both to the power source and to the control method for operation of the prosthesis. The prosthetic components are operated through the use of the body's own force and excursion. By harnessing the force and movement of joints more proximal to the amputation, it is possible to operate mechanical terminal devices, wrists, elbows, and humeral rotation devices. Joint movements produce excursion that is captured with the harness straps and transmitted through a cable system to cause movement of the prosthetic components. In this age of computers and miniaturized electronic components, it may be surprising that body power continues to be an effective form of prosthesis control for a select group of users. Several factors contribute to the continued utilization of body-powered control compared to externally powered devices including lower cost, generally more robust design, lower maintenance cost, generally lighter weight, and perhaps most importantly greater proprioceptive feedback through the harness and cable system. However, body-powered systems also have significant disadvantages. Some users may not be able to produce the high forces needed to optimally operate some body-powered systems, or these high forces may cause discomfort or pain during device activation [8]. Also some limb-deficient persons may not be able to produce sufficient excursion to effectively operate the system.

Prerequisites for Body-Powered Control

A fundamental requirement for the use of any body-powered component is the user's ability to produce adequate force and excursion with the controlling body motion. This means that the amputee must have the required strength and range of motion to operate the prosthetic component though its full range of operation. For terminal device operation alone, the user generally needs to produce forces at the harness between two and five times that of the grip force realized at the terminal device. This is due to the mechanical inefficiencies of the devices themselves and the friction loss of the cable system. In order to operate a typical hook through its full range of operation, the amputee must pull the cable approximately 50 mm. To operate a typical conventional elbow through its full range of motion, an additional 62 mm must be produced. The residual limb must also be able to comfortably withstand these high forces to operate a body-powered prosthesis effectively.

In order to transfer body motions to the prosthesis effectively, body motions must be captured by the harness with minimal lost motion;

J. Uellendahl (✉)
Hanger Prosthetics and Orthotics, Phoenix, AZ, USA

O. C. Aszmann, D. Farina (eds.), *Bionic Limb Reconstruction*,
https://doi.org/10.1007/978-3-030-60746-3_4

otherwise, excursion is lost. The user must be able to comfortably perform terminal device operation repeatedly throughout the day in order to accomplish their activities of daily living [8]. Persons with shorter residual limbs or more proximal amputations will have greater difficulty producing the needed excursion and force. When force and excursion are insufficient, alternative power sources and control methods are indicated.

Proprioceptive Feedback

Body-powered prostheses control provides users with a wealth of proprioceptive feedback through the harness and cable system [2]. Users of these devices can readily perceive the position and speed of movement of the prosthetic components [5]. Although the harness represents one of the main advantages of body-powered control, it is also one of the most disliked features of a body-powered prosthesis since it may cause discomfort and be difficult to don.

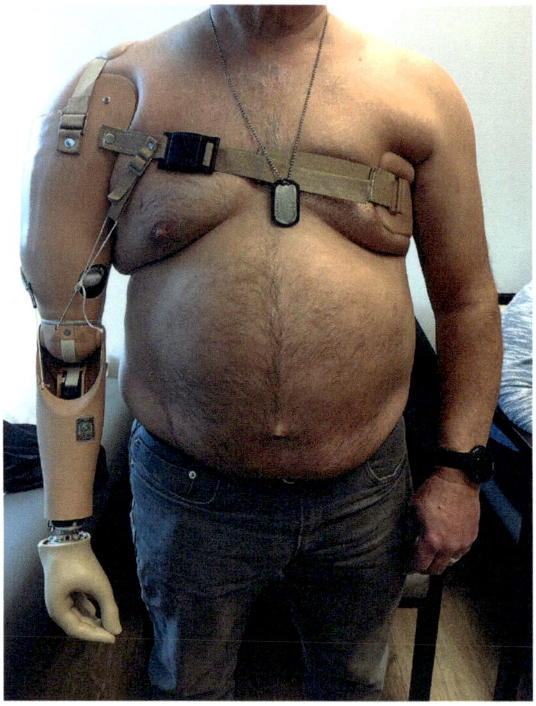

Fig. 4.1 Transhumeral hybrid prosthesis with dual-site myoelectric control of hand and cable-actuated body-powered control of elbow

Hybrid Control

Hybrid control refers to the use of two or more control methods: body-powered, myoelectric, switch, or linear transducer to name a few. But the term most commonly refers to combining body power with myoelectric control. Prostheses for amputation above the elbow often are provided with an elbow, wrist, and terminal device that need to be controlled. However, at this level, there are too few control sites to provide dedicated control inputs for each device. One common solution is to use a hybrid control system using a body-powered elbow with a myoelectrically controlled terminal device (Fig. 4.1). This type of system has the benefit of providing a separate and dedicated control input for elbow and terminal device and the potential to operate these simultaneously. Targeted muscle reinnervation (TMR) surgery as described by Kuiken [4], with or without pattern recognition control, can address the problem of limited control sites achievable with myoelectric signals. However, the author has also achieved good results using a hybrid control system combining body-powered control of the elbow with TMR control of the wrist and terminal device when sufficient force and excursion are available. This option provides a lighter weight prosthesis that is less expensive than a fully electronic system yet allows the user to realize the benefit of physiologically natural control of the hand and wrist, which is most important, while affording the user fast and smooth operation of the elbow with the proprioceptive feedback inherent with body-powered control.

Even without TMR surgery, hybrid control of a transhumeral prosthesis has been the chapter author's preferred method of fitting when a myoelectrically controlled prehensor is indicated. Compared to the use of an electric elbow, cable control of the elbow is faster and more accurate in positioning, and the user benefits from the proprioceptive feedback through the cable and harness regarding prosthesis velocity, force, and position. Electric elbows have the advantage of affording greater live lift capability. As mentioned above, this type of hybrid control offers the potential for simultaneous or seamless sequential control of the prosthetic elbow and hand. For some users, a hybrid transhumeral prosthesis with body-powered control of the elbow and myoelectric control of the terminal device is a less complicated control strategy compared to the conventional mode selection routines commonly used to allow two control sites to control two or more electric components when TMR surgery has not been performed. For a hybrid control strategy where the terminal device is myoelectrically controlled and the elbow is cable-actuated, the harness provides a similar function as that of a completely body-powered system; however, the excursion requirement is cut in half. This hybrid control strategy also eliminates the need to lock the elbow in order to operate the terminal device as would usually be required for a fully body-powered transhumeral system.

Hybrid control of a transhumeral prosthesis costs more than a fully body-powered system. Configuring and fitting a hybrid transhumeral prosthesis sometimes require a higher technical knowledge in order to combine components that the manufacturers may not have planned to be used in combination. Harnessing can be difficult especially when fitting short and very short transhumeral levels where the user may not have enough force and/or excursion to fully and efficiently operate the elbow. Elbows with a very strong spring lift assist mechanism have been a considerable improvement for this user group whenever this hybrid approach is implemented.

Terminal Devices

A terminal device (TD) is the component at the distal end of the prosthesis that is designed to substitute for the human hand. Because the human hand is an incredibly complex instrument of manipulation, expression, and sensation, a mechanical replacement is a poor substitution. Terminal devices may be classified as being either active or passive. An active TD is one that may be opened and closed with a cable and harness to grasp objects. A passive TD is one that does not open or close but may be used to stabilize objects. Hooks are typically active TDs although due to the hook shape they can be used with great success passively. For example, a passive device may stabilize paper, while the other hand writes, or it may hold a purse or wallet against the body while the sound side hand manipulates money. Hands are available as either active or passive versions. Most body-powered active TDs simply open and close in order to grasp and release objects. Shoulder and elbow motions provide the power for these movements. Even active terminal devices are often used passively to stabilize objects, while the sound side hand is used for object manipulation since these prosthetic grippers lack dexterity.

When considering body-powered terminal devices, it is important to make a distinction between body-powered hooks and hands. Body-powered hooks are very functional for a variety of activities. They are lightweight, relatively low cost, and simple in design and provide a less obstructed view of objects grasped. Hooks are especially well suited for heavy-duty activities. They are less prone to damage in excessively dirty, wet, and corrosive environments than other options. Hooks generally provide greater function than prosthetic hands as can be demonstrated by their preference by bilateral arm amputees. Their smaller profile provides a better line of sight, allowing users to better see the object being manipulated. The smaller size also allows hooks to more easily fit into small spaces such as wallets, purses, and pockets. Hooks are available in a wide variety of shapes and sizes. However, hooks do not restore the appearance of a hand and therefore are rejected by persons who desire a normal appearance [7].

Although there are a variety of body-powered hands available, few are used as active terminal devices [3]. Body-powered hands are mechanically inefficient; they are not lightweight and require a great amount of force to operate [8]. Smit et al. found that most body-powered voluntary opening hands do not provide enough prehensile force to accomplish activities of daily living [8]. These facts illustrate the point that not all body-powered options can be considered equal in their functional capabilities. Body-powered hooks can be very functional but do not provide a natural appearance. Body-powered hands, while more cosmetic, are often not functional and in those cases are little more than passive devices. Since most amputees seek to replace both the functions and appearance of the hand they lost, utilizing a hand prosthesis is appropriate and indicated for the majority of upper limb amputees. Traditionally this dilemma of function versus appearance has been addressed by providing users with a hook and interchangeable hand. Although quick disconnect wrists are common in body-powered systems allowing interchangeability of hook and hand, they are often difficult for the user to manage with their one intact hand due to the need to disconnect and reconnect the control cable. This problem can be further complicated when the glove of the prosthetic hand covers the control cable at the disconnection point. It is worth noting that an electric-powered hand provides both usable grip strength and a reasonably natural appearance and that exchanging electric hooks and hands is easier to perform absent the requisite control cable [7].

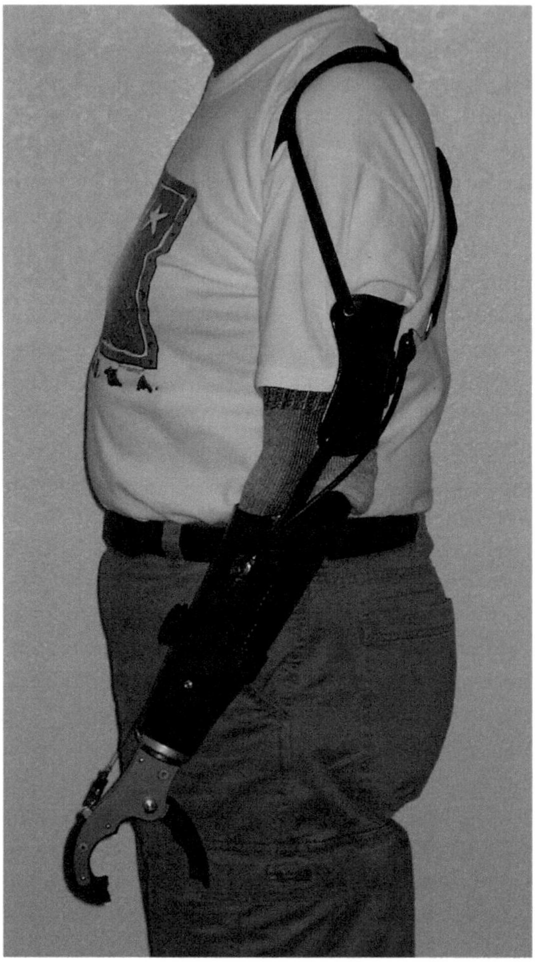

Fig. 4.2 Transradial prosthesis with voluntary closing terminal device and figure eight harness

Voluntary Opening Versus Voluntary Closing

There are two fundamentally different control methods for body-powered terminal devices, voluntary opening (VO) and voluntary closing (VC). Voluntary opening device are opened by pulling the control cable and close with either rubber bands or springs. The grip force is determined by the spring tension, and the user must maintain cable tension to apply less than maximum grip force. A voluntary closing terminal device uses springs to open the terminal device (Fig. 4.2). The hand is closed by applying tension through the harness and cable system. Maintaining tension on the cable allows the person to continue grasping the object; relaxing the tension on the cable opens the fingers again. The advantage of voluntary closing devices is that they allow for a wide range of prehension force. Also, much greater prehension forces are possible than with VO systems. However, in order to maintain grasp, the cable tension must be applied continuously or a locking mechanism employed. Voluntary opening devices have the advantage that no effort is required to maintain grasp assuming the object is not crushable. There are also a wider variety of voluntary opening devices available offering a large number of finger shapes and grip surfaces.

Some terminal devices are specially designed for use with tools having a knife holder, nail holder, and chisel holder. Some models have a "back lock" feature that locks the opening size of the fingers around the object and does not allow the fingers to spread open until the person applies tension through the cable and harnessing system.

A hook may be selected with or without a lining, such as nitrile. Lined surfaces provide more friction for grip, but the lining will need to be replaced when worn. Unlined hooks may have a serrated metal surface, but this can more easily damage items grasped.

Material choices for hooks include reinforced plastics, aluminum, titanium, and stainless steel. Aluminum and plastics are lightweight but are more easily broken and therefore best for light-duty usage and higher-level amputations. Stainless steel is durable but heavy (twice the weight of aluminum) and therefore for heavy-duty use. Titanium is lightweight and rigid but expensive.

Harnessing

Conventional harnessing serves the dual role of suspension and control of the body-powered prosthesis. In designing a harness system, it may be useful for the prosthetist to consider suspension and control separately. When socket design provides suspension, the harness design can be simpler and worn looser making the harness more comfortable [6].

In the absence of socket suspension, the harness must suspend the prosthesis securely from the shoulders in order to maintain the residual limb within the socket. The harness must be securely anchored and its straps precisely positioned to resist distraction forces. The contralateral axilla affords an anchor point for most harness designs.

For control, the harness must transmit power through the cable system to the distal components. For the transradial prosthesis where only the TD needs to be controlled, a single-function Bowden cable is employed. The cable passes through a continuous housing that is attached to the prosthesis to transmit shoulder and scapular motion to the TD without affecting the elbow. For transhumeral fittings where both the elbow and TD are to be controlled, a two-function "fairlead" housing is usually employed. When the elbow is unlocked, tension on the cable causes the two housing pieces to be pulled together, thereby flexing the elbow. With the elbow locked, the force is transmitted to the TD. With dual control, there is interaction with the TD and elbow such that maintenance of grip force is compromised when the elbow is actively flexed because force is exerted on the TD equal to the force required to flex the elbow. One solution to this problem is to design a control system where separate control cables are used for TD, elbow, and elbow lock. This triple control system uses Bowden cables for each component where shoulder flexion causes elbow flexion, biscapular abduction causes TD operation, and shoulder extension causes elbow lock operation. Many people find it difficult to achieve proper separation of the controls in operating the triple control system [2]. Triple control is used far less than hybrid fitting using an externally powered option for the TD or elbow.

The primary body motions used to control the prosthesis are glenohumeral flexion and biscapular abduction. In the case of the self-suspending transradial socket, it is desirable to attach the cross bar assembly tab to the posterior aspect of the socket which allows elbow flexion to contribute to terminal device control. Scapular or biscapular abduction is used for tasks near the midline of the body.

An alternative to the figure-of-eight harness is the shoulder saddle or thoracic suspension system. The shoulder saddle uses a pad located on the shoulder of the amputated side as the starting point for a similar array of straps as the figure-of-eight harness. Because the suspension straps are mounted to this pad on the shoulder, greater comfort is afforded for axial lifting. Also, these types of harnesses do not anchor in the sound side

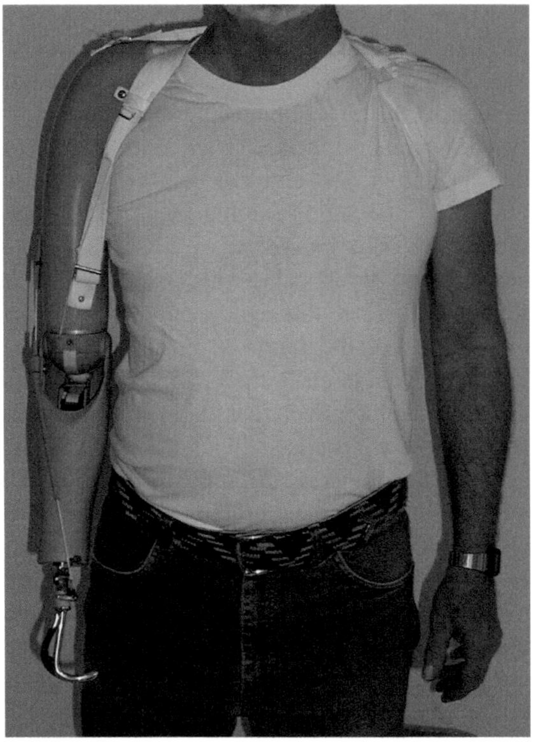

Fig. 4.3 Transhumeral prosthesis with figure 8 harness

axilla but rather against the sound side thorax below the axilla using a chest strap (Figs. 4.3 and 4.4). Thoracic suspension systems have the advantage over figure eight systems in that they offer more comfort in axial lifting and more comfort in the sound side axilla and are easier to don in many cases. Their disadvantages include being less cosmetic because the chest strap is visible in open neck clothing and they are less efficient in capturing excursion due to the less positive anchor and tend to be bulkier around the shoulder region.

When the transradial socket is self-suspending, a figure 9 harness is indicated. The figure 9 harness simply consists of an axilla loop leading to a control attachment strap. Because this type of harness is used for control only, the harness can be worn quite loosely, and it is generally well tolerated.

Cable efficiency is of critical importance to the success of a body-powered fitting. Careful attention should be devoted to producing the straightest line of pull using materials that offer the least amount of friction such as Spectra cable in a Teflon-lined housing [1]. Spectra is a string-like material with high tensile strength and a slippery surface.

Wrists

There are few wrist features that can presently be actively controlled with body power. Most body-powered prosthesis utilize wrists that are passively positioned. One notable exception is the N-Abler wrist from Texas Assistive Devices. This wrist affords the user the ability to flex and supinate the wrist using the control cable, while springs provide wrist extension and pronation. This wrist configuration is particularly useful for bilateral amputees where there is not sound side hand to passively position the wrist. This system has been used successfully on short transradial, transhumeral, and shoulder disarticulation prostheses.

Bilaterals

Without a sound side hand to perform more difficult manipulation, bilateral upper limb users often come to appreciate the functional capabilities of hook prostheses. This coupled with the proprioceptive feedback inherent in cable-driven prostheses makes body-powered hooks a well-accepted and often used prosthetic option for this user group.

Bilateral arm amputees will often benefit from the use of two different types of terminal devices. This will afford the user a wider variety of gripping options and greater versatility. There are a wide variety of body-powered hooks available, and the user will benefit from trying several options to see which best meet their specific needs. Another successful terminal device combination employs a body-powered hook on the dominant side and an electrically powered hook or hand on the nondominant side. This combination provides the fine manipulation capabilities of a body-powered split hook with the superior gripping forces available with an electric TD and

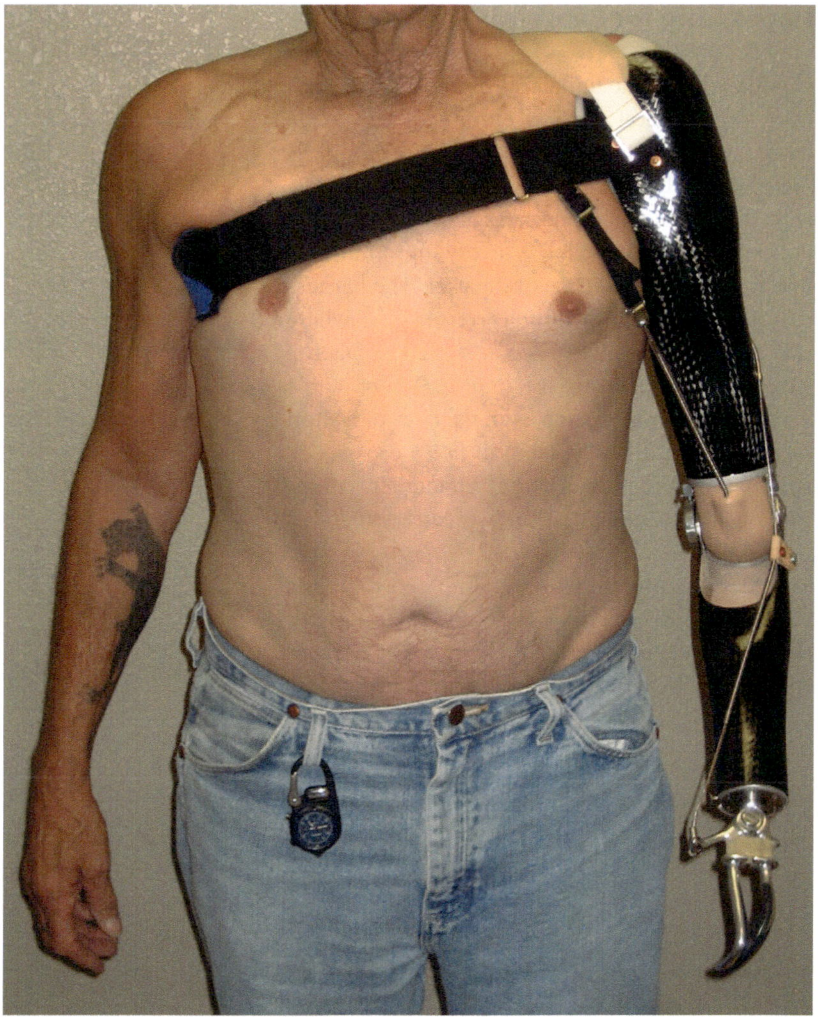

Fig. 4.4 Transhumeral prosthesis with shoulder saddle and chest strap harness. Note the line of pull of the harness on the sound side is directed against the thorax and not into the axilla

has been particularly well accepted by people with transhumeral/shoulder disarticulation amputations. Voluntary opening hooks are primarily used because they maintain grip without the need for continued cable tension. Voluntary closing hooks offer excellent feedback regarding prehension forces as well as higher grip force, however, at the transhumeral level when a single control cable operates the elbow and hook voluntary closing control complicates prosthesis use. Voluntary closing hooks also find more limited use for bilateral fittings due to the requisite continuous cable tension (or the need for a locking mechanism) to maintain grasp and the limited number of designs available.

Task-specific TDs should be considered for bilateral prostheses. These are available with quick disconnect wrist components that enable interchange between a utilitarian hook and specific use devices such as work tools and kitchen utensils.

When fitting any high-level bilateral amputee, it is advisable to start with as simple a system as possible. Due to the many components that need to be controlled, a hybrid approach regarding both power and control may be indicated. Issues of ease of control, prosthesis weight, and control reliability along with donning ease become even more important for bilateral users than unilateral fittings.

Harnessing for Bilaterals

Compared with the unilateral figure-of-eight harness, the bilateral version eliminates the axilla loop, a frequent area for discomfort, and is therefore well tolerated by almost all patients (Fig. 4.5). This type of harness is easy to don and doff independently which is a very important feature for the independence of the user. The harness may have a single or double ring design where two rings are fixed to each other by a Dacron strap, one inferior to the other. Where cable excursion is limited, such as in the case of a short residual limb, it is advisable to use a cross-back strap which keeps the control attachment straps low on the scapulae yielding greater excursion (Fig. 4.6).

When the prostheses are harnessed together, each prosthesis serves as the anchor point for the other. When both prostheses use the harness for control, inadvertent cross control becomes a potential problem. One solution to cross control is to provide a fully body-powered prosthesis on one side and a fully electric prosthesis on the other so the control motions cause no interaction with the control of any component other than the one intended. In some cases when one side is amputated at a very proximal level such as the interscapulothoracic amputation level and the contralateral side is a transhumeral or longer, it may be advisable to provide a passive prosthesis to serve as a firm anchor point for the harness of that more functional contralateral prosthesis.

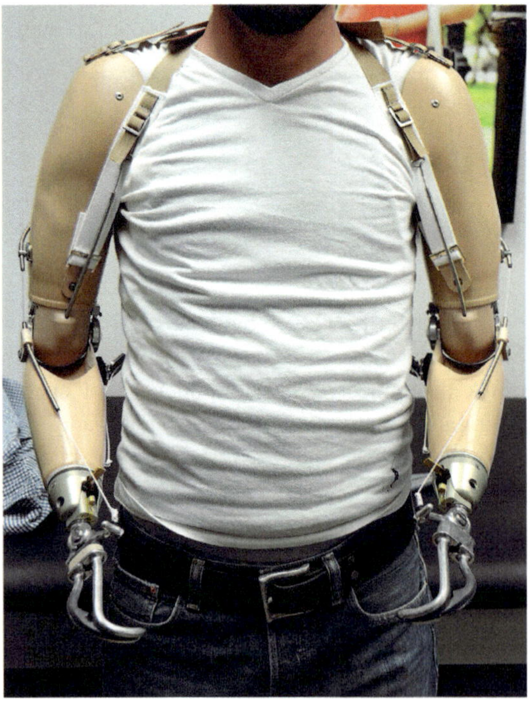

Fig. 4.5 Typical configuration of body-powered transhumeral prostheses for the bilateral amputee. Wrist rotation, wrist flexion, terminal device opening, and elbow flexion are controlled by a single control cable

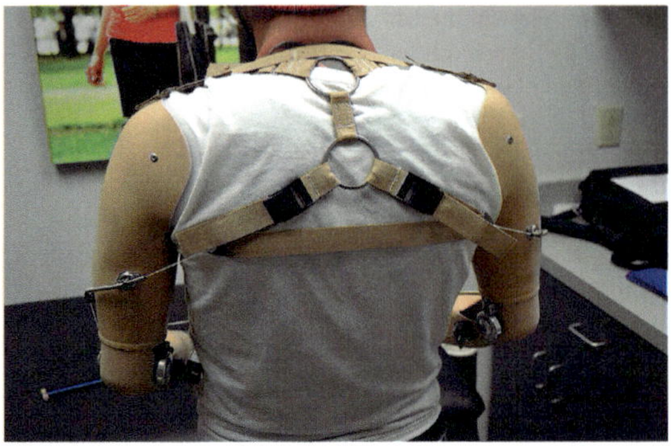

Fig. 4.6 The double ring with cross-back strap afford an optimal position of the control strap on the inferior aspect of the scapula. This improves the amputee's ability to operate the elbow and terminal device

Conclusion

Body-powered prostheses continue to be a viable option for some amputees. Their simple and often robust design makes them well suited for use by manual laborers. Also owing to their simple design, they can often be maintained and repaired by individuals without specialized knowledge and training, making them appropriate for use in rural and remote areas where access to trained professionals may not exist. Whenever funding is limited, body-powered prostheses may be an option due to their lower cost compared to electronic alternatives. Determination of the most appropriate upper limb prosthesis must take into account the functional possibilities of the prosthetic components as well as comfort, cosmesis, and the life situation of the user. It is incumbent upon the rehabilitation team to identify those prosthetic design options that will be most beneficial to the individual user.

References

1. Carlson L, Veatch B, Frey D. Efficiency of prosthetic cable and housing. J Prosthet Orthot. 1995;7(3):96-99.
2. Heckathorne CW. Manipulation in unstructured environments: extended physiological proprioception, position control, and arm prostheses. Proc Int Conf Rehabil Robot. 1990;1990:25–40.
3. Hess A. Upper limb body-powered components. In: Krajbich JI, Pinzur MS, Bk P, Stevens PM, editors. Atlas of amputations and limb deficiencies: surgical, prosthetic, and rehabilitation principles. Rosemont, IL: American Academy of Orthopedic Surgeons; 2016. p. 139–58.
4. Kuiken TA, Li G, Lock BA, et al. Targeted muscle reinnervation for real-time myoelectric control of multifunctional artificial arms. JAMA. 2009;301:619–28.
5. Uellendahl JE, Heckathorne CW. Creative prosthetic solutions for the person with bilateral upper extremity amputations. In: Atkins D, Meier R, editors. Functional restoration of adults and children with upper extremity amputation. New York, NY: Demos Medical Publishing; 2004. p. 225–37.
6. Uellendahl JE. Bilateral upper limb prostheses. In: Smith DG, Michael JW, Bowker JH, editors. Atlas of amputations and limb deficiencies: surgical, prosthetic, and rehabilitation principles. Rosemont, IL: American Academy of Orthopedic Surgeons; 2004. p. 311–25.
7. Uellendahl J. Myoelectric versus body-powered upper-limb prostheses: a clinical perspective. J Prosthet Orthot. 2017;29(4S):25–9.
8. Smit G, Bongers RM, Van der Sluis CK, et al. Efficiency of voluntary opening hand and hook prosthetic devices: 24 years of development? JRRD. 2012;49:523–34.

Mechatronic Design of Functional Prosthetic Systems

5

Ivan Vujaklija and Dario Farina

This chapter focuses on the mechatronic components of active prosthetics and will give an overview of various design approaches, hardware interfacing solutions, and end-effector options. It describes the hardware commonly found in both body-powered and externally powered devices, and it analyzes the components of current clinical state-of-the-art prosthetics. Furthermore, the chapter provides insights into mechatronic designs of established research devices as well as more pioneering solutions.

In 200 BC, a Roman general lost his hand in combat and became the first documented case of a prosthetic limb fitting [26]. This prosthetic device was passive and rather bulky since it was entirely made of cast iron. The prosthetic designs have since been evolving by continuously adapting the latest engineering advances and materials. Nevertheless, the first functional upper limb prosthesis has only been available to patients with limb impairments in late 1912. By that time, Bowden cables had become widely available, and the first split hook prosthesis was patented [13]. This device featured a voluntary opening of the prehensor that was coupled proportionally to the level of elbow flexion through a set of pulleys and cables. This core design has been so efficient that it is still widely present on the market [2].

As the electrical motors became more compact and efficient, electrically actuated hand prostheses were introduced into the market [9]. The initial solutions, however, were heavy and functionality limited. The next big step in prosthetic design occurred when the transistors came to mass production. The Russian Hand was the first to fully embrace this technology in 1960 [9]. Shortly after, in 1965, the Belgrade Hand [41] was developed as the first multi-articulated fully wearable hand prosthesis. It represented the state-of-the-art system that thrived on the most recent advances in electronics and actuator miniaturization.

Since then, the prosthetic design has seen steady incremental improvements associated with the advances in manufacturing and materials technology. As of recent, additive manufacturing has been introduced in prosthetic design, and the first fully 3D printed system has been CE marked [53]. Moreover, the general principles of robotic manipulator design are slowly being re-evaluated in light of bioinspired approaches [4].

Body-Powered Prosthetic Systems

Body-powered or "cable-driven" prostheses are those for which the power to operate certain functions of the device comes from the user's

I. Vujaklija (✉)
Department of Electrical Engineering and Automation, Aalto University, Espoo, Finland
e-mail: ivan.vujaklija@aalto.fi

D. Farina
Department of Bioengineering, Imperial College London, London, UK

O. C. Aszmann, D. Farina (eds.), *Bionic Limb Reconstruction*,
https://doi.org/10.1007/978-3-030-60746-3_5

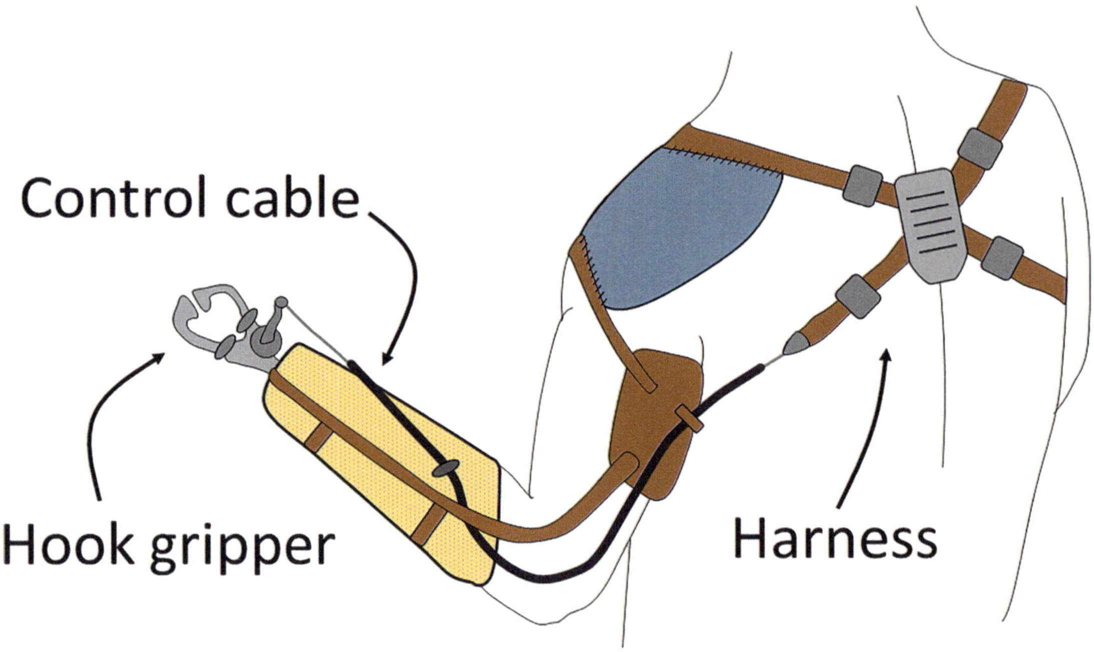

Fig. 5.1 Body-powered hook prosthesis. Opening and closing of the hook gripper are controlled by shoulder shrugging using a harness and a system of pulleys

own body motion. Namely, as described in details in Chap. 4 an amputee would wear a harness that translates a dedicated intact joint motion into a prosthetic joint movement through a system of pulleys (Fig. 5.1) [12]. These simple yet very effective devices deliver much-needed support in everyday life of an amputee in a fairly straightforward and reliable manner. Their effectiveness has proven so significant that some of their fundamental principles can still be found in modern prosthetics. Indeed, most users that are conducting heavy lifting or intense manual activities on a daily basis rely on body-powered grippers [17, 45]. While the cost of these devices is lower than that of externally powered systems, the level of comfort as well as potential functionality is somewhat limited. However, body-powered systems tend to be lighter in weight and as such preferred by users of smaller stature. However, the actuation requires exertion of larger forces due to the mechanical disadvantage. This further limits the loading capabilities of body-powered prosthetics and as such their usability.

Moreover, body-powered devices suffer from a higher incidence of breakages than other systems [14]. On the other hand, due to their simple design, the repairs are easy to perform and mostly involve a replacement of the cable system or the load springs. Yet, the downtime per incident ranges from a few days to over a week, which is a source of dissatisfaction for frequent users [45].

Clinical State of the Art

The market of electrically actuated upper limb prosthetics can be divided into three major groups depending on the level of disability that the systems aim to address: transradial and transcarpal solutions, transhumeral devices, and systems for high-level shoulder disabilities. Requirements for each group are different, and the corresponding devices are equipped with impairment-specific features.

Transradial/Transcarpal Solutions

Human hands feature complex yet very efficient design and play a crucial role in daily interaction with the environment, own body, and other individuals. Therefore, their substitution is a challenge that is still to be matched by a comparably versatile hardware. To the day, simple single degree of freedom (DoF) grippers dominate the market. They essentially rely on a single motor that adjusts the aperture of the prehensor according to the input signal. Centi Child Myoelectric Hand, Ottobock Healthcare Transcarpal-Hand, and Liberating Technologies Select Electric are some of the commonly distributed prosthetic grippers. Their characteristics are listed in Table 5.1.

Over the last 5 years, all major prosthetic companies developed multi-actuated prosthetic hands capable of performing various grips or even moving individual digits. Similar to grippers, the body of these prosthetic hands hosts the actuators responsible for proportional control over the dedicated DoFs. However, the raise in number of motors leads to an increase in the weight, as seen in Table 5.2.

In addition to these clinical devices, Vincent Systems' Evolution 2/3, DARPA-funded DEKA Arm RC, and Open Bionics Hero Arm are examples of uniquely designed systems, which are partly clinically present. Vincent Systems offers one of the lightest solutions of its kind on the market (~400 g) [52] that additionally combines sensory feedback information with individually motorized digits and fully actuated thumb. In contrast, the DEKA Arm RC is one of the heaviest prosthetic hands (1270 g) [42], but it provides a fully actuated 3 DoF wrist. Finally, Hero Arm by Open Bionics has been the first fully 3D printed prosthetic hand to receive a CE certificate [34] and comprises individually actuated digits.

Compared to standard grippers, all advanced prosthetic hands are fitted with comprehensive

Table 5.1 Overview of representative commercially available myoelectric prosthetic grippers

Model	Weight (g)	Size (mm)	Full closing time (s)	Maximal grip force (N)	Thumb rotation properties	Digit dexterity	Wrist options
Child Myoelectric Hand by Centri Hosmer Dorrance Corp [18, 20]	238	171	0.35	63	Static	First two digits coupled	Passive rotation
Transcarpal-Hand Ottobock Healthcare [36]	308	184–210	0.91	90	Static	First two digits coupled	Passive flexion and/or active rotation
Select Electric Hand Liberating Technologies [27, 28]	470–520	184–210	0.90	–	Static	First digit active	Passive rotation

Table 5.2 Overview of representative commercially available multi-articulated prosthetic hands

Model	Weight (g)	Size (mm)	Full closing time (s)	Maximal grip force (N)	Thumb rotation properties	Digit dexterity	Wrist options
i-Limb Quantum Touch Bionics [46–49]	474	154–182	0.80	136	Passive or motorized	Four individually motorized	Active or passive rotation and passive flexion
BeBionic3 Ottobock Healthcare (RSL Steeper) [44]	570–590	190–200	1.00	140.1	Passive	Four individually motorized	Active or passive in all directions
Michelangelo Ottobock Healthcare [38]	420–510	177–210	0.37	70	Motorized	First two digits coupled	Active and passive rotation and passive flexion

wrist solutions. These range from purely passive ones to a combination of passive and active wrist systems. Commonly, the user can manually adjust the position of passive prosthetic wrists, which often features two modes of operation. The first mode enables them to lock the hand in one of the predefined positions in flexion/extension direction (Ottobock AxonWrist, Touch Bionics Flexion Wrist, BeBionic FlexionWrist) or along the rotation axis (Ottobock AxonRotation, Touch Bionics QWD, BeBionic Short Wrist). The second mode leaves the passive wrist unhinged so that it can remain compliant as the hand interacts with the environment. Besides these two DoFs, a few passive wrists allow prosthetic hands to be positioned in virtually any direction covering the full range of motion of the natural hand (MyolinoWrist 2000, BeBionic Multi-Flex). Most of the commercially available active wrists are focused solely on the wrist rotation (MC Wrist Rotator, Ottobock Electric Wrist Rotator).

Myoelectric devices experience fewer malfunctions per year then their body-powered counterparts [14]. However, these repairs do tend to take longer and be more costly.

Transhumeral and Shoulder Disarticulation Solutions

The difficulty of functional limb restoration increases with each additional missing joint. For instance, in cases of transhumeral amputations, even though the wrist rotation unit can substitute the forearm rotation, the larger absence of the anatomy poses great challenges in designing a balanced, lightweight system that can handle an additional elbow DoF in parallel to those of the hand and wrist. Similar to the prosthetic wrist, elbows can too be passive or active, with passive systems dominating the market. Still, there are a number of actuated elbows, out of which a few representative ones are listed in Table 5.3, offering added functionality. All these support a free swing mode that allows users to have a spontaneous arm swing during walking. The lack of this function on the motorized elbow of DEKA Arm HC has been highly criticized by the users [43].

Table 5.3 Overview of representative commercially available active prosthetic elbows

Model	Weight (g)	Max. lift capacity (Nm)	Terminal device compatibility
Utah Arm 3+ Fillauer [16, 21]	900–1000	4.3	iLimb, MC TDs, Sensor Speed
Arm System Boston Digital [5]	965	14.2	BeBionic, iLimb, MC TDs, Sensor Speed, El. Greifer, Select El. Hand
Dynamic Arm+ Ottobock Healthcare [37]	680–710	18.0	SensorHand Speed, MyoHand VariPlus Speed, El. Greifer
NY Electric Elbow Hosmer Dorrance Corp [19, 21].	439–453	3.4	Michigan Electric Hook, NY-Greifer, NU-VA Synergetic Claw

The shoulder is the most complex of the arm joints. For its replacement, the DEKA Arm HC is the current only system with an FDA approval [15]. This particular prosthetic offers four movement directions across the joint [43].

Research-Oriented Systems

Since the human hand is the gold standard of dexterous manipulators, roboticists have been trying to replicate its design and function for a very long time. Although artificial hands may be built so that they are both stronger and faster than their biological counterparts, demands for versatility, robustness, and lightweight design are largely unmet by many of the proposed solutions. For this reason, only a few of these systems obtained certification and can be found in clinical settings. However, though currently not available for clinical applications, some research-oriented systems may be valuable test platforms for the development of the next generation of prosthetic hardware and interfaces.

The SmartHand transradial prosthesis [10] was a research platform which offered 16 DoFs and a maximum power grasp of 3.6 kg. This system was commercialized and is now available for research purposes as the IH2 hand [39]. With a weight of 640 g, it has three independently controlled fingers and two compliant digits. It features an independent controlled in abduction/adduction by brushed DC motor with nonbackdrivable mechanism. Its versatility and ease of interfacing make it a solid platform for further investigation into prosthetic design, control, and interfacing.

Unique in its actuation design, the Pisa/IIT SoftHand is a wearable research prosthetic hand with 19 joints [8]. It has anthropomorphic features and is able to grasp objects of different shapes by using only one motor. This neuroscience research-inspired system is also commercially available as qb SoftHand [40]. With only one DC brushed motor, it weights 500 g and delivers up to 84 N of force during a power grasp. In comparison to other systems, SoftHand's compliant design provides a significantly different test bench for research.

Open Bionics Brunel Hand is a fully 3D printed research-oriented prosthetic hand [35]. It has a power grasp of only 2.2 kg, but it weighs only 340 g and offers 4 degrees of actuation. Being fully an open source and with a simple built, it has a strong community of users, as a unique environment for research and development.

Similar to commercial systems, academia has offered fewer solutions when it comes to more proximal prosthetic joints. There are some passive [11, 50] and active [51] elbow solutions that are promising, but are still not at the stage to be offered as solid development platforms. One of the most interesting concepts is a modification of the LTI-Collier shoulder joint to which an electronic lock/unlock feature was added [29]. With some further improvements, this could be a solid start for devising a much-needed active shoulder development bench.

Beyond Commercial Systems

While both clinical and commercially available academic prosthetic systems adhere, even though sometimes loosely, to a set of standards and standardized design practices, there is a range of prosthetic hardware designs that go beyond these conventions. These devices do not aim to become clinically viable but rather to test the boundaries of design and to provide insights and inputs for developing the next generation of bionic limbs.

A number of these unconventional systems come from a growing research area that exploits the capabilities of soft structures and their potential for developing highly adaptive robots. These robots rely on materials and actuation methods which are soft, flexible, and compliant [23]. While there is a number of soft robotic manipulators [22, 24, 30, 31], a truly wearable solution is yet to be designed. Actuation is the primary challenge, preventing these manipulators to become viable prosthetic solutions. Namely, a suitable compromise between the size of the actuation units and the delivered grip strength has not been found yet. To the day, the majority of soft manipulators rely on large and noisy pneumatic systems to operate [31, 32]. As an alternative, cable-driven actuation [7, 33] and shape memory alloys [1, 25] have been considered. However, in case of soft manipulators, both approaches seem to fail in delivering a sufficient grip strength and/or are very slow.

Designing actuators with anthropomorphic characteristics is another alternative approach that goes beyond the classic prosthetic design. By combining belt and cable drive transmission with a brushless DC motor, a prosthetic elbow with matching output torque, mass, and size has been proposed [3]. However, this and similarly complex prosthetic systems [6] are yet to be tested and proven robust in a clinical setting.

Conclusions

Commercially available and clinically viable prosthetic hardware has seen significant improvements in the last decade. Most of the changes were made possible due the advances in technology and manufacturing techniques. This allowed more actuators to be embedded in a smaller anthropomorphic packaging resulting in more appealing, dexterous devices. However, further improvements are still needed. With an increasing number of actuated DoFs, the problem of robustness has become central [14]. In addition, the weight of multi-articulated systems has become substantial, making some of them suitable for only selected users [54].

The current prosthetic hardware development follows two directions. The first includes approaches that aim to deliver commercially available, but due to certain trade-offs, not clinically certified systems. These devices build up on the concepts of the current clinical solutions, by implementing novel technologies and scientific methodologies. They offer high flexibility and new features but are usually not sufficiently robust or aligned with the strict clinical requirements. Still, these systems are important as development and testing tools since they are easily accessible by researchers and practitioners. The second direction comprises solutions that do not adhere to the conventional design approaches, but rather serve to critically question the user needs and put the technology to test in order to pave the way for the next generation of prosthetic devices. Such systems are commonly a result of in-lab developments and usually do not end up being translated into commercial products. However, the knowledge and the experience obtained from these designs likely end up driving future commercial developments.

The next generation of prosthetic hardware will certainly be more compliant than current systems and will likely incorporate neuromorphic features via a combination of manufacturing techniques, redesign of actuators, and their low-level controllers. Through the combination of new materials and additive manufacturing, prosthetic systems targeting more proximal limb deficiencies might also develop further. This would fill an important gap in the market that lacks functional, light, and tailored solutions.

References

1. Andrianesis K, Tzes A. Development and control of a multifunctional prosthetic hand with shape memory alloy actuators. J Intell Robot Syst. 2015;78:257–89. https://doi.org/10.1007/s10846-014-0061-6.
2. Belter JT, Reynolds BC, Dollar AM. Grasp and force based taxonomy of split-hook prosthetic terminal devices. In: 2014 36th annual international conference of the IEEE Engineering in Medicine and Biology Society. Piscataway, NJ: IEEE; 2014. p. 6613–8.
3. Bennett DA, Mitchell J, Goldfarb M. Design and characterization of a powered elbow prosthesis. In: Proceedings of the annual international conference of the IEEE Engineering in Medicine and Biology Society, EMBS IEEE. Piscataway, NJ: IEEE; 2015. p. 2458–61.
4. Bicchi A. Hands for dexterous manipulation and robust grasping: a difficult road toward simplicity. IEEE Trans Robot Autom. 2000;16:652–62. https://doi.org/10.1109/70.897777.
5. Boston Digital Boston Digital Arm System Catalog.
6. Braun DJ, Chalvet V, Dahiya A. Positive–negative stiffness actuators. IEEE Trans Robot. 2018;2018:1–12. https://doi.org/10.1109/TRO.2018.2872284.
7. Camarillo DB, Carlson CR, Salisbury JK. Configuration tracking for continuum manipulators with coupled tendon drive. IEEE Trans Robot. 2009;25:798–808. https://doi.org/10.1109/TRO.2009.2022426.
8. Catalano MG, Grioli G, Farnioli E, Serio A, Piazza C, Bicchi A. Adaptive synergies for the design and control of the Pisa/IIT SoftHand. Int J Robot Res. 2014;33:768–82. https://doi.org/10.1177/0278364913518998.
9. Childress D. Historical aspects of powered limb prostheses. Clin Prosthet Orthot. 1985;9(1):2–13.
10. Cipriani C, Controzzi M, Carrozza MC. The SmartHand transradial prosthesis. J Neuroeng Rehabil. 2011;8:29. https://doi.org/10.1186/1743-0003-8-29.
11. Create O&P. The create arm; 2017.
12. Doeringer JA, Hogan N. Performance of above elbow body-powered prostheses in visually guided unconstrained motion tasks. IEEE Trans Biomed Eng. 1995;42:621–31. https://doi.org/10.1109/10.387202.
13. Dorrance DW. Artificial hand; 1912.
14. Etter K, Borgia M, Resnik L. Prescription and repair rates of prosthetic limbs in the VA healthcare system: implications for national prosthetic parity. Disabil Rehabil Assist Technol. 2015;10:493–500. https://doi.org/10.3109/17483107.2014.921246.
15. FDA FDA allows marketing of first prosthetic arm that translates signals from person's muscles to perform complex tasks. https://www.medicaldesignandoutsourcing.com/fda-allows-marketing-of-prosthetic-arm-that-translates-signals-from-persons-muscles/. Accessed 14 Feb 2020.
16. Fillauer. Utah arm 3 plus sales sheet; 2012.
17. Gaine WJ, Smart C, Bransby-Zachary M. Upper limb traumatic amputees. J Hand Surg Am. 1997;22:73–6. https://doi.org/10.1016/S0266-7681(97)80023-X.
18. Hosmer Dorrance Corporation. Hooks, hands, and wrists – product brochure. Campbell, CA: Hosmer Dorrance Corporation; 2003. p. 1–58.
19. Hosmer Dorrance Corporation. Externally powered systems product catalog. Campbell, CA: Hosmer Dorrance Corporation; 2011.

20. Hosmer Dorrance Corporation. Hosmer upper extremity product catalog. Campbell, CA: Hosmer Dorrance Corporation; 2012.
21. Hussain I, Masood A, Iqbal J, Khan US, Elbow BNYE. Design of high torque elbow joint for above elbow prosthesis. Conference. 2011;5:318–21.
22. Ilievski F, Mazzeo AD, Shepherd RF, Chen X, Whitesides GM. Soft robotics for chemists. Angew Chemie. 2011;123:1930–5. https://doi.org/10.1002/ange.201006464.
23. Laschi C, Cianchetti M. Soft robotics: new perspectives for robot bodyware and control. Front Bioeng Biotechnol. 2014;2:3. https://doi.org/10.3389/fbioe.2014.00003.
24. Laschi C, Cianchetti M, Mazzolai B, Margheri L, Follador M, Dario P. Soft robot arm inspired by the octopus. Adv Robot. 2012;26:709–27. https://doi.org/10.1163/156855312X626343.
25. De Laurentis KJ, Mavroidis C. Mechanical design of a shape memory alloy actuated prosthetic hand. Technol Health Care. 2002;10:91–106. https://doi.org/10.1063/1.4983336.
26. Leblanc M. Use of prosthetic prehensors. Prosthetics Orthot Int. 1988;12:152–4.
27. Liberating Technologies Inc. Select electric hands & gloves data sheet. Holliston, MA: Liberating Technologies Inc; 2007.
28. Liberating Technologies Inc. Select electric hands for adults product catalog. Holliston, MA: Liberating Technologies Inc; 2013.
29. Lipschutz RD, Kuiken TA, Miller LA, Dumanian GA, Stubblefield KA, Otr L. Shoulder disarticulation externally powered prosthetic fitting following targeted muscle reinnervation for improved myoelectric control. Prosthetics Orthot Int. 2006;18:245–53. https://doi.org/10.1097/00008526-200604000-00002.
30. Manti M, Hassan T, Passetti G, D'Elia N, Laschi C, Cianchetti M. A bioinspired soft robotic gripper for adaptable and effective grasping. Soft Robot. 2015;2:107–16. https://doi.org/10.1089/soro.2015.0009.
31. Mikey77 3d print an artificial muscle robot hand: 10 steps (with pictures). http://www.instructables.com/id/3d-Print-An-Artificial-Muscle-Robot-Hand/. Accessed 18 Feb 2018.
32. Mosadegh B, Polygerinos P, Keplinger C, Wennstedt S, Shepherd RF, Gupta U, Shim J, Bertoldi K, Walsh CJ, Whitesides GM. Pneumatic networks for soft robotics that actuate rapidly. Adv Funct Mater. 2014;24:2163–70. https://doi.org/10.1002/adfm.201303288.
33. Nguyen T-D, Burgner-Kahrs J. A tendon-driven continuum robot with extensible sections. In: 2015 IEEE/RSJ International Conference on Intelligent Robots and Systems (IROS) IEEE. Piscataway, NJ: IEEE; 2015. p. 2130–5.
34. OpenBionics. Hero arm user manual. Bristol: OpenBionics; 2018.
35. OpenBionics. Humanoid robotic hand brunel V1.0-datasheet. Bristol: OpenBionics; 2017.
36. Otto Bock Healthcare GmbH Upper Limb Prosthetics Catalog.
37. Otto Bock Healthcare GmbH Elbow Components – at a Glance.
38. Otto Bock Healthcare GmbH. Michelangelo technician product brochure; 2013
39. Prensila s.r.l. IH2 azzura series data sheet; 2014.
40. qb SoftRobotics. (2018) qb softhand technical sheet.
41. Rakic M. The "belgrade hand prosthesis". Inst Mech Eng Proc. 1968;183:60–7.
42. Resnik L, Klinger SL, Etter K. The DEKA Arm: its features, functionality, and evolution during the veterans affairs study to optimize the DEKA Arm. Prosthetics Orthot Int. 2014;38:492–504. https://doi.org/10.1177/0309364613506913.
43. Resnik L, Klinger SL, Etter K. User and clinician perspectives on DEKA arm: results of VA study to optimize DEKA arm. J Rehabil Res Dev. 2014;51:27–38. https://doi.org/10.1682/JRRD.2013.03.0068.
44. Steeper RSL. BeBionic v3 product brochure; 2013.
45. Schweitzer W, Thali MJ, Egger D. Case-study of a user-driven prosthetic arm design: bionic hand versus customized body-powered technology in a highly demanding work environment. J Neuroeng Rehabil. 2018;15:1. https://doi.org/10.1186/s12984-017-0340-0.
46. Touch Bionics. i-limb quantum clinician manual; 2015.
47. Touch Bionics. i-limb quantum product sheet; 2015. pp. 1–2.
48. Touch Bionics. i-Limb quantum user manual. 1; 2015. p. 1–21.
49. Touch Bionics. Flexion wrist product sheet; 2015. pp. 1–2.
50. UCFArmory Limbitless Arm for Alex (V1) – Thingiverse. https://www.thingiverse.com/thing:408641/#comments. Accessed 20 Feb 2018.
51. Unlimited Tomorrow Robotic Arm. http://theroboarm.com/index.html. Accessed 20 Feb 2018.
52. Vincent Systems (2016) Vincent evolution 2. http://vincentsystems.de/en/prosthetics/vincent-evolution-2/. Accessed 24 Feb 2016.
53. Vujaklija I, Farina D. 3D printed upper limb prosthetics. Expert Rev Med Devices. 2018;17434440(2018):1494568. https://doi.org/10.1080/17434440.2018.1494568.
54. Vujaklija I, Farina D, Aszmann O. New developments in prosthetic arm systems. Orthop Res Rev. 2016;8:31–9. https://doi.org/10.2147/ORR.S71468.

6 Osseointegrated Amputation Prostheses and Implanted Electrodes

Yan Li, Max Ortiz-Catalan, and Rickard Brånemark

The load transfer from the external prosthesis to the residual limb via the socket can cause significant stress on the soft tissues, leading to irritation and skin ulcers. Osseointegrated bone-anchored prostheses systems create a direct structural and functional connection between the prosthesis and residual skeleton. Up to date, standardized implant systems, surgical techniques, and post-operative rehabilitation protocols have been developed for osseointegrated prostheses for the rehabilitation of amputees (OPRA), which has resulted in better functionality, fewer complications, and a better quality of life for implant recipients. The OPRA implant systems can now incorporate neuromuscular electrodes to facilitate myoelectric control and sensory feedback, which is especially important for upper extremity amputees. The latest development, called the osseointegrated human-machine gateway, allows for permanent implantation of neuromuscular electrodes, which provide long-term stable signals for myoelectric control, independent of limb position or environmental conditions, as well as artificial sensory feedback. In addition, the modular design of this system allows any part to be upgraded or replaced with minimal disturbance to the other components. The osseointegrated implants and the human-machine gateway represent frontiers in amputee rehabilitation.

Y. Li
Division of Orthopaedics and Biotechnology, Department of Clinical Intervention and Technology (CLINTEC), Karolinska Institutet, Stockholm, Sweden

M. Ortiz-Catalan
Biomechatronics and Neurorehabilitation Laboratory, Department of Electrical Engineering, Chalmers University of Technology, Gothenburg, Sweden

Integrum AB, Mölndal, Sweden

R. Brånemark (✉)
Department of Orthopedics, Sahlgrenska Academy, Gothenburg University, Gothenburg, Sweden

Center for Extreme Bionics, Biomechatronics Group, MIT Media Lab, Massachusetts Institute of Technology, Cambridge, MA, USA
e-mail: rickard.branemark@orthop.gu.se

Background

Despite improvements in medical and surgical techniques for limb salvage interventions, amputation remains a common surgical procedure with a prevalence between 33 and 39/100,000 inhabitants in Sweden [1]. In 2005, there were approximately 1.6 million amputees in the United States, and that number is expected to be double by 2050 [2, 3]. The current rehabilitation standard following amputation is mainly through a socket-mounted prosthesis [4]. Unfortunately, the load transfer from the external prosthesis to the residual limb via the socket connection causes significant stress on the skin and the underlying soft tissues, leading to irritation and ulcers [5].

O. C. Aszmann, D. Farina (eds.), *Bionic Limb Reconstruction*,
https://doi.org/10.1007/978-3-030-60746-3_6

One solution to these load transfer issues is to use an osseointegrated bone-anchored implant. Because the implant directly attaches to the skeleton, it avoids the inherent problems of socket suspension. Studies have shown that patients with osseointegrated prostheses receive more sensory feedback from the environment (osseoperception), use their prostheses more often, experience fewer problems, and have a higher quality of life [6–9]. However, early attempts to create percutaneous bone-anchored prostheses in the 1960s and 1970s were unsuccessful due to a high incidence of infections at the skin-implant interface and inadequate bone anchorage. The development of the OPRA implant system presents the histological aspects for refining a novel concept based on the laboratory study and clinical experience. Currently, multiple percutaneous implant systems exist for clinical use internationally, each attempting to create a stable bone-implant interface while avoiding complications such as infection and loosening [10].

Development of Osseointegration

In the early 1960s, the Swedish researcher Per-Ingvar Brånemark accidentally discovered that an implant made of titanium, a new biomaterial at that time, remained firmly attached to a rabbit tibia without loosening. Brånemark used the implant to study microcirculation in bone tissue and came to realize that the excellent anchorage using titanium implants could spawn a whole new field of research. After thorough animal testing, P.I. Brånemark implanted the first titanium dental implant in a patient in 1965. After 17 years of clinical work, this revolutionary treatment finally garnered international acceptance at a dental meeting in Toronto in 1982. A vast number of clinical trials have since confirmed that the functional and structural connection between living bone and a titanium implant can be maintained for more than 50 years. P.I. Brånemark coined the term "osseointegration" in1976 to describe the interaction between the biomaterial titanium and living bone tissue [4, 11].

The clinical application of osseointegration for amputee rehabilitation began in 1990, based on dental and craniofacial osseointegration experience and extensive basic research, including experimental biomechanical studies of osseointegration in animals and humans [12–14]. The results of the biomechanical experiments became the foundation for designing implant components and rehabilitation protocols for percutaneous osseointegrated amputation prostheses.

An osseointegrated prosthesis was first implanted in an amputee patient on May 15, 1990, by P.I. Brånemark and Professor B. Rydevik. The recipient was a 25-year-old woman who had suffered bilateral transfemoral amputations at the age of 15 due to a tram accident. She used her implants for more than 20 years; however, her implants were revised in 2014 due to skin infections and bone resorption.

The Initiation of the Osseointegrated Prostheses for the Rehabilitation of Amputees (OPRA) Program

Based on the results of early clinical pilot trials, P.I. Brånemark's son, Dr. Rickard Brånemark, worked to standardize the implant system, surgical technique, and postoperative rehabilitation protocol. This programmed approach was named Osseointegrated Prostheses for the Rehabilitation of Amputees (OPRA). The first implant systems to be standardized were femoral implants (1998), followed by humeral (2003) and thumb implants (2005). The OPRA program currently provides standard operating procedures and rehabilitation protocols for patients with amputations of the femur, forearm, humerus, and thumb.

The major components of the OPRA™ Implant System (Integrum AB, Molndal, Sweden) are the fixture, the abutment, and the abutment screw (Fig. 6.1). The fixture has external threads that engage with the endosteal surface of the bone. The distal part of the fixture contains an internal fitting, which press-fits the skin-penetrating abutment and is secured by the abutment screw. The outer end of the abutment is attached to the external prosthetic components (Fig. 6.2).

The original surgical osseointegration protocol used a two-stage surgery, whether for dental, craniomaxillofacial, or orthopedic procedures.

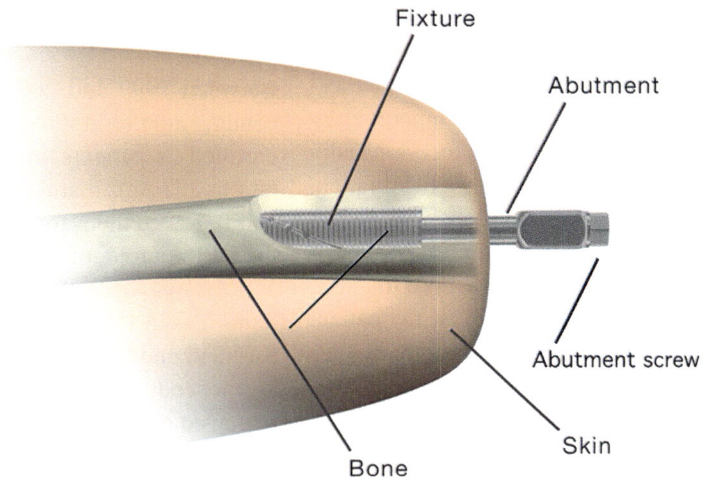

Fig. 6.1 The design of the osseointegrated prostheses for the rehabilitation of amputees (OPRA) implant system. Three major components are used: the fixture, the abutment, and the abutment screw. Reproduced with permission from The British Editorial Society of Bone and Joint Surgery, from A novel osseointegrated percutaneous prosthetic system for the treatment of patients with transfemoral amputation, R. Brånemark, Ö. Berlin, K. Hagberg, P. Bergh, B. Gunterberg and B. Rydevik, 96-B, 1, 2014 [18]

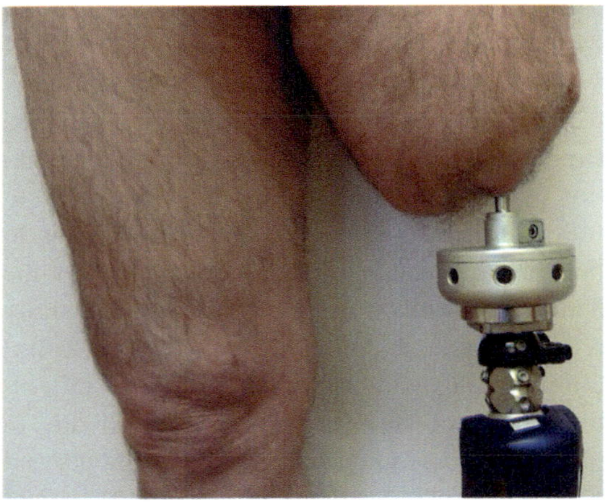

Fig. 6.2 Transfemoral amputee with a prosthesis attached to the skin-penetrating abutment. Reproduced with permission from The British Editorial Society of Bone and Joint Surgery, from A novel osseointegrated percutaneous prosthetic system for the treatment of patients with transfemoral amputation, R. Brånemark, Ö. Berlin, K. Hagberg, P. Bergh, B. Gunterberg and B. Rydevik, 96-B, 1, 2014 [18]

During the first surgery (S1), the fixture was anchored to the bone. During the second surgery (S2), which generally took place 6 months after the S1 procedure, the abutment was attached to the fixture, and the soft tissues were refashioned using a meticulous surgical technique to avoid skin mobility in the abutment-skin interface. The 6-month healing period between the two surgeries was considered adequate, even for cases with an undesirable bone situation (e.g., osteoporosis) and non-optimal primary stability. Moreover, the technique used to refashion the soft tissue is based on extensive clinical experience from implanting bone-anchored hearing aids (based on the osseointegration concept) [15], which has been performed for more than 100,000 patients. The primary lesson learned from implanting bone-anchored hearing aids is that immobilizing the abutment-skin interface drastically reduces the incidence of skin and soft tissue inflammation and subsequent infection.

During the healing period, patients were allowed to use their socket prosthesis, but no load should be directly transferred to the distal end of the bone and the fixture, which means that the socket should not be "end-bearing." The postoperative rehabilitation for transfemoral amputees started about 2 weeks after the S2 procedure by performing gentle exercises (i.e., range of motion exercises without full voluntary muscle contraction). More active training usually began with loading the skeleton around 4–6 weeks after the S2 procedure, once the skin penetration area and soft tissues were adequately healed.

During the early load training process, the patient used a short training prosthesis to apply load onto a bathroom scale. He or she began by bearing 10 kg of weight on the prosthesis and then increased the load by 10 kg each week until his or her full body weight can be painlessly (as measured with a visual analogue scale) supported by the prosthesis. The load training process usually took 6–8 weeks. The next phase of rehabilitation, prosthetic gait training, usually began 12 weeks after the S2 procedure. For gait training, the patient initially used the prosthesis for no more than 2 h per day and only indoors. In addition, the patient used the support of two crutches to limit weight-bearing on the prosthetic foot. The prosthesis wearing time, prosthetic activity, and weight-bearing were then gradually increased over a period of weeks. Patients typically achieve full-day prosthetic use after 4–6 weeks of gait training.

Patients were encouraged to walk with the support of two crutches or canes during the first 3 months of prosthetic use. About 6 months after the S2 procedure, the treatment team (surgeon, prosthetist, physiotherapist, and occupational therapist) reviewed the patient's radiographic findings and clinical status to determine whether the patient can stop using a walking support and whether he or she was ready to begin walking outdoors.

During both the load training and gait training processes, pain is an especially useful indicator. A pain score of 5 or higher using the visual analogue scale (VAS) is considered a sign that the bone-implant interface has been overloaded. If needed, the training progress should be slowed to maintain a VAS score below 5. The goal is to prevent overloading the ongoing integration of the bone-implant interface, which might lead to loosening of the integration and failure of the implant system [16].

Generally, these treatment principles apply to all amputation levels; however, some modifications are needed for certain types of amputations. For transradial amputations, a fixture is inserted in both the radius and ulna. With upper extremity prostheses, the rehabilitation is generally faster, and the healing period can be shorter because the patient does not need to bear bodyweight on the upper extremity. Moreover, the surgery can usually be performed in a single-stage when there is adequate bone tissue [17].

Transfemoral Amputations

Between 1999 and 2007, a prospective, single-center, non-randomized study consecutively enrolled 51 patients with 55 transfemoral amputations (TFAs), mainly due to trauma or tumor, and followed them for 2 years [18]. In this study, the OPRA protocol was strictly followed for each patient. All operations were performed at the Sahlgrenska University Hospital in Gothenburg, Sweden, and removal of the fixture was regarded as the endpoint for failure. The patients were reviewed at 3, 6, 12, and 24 months after the S2 procedure, and any complications were recorded. Two validated, self-reported questionnaires, the Questionnaire for Persons with a Transfemoral Amputation (Q-TFA) [19] and the Short Form 36 Health Survey (SF-36) [20], were used to assess the functional outcome and the patients' health-related quality of life. Both questionnaires were completed before the S1 procedure and at 12 and 24 months after the S2 procedure.

Three patients were withdrawn from the study for reasons unrelated to the implant (one died from an unrelated cause, one had severe dysfunction of the contralateral knee, and one was lost to follow-up). Four patients had falls leading to bending or fracture of the abutment and/or the abutment screw, which were successfully replaced without the need to exchange the fixture. Three patients had their entire implant systems removed during the study period due to inadequate osseointegration in two patients and deep infection in one patient, and a fourth patient had the implant system removed shortly after the study ended due to insufficient integration and infection. Therefore, the cumulative fixture survival rate was 92% at the 2-year follow-up. All Q-TFA scores significantly improved ($p < 0.0001$), indicating that patients experienced better prosthetic mobility, fewer problems, and an overall clinical improvement. The mean prosthetic use score (0–100) improved from 47 before the S1 procedure to 79, 2 years after the S2 procedure ($p < 0.0001$). In addition, the SF-36 physical function scores showed that the patients' general quality of life also improved ($p < 0.0001$).

Superficial infection was the most frequent complication, occurring 41 times in 28 patients. On average, each patient experienced one superficial skin infection every 2 years. Most infections were effectively treated with short-term oral antibiotics; however, longer-term antibiotics, including intravenous antibiotics, were sometimes required. In all, nine mechanical complications in four patients were reported with the abutment and/or the abutment screw, resulting in fracture or bending of the abutment and/or the abutment screw. All patients regained normal functioning after the damaged components were replaced. No mechanical problems of the fixture were reported.

Transhumeral, Transradial, and Thumb Amputations

The clinical experience with osseointegrated implants at other amputation levels is more limited. Between 1995 and 2010, there were 18 primary percutaneous osseointegrated implants and two implant revisions performed in 18 transhumeral amputees; of those, 16 patients were available for follow-up at a minimum of 2 years (median, 8 years; range, 2–19 years) [21]. Two primary and one revised implants failed due to early loosening, and the complete implant systems were removed. A fourth implant system was partially removed (the abutment and abutment screw were removed, but the fixture was left in place, and the skin was closed) due to ipsilateral shoulder osteoarthritis and subsequent arthrodesis of the glenohumeral joint. The overall implant success rate was 83% at 5 years, including the revised cases. The most common adverse event was skin infection around the abutment. One deep infection occurred and was successfully treated with antibiotics without requiring removal of the implant system.

Ten patients (nine men and one woman) in Sweden received percutaneous osseointegrated prostheses for transradial amputations [17] (Fig. 6.3), including two patients with dysmelia and one patient with traumatic bilateral transradial amputation who was treated bilaterally. All patients had custom-designed implant systems installed in both the radius and ulna. Two patients followed the standard OPRA treatment protocol, which was introduced in 2003. Fixture fractures occurred in three patients, all of whom had received the older implant design before the OPRA program began. Notably, no mechanical problems have occurred since the introduction of the OPRA program in 2003.

Fig. 6.3 Transradial amputee with an osseointegrated myoelectric prosthesis

Between 1990 and 2014, a total of 13 patients (ten men and three women) in Sweden have received percutaneous osseointegrated prostheses for unilateral thumb amputations [22] (Fig. 6.4). Eleven of these patients underwent amputation due to trauma, and two patients underwent amputation due to tumors. Seven patients (including all six after the introduction of the standardized OPRA protocol) performed a grip strength test (Jamar) with an average of 28.3 kg on the operated side versus 40.4 kg in the unaffected hand (70%), and key grip strength was 6 vs. 9.1 kg. Hand function was 94% of the normal hand using the Sollerman test. The most common complications were mechanical failures necessitating changes of components (eight times in three patients) and superficial infections (seven times in five patients). Three patients in the early group lost their implants due to loosening. Five patients had no complications. The improved implant design and the standardized OPRA treatment protocol achieved a 100% cumulative success rate, with, on average, 9.5 years of follow-up.

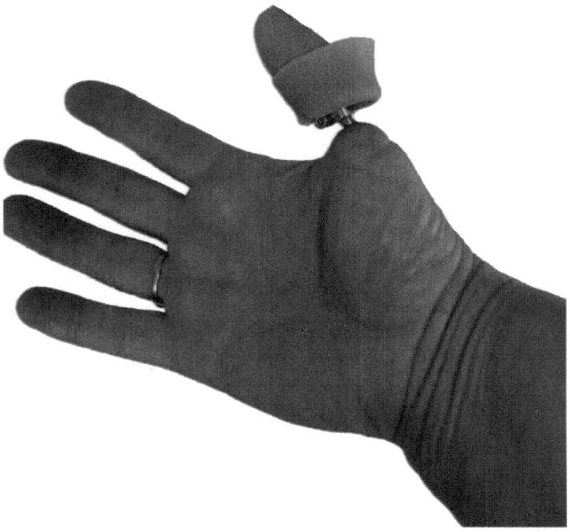

Fig. 6.4 Thumb amputee with an osseointegrated prosthesis with permission of Li Y, Kulbacka-Ortiz K, Caine-Winterberger K, Brånemark R. Thumb amputations treated with osseointegrated percutaneous prostheses with up to 25 years of follow-up. J Am Acad Orthop Surg Glob Res Rev. 2019;3(1):e097. https://doi.org/10.5435/JAAOSGlobal-D-18-00097. eCollection 2019 Jan

The current OPRA protocol has undergone some significant changes. A second generation of the OPRA implant system, which has up to a 40% better mechanical endurance for long-term usage, has been developed and clinically implemented since 2016. The new implant system also has a part of the surface laser-modified and nanostructured for enhanced osseointegration [23], which provides improved primary stability and allows for an accelerated rehabilitation process, including one-step surgery and early weight-bearing.

Other Currently Active Osseointegration Implant Systems

The press-fit systems present another osseointegration design, which was mainly used for lower extremity amputees. The Integral Leg Prosthesis (ILP) (Orthodynamics) evolved from the Endo-Exo implant (ESKA Orthopaedic Handels), which was introduced by Hans Grundei in Germany. The implants have a cast stem of cobalt-chrome-molybdenum alloy, and the implant stem is 140–180 mm long and slightly curved to prevent rotation in the intramedullary cavity and to fit the normal curvature of the femur. Since its introduction, the system has gone through several design changes. The models achieve immediate implant retention via the press-fit implantation, analogous to hip arthroplasty, and the external prosthetic limb is mounted via a multi-component dual cone and screw system [24].

Based on similar design principles, the OPL was introduced in Australia in 2013 and the Netherlands in 2015. Standardized implants are used for transfemoral amputees with sufficient stump length (≥160 mm). In contrast, custom-made implants are available for transtibial amputees and transfemoral amputees with very short stumps [25].

Unlike the ILP and OPRA, the COMPRESS system was first developed as an endo-prosthetic system for oncologic limb salvage reconstruction by Biomet Corporation (now Zimmer Biomet, Warsaw, USA). The intramedullary part of the implant is attached to the bone by transverse pins in a bone-anchor plug. A porous-coated collar designed to promote osseointegration is located at the distal interface of the amputated bone. To enhance osseointegration and to prevent stress-shielding of the bone, the concept of compliant pre-stress is utilized, exposing the bone-collar interface to a compressive force. Under a FDA custom device exemption, a percutaneous version of this system enabling attachment of an external limb prosthesis has been developed and implanted in ten transfemoral amputees and one transhumeral amputee. Both single-stage and two-stage surgeries have been used for implantation of the system. Two cases of peri-prosthetic fractures caused by falls have been reported among the transfemoral subjects [26].

Myoelectric Control and Sensory Feedback Using a Percutaneous Osseointegrated Implant

Percutaneous bone-anchored prostheses allow for full range of motion of the remaining joints. However, myoelectric control is compromised in certain limb positions when utilizing skin surface electrodes due to myoelectric interference from adjacent muscles (Fig. 6.5). Such interfer-

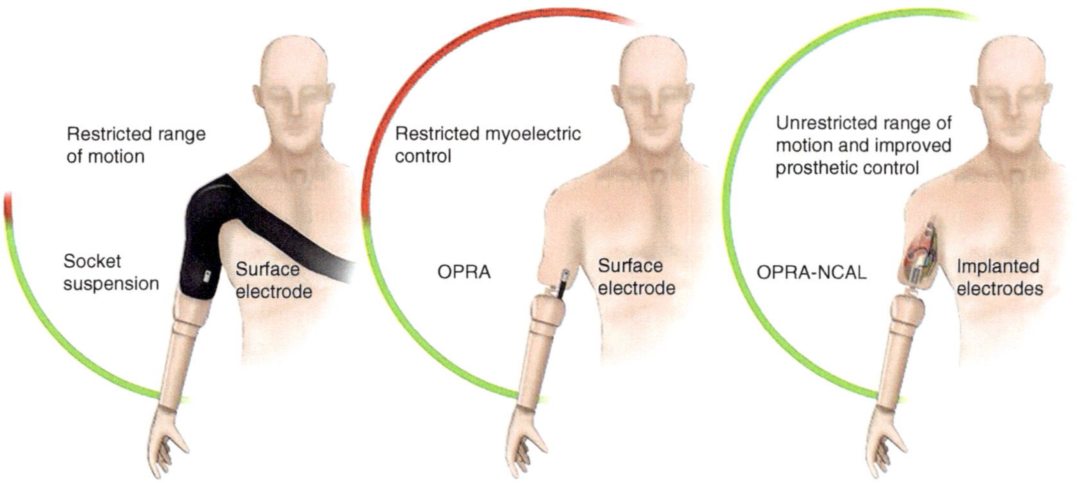

Fig. 6.5 Illustration of a common socket fitting (left inset) for a high transhumeral amputee, which limits the range of motion (green line) and prosthetic control at certain heights (red line). The osseointegrated prostheses for the rehabilitation of amputees (OPRA) implant system releases the adjacent joint and allows for full range of motion (central inset). However, the controllability of the prosthesis is compromised in certain limb positions due to myoelectric crosstalk from adjacent muscles (red line). Enhancing the OPRA implant system to become an osseointegrated human-machine gateway (OHMG) allows for permanent implantation of neuromuscular electrodes, which provide long-term, stable signals for myoelectric control, independent of limb position or environmental conditions. Illustration reproduced with permission from M. Ortiz-Catalan [27]

ence limits the advantages of treatment with a bone-anchored implant. Moreover, restricted myoelectric control can limit professional activities and activities of daily life. In addition, myoelectric control can be affected by environmental conditions, such as cold weather, during outdoor activities. In cold weather, the impedance between the skin and surface electrodes changes, reducing conduction. The use of implanted neuromuscular electrodes eliminates these skin issues, and the controllability of the prosthesis depends less on limb position or environmental conditions [28]. Furthermore, since the electrode is closer to the source, the control signals do not need to travel through soft tissue and skin before reaching the electrodes. Thus, compared with skin electrodes, a lower muscular effort is required to activate the prosthesis when implanted electrodes are used. This, in turn, increases grip resolution and improves proportional control, as would be expected from an increased signal bandwidth [28].

Extensive work conducted in neuroprosthetics has shown that epimysial and cuff electrodes are safe, reliable, and well-characterized neuromuscular interfaces [29]. Epimysial electrodes have been used extensively in humans for recording and stimulation as part of functional electrical stimulation (FES) neuroprostheses [30]. These devices have remained implanted for over 20 years and have been successfully used for activities of daily living [31]. The clinical trial of the Freehand system reported 408 electrodes implanted in 51 patients, of which only three patients were deemed treatment failures and only one of which was presumably due to mechanical fatigue [32]. In a related study, only 2 out of 204 epimysial electrodes in 27 patients failed with an average follow-up of 7.1 years (range, 3.2–16.4 years) [33]. In both studies, the electrode leads (lengths, 28–83 cm in the latter study) crossed up to three joints, which increases the mechanical stress to which they are exposed. Nevertheless, these systems demonstrated high survival rates. Nerve cuff electrodes have also been extensively used in humans to treat chronic pain [34], tinnitus [35], epilepsy [36, 37], sleep apnea [38], and blindness [39, 40], as well as for restoring upper [41–43] and lower [41, 44] limb function in patients with tetraplegia or hemiplegia, respectively.

The OPRA implant system for the transhumeral level has recently been redesigned to allow for bidirectional communication with implanted neuromuscular interfaces for closed-loop control of the amputation prostheses. This novel OHMG system makes it possible to connect an arm prosthesis to the patient's bone, nerves, and muscles [28]. To the best of our knowledge, this system, which is currently in a clinical trial, is the first neuroprosthesis to allow patients to operate an artificial arm using implanted electrodes for activities of daily living while also receiving direct neural feedback for closed-loop control [45].

Following the design philosophy of the OPRA implant system, the OHMG system has a modular design (Fig. 6.6); therefore, individual components can be easily exchanged without the need to replace the entire implant system. The percutaneous and osseointegrated components (abutment and fixture, respectively) were kept intact to maintain the mechanical stability required for load transfer between the prosthesis and the bone. Thus, patients who have already received the OPRA implant system can be upgraded to the OHMG system without needing to replace the implant system. The abutment screw can be modified to embed two feedthrough connectors, one parallel at the distal end and one in-line at the proximal end. An in-line pin extends from the central sealing component to the interface with the proximal connector of the abutment screw. From there, signals are transferred via the feedthrough sealing component to leads that extend intramedullary and then through a hole in the bone cortex to a connector unit positioned in soft tissues. Neuromuscular interfaces can then be connected to this latter connector unit and placed on muscles using epimysial electrodes and around peripheral nerves using nerve cuffs. This modular design allows any component to be upgraded or replaced with minimal disturbance to the other components (Fig. 6.6).

In this first generation of OHMG systems, no active components are implanted. Instead, the biopotential amplifiers, control, and neurostimu-

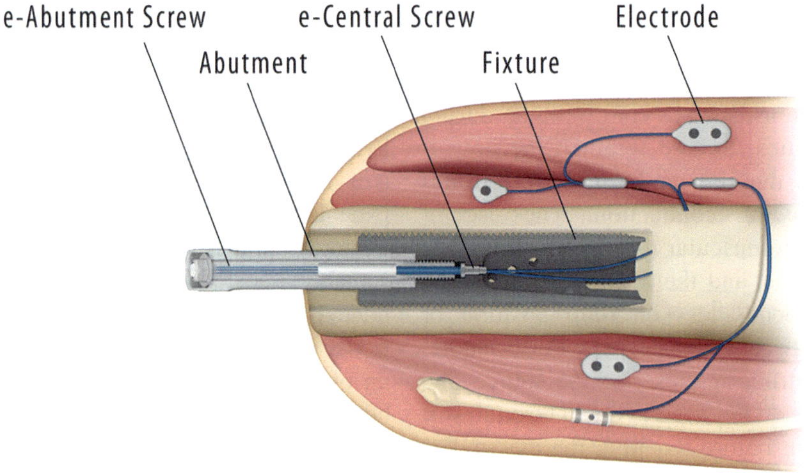

Fig. 6.6 Illustration of the osseointegrated human-machine gateway (OHMG) system [28]. Loads are transferred from the artificial limb to the abutment, from the abutment to the fixture, and then from the fixture to the bone. The e-abutment screw (eAS), which goes through the abutment to the fixture, is designed to keep the abutment in place. A parallel connector is embedded in the eAS's distal end (1) to electrically interface with the artificial limb. This feedthrough is electrically connected to an in-line connector embedded in the eAS's proximal end (2). The in-line connector interfaces with the corresponding in-line pin, extending from the central sealing component (3). From there, leads extend intramedullary to a connector unit located in the soft tissues, and neuromuscular electrodes are then mated to this latter connector. Reproduced with permission from Ortiz-Catalan M, Håkansson B, Brånemark R. An osseointegrated human-machine gateway for longterm sensory feedback and motor control of artificial limbs. Sci Transl Med. 2014

lation electronics are placed within the arm prosthesis [46]. This embedded system is capable of computing signals as well as processing and decoding algorithms [47, 48] to infer simultaneous movements from different degrees of freedom in real time [46]. The resulting system is self-contained, and no components over the skin are required. The patient needs only to couple the prosthetic arm to the abutment (which can be done single-handedly as shown in Video 1 of Ortiz-Catalan et al. [28]), and all electrical connections and functionality are made available automatically.

One crucial aspect of the clinical implementation of novel medical technologies is the regulatory framework in which they are developed and manufactured. The OHMG system was developed within the ISO 13485:2003 standards (Med Dev—Quality Management) and by following a variety of international and European standards, such as the 93/42/EEC (Med Dev), 90/385/EEC (Active Imp Med Dev), ISO 14708-1 (Active Imp Med Dev), ISO 14708-3 (Active Imp Med Dev, Neurostimulators), and IEC 60601-1 (Med Elect Equip). Relevant sections of these various standards were considered so that this system could be used outside the laboratory as a medical device.

We recently reported the use of a bone-anchored, self-contained robotic arm with both sensory and motor components over 3–7 years in four patients after transhumeral amputation. The implant allowed for bidirectional communication between a prosthetic hand and electrodes implanted in the nerves and muscles of the upper arm and was anchored to the humerus through osseointegration, the process in which bone cells attach to an artificial surface without formation of fibrous tissue. The use of the device did not require formal training and depended on the intuitive intent of the user to activate movement and sensory feedback from the prosthesis. Daily use resulted in increasing sensory acuity and effectiveness in work and other activities of daily life [49].

Further work is currently ongoing to develop an analogous OHMG system for transradial and transfemoral amputees.

Conclusion

The OPRA program's standardization of implant systems, surgical techniques, and postoperative rehabilitation protocols has resulted in better functionality, fewer complications, and a better quality of life for implant recipients. Moreover, the ability to integrate human-machine gateways into these osseointegrated systems can provide implant recipients with myoelectric control and sensory feedback. Together, osseointegrated implant systems and osseointegrated human-machine gateways represent a new frontier in amputee rehabilitation.

References

1. Kamrad I, Soderberg B, Orneholm H, Hagberg K. SwedeAmp-the Swedish Amputation and Prosthetics Registry: 8-year data on 5762 patients with lower limb amputation show sex differences in amputation level and in patient-reported outcome. Acta Orthop. 2020;91:464–70. https://doi.org/10.1080/17453674.2020.1756101.
2. Ephraim PL, Dillingham TR, Sector M, Pezzin LE, Mackenzie EJ. Epidemiology of limb loss and congenital limb deficiency: a review of the literature. Arch Phys Med Rehabil. 2003;84:747–61. http://www.ncbi.nlm.nih.gov/pubmed/12736892. Accessed 12 Feb 2018.
3. Sobaci G, Akýn T, Mutlu FM, Karagül S, Bayraktar MZ. Terror-related open-globe injuries: a 10-year review. Am J Ophthalmol. 2005;139:937–9. http://linkinghub.elsevier.com/retrieve/pii/S0002939404013728. Accessed 12 Feb 2018.
4. Millstein S, Bain D, Hunter GA. A review of employment patterns of industrial amputees—factors influencing rehabilitation. Prosthetics Orthot Int. 1985;9:69–78. http://www.ncbi.nlm.nih.gov/pubmed/4047922. Accessed 12 Feb 2018.
5. Reiber GE. Who is at risk of limb loss and what to do about it? J Rehabil Res Dev. 1994;31:357–62. http://www.ncbi.nlm.nih.gov/pubmed/7869284. Accessed 12 Feb 2018.
6. Lundberg M, Hagberg K, Bullington J. My prosthesis as a part of me: a qualitative analysis of living with an osseointegrated prosthetic limb. Prosthetics Orthot Int. 2011;35:207–14. http://journals.sagepub.com/doi/10.1177/0309364611409795. Accessed 12 Feb 2018.
7. Häggström E, Hagberg K, Rydevik B, Brånemark R. Vibrotactile evaluation: osseointegrated versus socket-suspended transfemoral prostheses. J Rehabil Res Dev. 2013;50:1423–34. http://www.rehab.

research.va.gov/jour/2013/5010/pdf/JRRD-2012-08-0135.pdf. Accessed 12 Feb 2018.
8. Rusaw D, Hagberg K, Nolan L, Ramstrand N. Can vibratory feedback be used to improve postural stability in persons with transtibial limb loss? J Rehabil Res Dev. 2012;49:1239–54. http://www.ncbi.nlm.nih.gov/pubmed/23341316. Accessed 12 Feb 2018.
9. Hagberg K, Brånemark R, Gunterberg B, Rydevik B. Osseointegrated trans-femoral amputation prostheses: prospective results of general and condition-specific quality of life in 18 patients at 2-year follow-up. Prosthetics Orthot Int. 2008;32:29–41. http://journals.sagepub.com/doi/10.1080/03093640701553922. Accessed 12 Feb 2018.
10. Zaid MB, O'Donnell RJ, Potter BK, Forsberg JA. Orthopaedic osseointegration: state of the art. J Am Acad Orthop Surg. 2019;27(22):e977–85. https://doi.org/10.5435/JAAOS-D-19-00016.
11. Brånemark PI, Hansson BO, Adell R, Breine U, Lindström J, Hallén O, et al. Osseointegrated implants in the treatment of the edentulous jaw. Experience from a 10-year period. Scand J Plast Reconstr Surg Suppl. 1977;16:1–132. http://www.ncbi.nlm.nih.gov/pubmed/356184. Accessed 12 Feb 2018.
12. Brånemark PI, Adell R, Breine U, Hansson BO, Lindström J, Ohlsson A. Intra-osseous anchorage of dental prostheses. I. Experimental studies. Scand J Plast Reconstr Surg. 1969;3:81–100. http://www.ncbi.nlm.nih.gov/pubmed/4924041. Accessed 12 Feb 2018.
13. Branemark R. 1960-R. A biomechanical study of osseointegration : in-vivo measurements in rat: rabbit: dog and man. Goteborg: Goteborg University; 1996. https://gupea.ub.gu.se/handle/2077/17036. Accessed 12 Feb 2018.
14. Brånemark R, Thomsen P. Biomechanical and morphological studies on osseointegration in immunological arthritis in rabbits. Scand J Plast Reconstr Surg Hand Surg. 1997;31:185–95. http://www.ncbi.nlm.nih.gov/pubmed/9299679. Accessed 12 Feb 2018.
15. Tjellström A, Lindström J, Hallén O, Albrektsson T, Brånemark PI. Osseointegrated titanium implants in the temporal bone. A clinical study on bone-anchored hearing aids. Am J Otol. 1981;2:304–10. http://www.ncbi.nlm.nih.gov/pubmed/6894824. Accessed 12 Feb 2018.
16. Hagberg K, Brånemark R. One hundred patients treated with osseointegrated transfemoral amputation prostheses—rehabilitation perspective. J Rehabil Res Dev. 2009;46:331–44. http://www.ncbi.nlm.nih.gov/pubmed/19675986. Accessed 12 Feb 2018.
17. Jönsson S, Caine-Winterberger K, Brånemark R. Osseointegration amputation prostheses on the upper limbs: methods, prosthetics and rehabilitation. Prosthet Orthot Int. 2011;35(2):190–200. https://doi.org/10.1177/0309364611409003. PMID: 21697201. http://www.ncbi.nlm.nih.gov/pubmed/21697201. Accessed 12 Feb 2018.
18. Brånemark R, Berlin O, Hagberg K, Bergh P, Gunterberg B, Rydevik B. A novel osseointegrated percutaneous prosthetic system for the treatment of patients with transfemoral amputation: a prospective study of 51 patients. Bone Joint J. 2014;96B:106–13. http://www.bjj.boneandjoint.org.uk/cgi/doi/10.1302/0301-620X.96B1.31905. Accessed 12 Feb 2018.
19. Hagberg K, Brånemark R, Hägg O. Questionnaire for persons with a transfemoral amputation (Q-TFA): initial validity and reliability of a new outcome measure. J Rehabil Res Dev. 2004;41:695–706. http://www.ncbi.nlm.nih.gov/pubmed/15558399. Accessed 12 Feb 2018.
20. Ware JE, Sherbourne CD. The MOS 36-item short-form health survey (SF-36). I. Conceptual framework and item selection. Med Care. 1992;30:473–83. http://www.ncbi.nlm.nih.gov/pubmed/1593914. Accessed 12 Feb 2018.
21. Ibáñez-Gimeno P, Galtés I, Jordana X, Malgosa A, Manyosa J. Biomechanics of forearm rotation: force and efficiency of pronator teres. PLoS One. 2014;9:e90319. http://dx.plos.org/10.1371/journal.pone.0090319. Accessed 12 Feb 2018.
22. Li Y, Kulbacka-Ortiz K, Caine-Winterberger K, Brånemark R. Thumb amputations treated with osseointegrated percutaneous prostheses with up to 25 years of follow-up. J Am Acad Orthop Surg Glob Res Rev. 2019;3(1):e097. https://doi.org/10.5435/JAAOSGlobal-D-18-00097. eCollection 2019 Jan.
23. Brånemark R, Emanuelsson L, Palmquist A, Thomsen P. Bone response to laser-induced micro- and nano-size titanium surface features. Nanomed Nanotechnol Biol Med. 2011;7:220–7.
24. Thesleff A, Branemark R, Hakansson B, Ortiz-Catalan M. Biomechanical characterisation of bone-anchored implant systems for amputation limb prostheses: a systematic review. Ann Biomed Eng. 2018;46(3):377–91. https://doi.org/10.1007/s10439-017-1976-4.
25. Hoellwarth JS, Tetsworth K, Rozbruch SR, Handal MB, Coughlan A, Al Muderis M. Osseointegration for amputees: current implants, techniques, and future directions. JBJS Rev. 2020;8(3):e0043. https://doi.org/10.2106/JBJS.RVW.19.00043.
26. McGough RL, Goodman MA, Randall RL, Forsberg JA, Potter BK, Lindsey B. The compress(R) transcutaneous implant for rehabilitation following limb amputation. Unfallchirurg. 2017;120(4):300–5. https://doi.org/10.1007/s00113-017-0339-9.
27. Ortiz-Catalan M. Towards natural control of artificial limbs: a novel osseointegrated human-machine gateway, neuromuscular electrodes, and pattern recognition. Gothenburg: Chalmers University of Technology; 2014.
28. Ortiz-Catalan M, Håkansson B, Brånemark R. An osseointegrated human-machine gateway for long-term sensory feedback and motor control of artificial limbs. Sci Transl Med. 2014;6:257re6.
29. Ortiz-Catalan M, Brånemark R, Håkansson B, Delbeke J. On the viability of implantable electrodes for the natural control of artificial limbs: review and discussion. Biomed Eng Online. 2012;11:33.

30. Kilgore KL, Peckham PH, Montague FW, Hart RL, Bryden AM, Keith MW, et al. An implanted upper extremity neuroprosthesis utilizing myoelectric control. Proceedings of the 2nd international IEEE EMBS, conference on neural engineering. Arlington, VA: IEEE; 2005. p. 368–71.
31. Kilgore KL, H a H, Bryden AM, Hart RL, Keith MW, Peckham PH. An implanted upper-extremity neuroprosthesis using myoelectric control. J Hand Surg Am. 2008;33:539–50.
32. Peckham PH, Keith MW, Kilgore KL, Grill JH, Wuolle KS, Thrope GB, et al. Efficacy of an implanted neuroprosthesis for restoring hand grasp in tetraplegia: a multicenter study. Arch Phys Med Rehabil. 2001;82:1380–8.
33. Kilgore KL, Peckham PH, Keith MW, Montague FW, Hart RL, Gazdik MM, et al. Durability of implanted electrodes and leads in an upper-limb neuroprosthesis. J Rehabil Res Dev. 2003;40:457–68.
34. Nashold BS, Goldner JL, Mullen JB, Bright DS. Long-term pain control by direct peripheral-nerve stimulation the of joint long-term by direct. J Bone Jt Surg Am. 1982;64:1–10.
35. De RD, Vanneste S, Engineer ND, Kilgard MP. Safety and efficacy of vagus nerve stimulation paired with tones for the treatment of tinnitus: a case series. Neuromodulation Technol Neural Interface 2013;17:170–9.
36. El TR, Raedt R, Mollet L, De Herdt V, Wyckhuys T, Wyckuys T, et al. A novel implantable vagus nerve stimulation system (ADNS-300) for combined stimulation and recording of the vagus nerve: pilot trial at Ghent University Hospital. Epilepsy Res 2010;92:231–9.
37. Ben-Menachem E, Rydenhag B, Silander H. Preliminary experience with a new system for vagus nerve stimulation for the treatment of refractory focal onset seizures. Epilepsy Behav. 2013;29:416–9.
38. Schwartz AR, Bennett ML, Smith PL, De Backer W, Hedner J, Boudewyns A, et al. Therapeutic electrical stimulation of the hypoglossal nerve in obstructive sleep apnea. Arch Otolaryngol Head Neck Surg. 2001;127:1216–23.
39. Veraart C, Raftopoulos C, Mortimer JT, Delbeke J, Pins D, Michaux G, et al. Visual sensations produced by optic nerve stimulation using an implanted self-sizing spiral cuff electrode. Brain Res. 1998;813:181–6.
40. Delbeke J. Electrodes and chronic optic nerve stimulation. Biocybern Biomed Eng. 2011;31:81–94.
41. Haugland M, Sinkjaer T. Cutaneous whole nerve recordings used for correction of footdrop in hemiplegic man. IEEE Trans Rehabil Eng. 1995;3:307–17.
42. Polasek KH, Hoyen HA, Keith MW, Kirsch RF, Tyler DJ. Stimulation stability and selectivity of chronically implanted multicontact nerve cuff electrodes in the human upper extremity. IEEE Trans Neural Syst Rehabil Eng. 2009;17:428–37.
43. Memberg WD, Polasek KH, Hart RL, Bryden AM, Kilgore KL, G a N, et al. An implanted neuroprosthesis for restoring arm and hand function in people with high level tetraplegia. Arch Phys Med Rehabil. 2014;95(6):1201–1211.e1.
44. Waters RL, McNeal DR, Faloon W, Clifford B, California D. Functional electrical stimulation of the peroneal nerve for hemiplegia. Long-term clinical follow-up. J Bone Jt Surg Am. 1985;67:792–3.
45. Ortiz-Catalan M, Mastinu E, Brånemark R, Håkansson B. Direct neural sensory feedback and control via osseointegration. Cape Town: XVI World Congress of the International Society for Prosthetics and Orthotics; 2017.
46. Mastinu E, Doguet P, Botquin Y, Hakansson B, Ortiz-Catalan M. Embedded system for prosthetic control using implanted neuromuscular interfaces accessed via an osseointegrated implant. IEEE Trans Biomed Circuits Syst. 2017;11:867–77.
47. Ortiz-Catalan M, Brånemark R. BioPatRec: a modular research platform for prosthetic control algorithms based on bioelectric pattern recognition. Hyderabad: XIV World Congress of the International Society for Prosthetics and Orthotics; 2013.
48. Ortiz-Catalan M, Håkansson B, Brånemark R. Real-time and simultaneous control of artificial limbs based on pattern recognition algorithms. IEEE Trans Neural Syst Rehabil Eng. 2014;22:756–64.
49. Ortiz-Catalan M, Mastinu E, Sassu P, Aszmann O, Branemark R. Self-contained neuromusculoskeletal arm prostheses. N Engl J Med. 2020;382(18):1732–8. https://doi.org/10.1056/NEJMoa1917537.

7 Outcome Measures

Peter J. Kyberd

The ability to measure the performance of the user, or their prosthetic limb, is key to being able to track the progress of an individual, the result of an intervention or the impact of a change in the design of the device. Properly designed, such tools can be useful both for treatment and in research. However, to ensure a meaningful result, the aim of a particular measurement must direct the form of the assessment.

Introduction

Historically, there was little consensus over what to measure or how to measure in upper limb replacement [1]. However, over the last 15 years, the emphasis on medical interventions based on objective measurement has swept across the entire medical field, (along with similar changes in attitudes in other professions). This is a good idea, but it is not a straightforward process to measure outcomes as different stakeholders have different perspectives on what is the purpose of a measurement. For example, the payers (e.g. insurance companies and states) will wish to have a record of what they see as the best value for money, the clinical providers are more concerned with getting the best prescription for the user and then ensuring each user gets the most out of the fitting, while the users often want to be able to get on with their lives, performing those tasks that are meaningful to them. This difference creates different needs and so different requirements from any measurement system. From a technical perspective, whatever aspect is of interest, a valid and objective measurement is important. This requires the tool to be designed to measure the factors of interest, and then the tool must be tested to ensure that its psychometric properties are adequate for the job. So it must actually measure what the designer thinks it measures. It must do so repeatedly with different observers and on different days. Also users should be able to know how sensitive the tool is to changes in the performance of the observed party [2, 3].

Once obtained, the information about a device or a person needs to be communicated in an objective and precise way. So it is necessary to ensure the language used is precisely defined. One systematic way to approach the problem is to define what needs to be measured using the WHO ICF model [4, 5] as a framework to study the different aspects of the problem [2]. The area is divided into three domains:

- **Function**: Which are the properties of the device itself: its speed, size, mass, grip force and range of things it can grip. These can be measured using simple, objective tools (rulers

P. J. Kyberd (✉)
School of Energy and Electronic Engineering, University of Portsmouth, Portsmouth, UK
e-mail: peter.kyberd@port.ac.uk

O. C. Aszmann, D. Farina (eds.), *Bionic Limb Reconstruction*,
https://doi.org/10.1007/978-3-030-60746-3_7

to measure size or hand gape, gauges to measure forces or ranges and stopwatches to measure speed).

- **Activity**: What the device is capable of doing with a competent user in control. This can be measured with simulated activities of daily living (ADLs). The motion of the subject can be measured by simple observation or more precise measurements.
- **Participation**: What the actual wearer uses the device for. Historically, this was learned through surveys, asking the person what they did with their devices. More recently, technology has allowed this to change to newer measures (see below).

Each domain has a different emphasis, and so the way that the areas are measured has to be different. No one system could possibly measure all of the aspects of the three domains. Any more than a single device could measure the temperature of a meal cooked by a chef *and* the satisfaction of the customer eating it. Some tests can straddle two, but they cannot cover the full range. Hence any tool is restricted to a particular domain, and any wish to cover the entire range of experience and prosthetic performance with a single measure is a forlorn hope.

The Measurement/Analysis Paradox

There is a trade-off between the complexity of the measurement and the level of information created (see Fig. 7.1). So simple measurements will give simple, straightforward unequivocal results, but their use and range is limited to statements such as 'this is better than that by this much, under these specific conditions' (e.g. *this hand opens faster than that one*). On the other end of the scale are highly complex measurements that could be sensitive enough to detect subtle differences in form on the overall function (e.g. change the length of a finger, or a change in the threshold of an EMG command switch), makes performing an operation easier or harder.

The scale also matches the transition from design to application or *Function* to *Participation* in the ICF model [5]. At one end something specific about the properties of a prosthesis can be measured; at the other, the requirement is to gauge how the device is used in the field by a wearer.

A complex measurement attempts to capture something subtle and meaningful about the item in question. However to get that sort of quality of information, the measurement will take longer: longer to set up, longer to run and longer to be conducted. The horizontal scale is the complexity of the measure and the ICF domains; indicated on this scale is the approximate position of different functional tests mentioned in the chapter. (c) Peter Kyberd

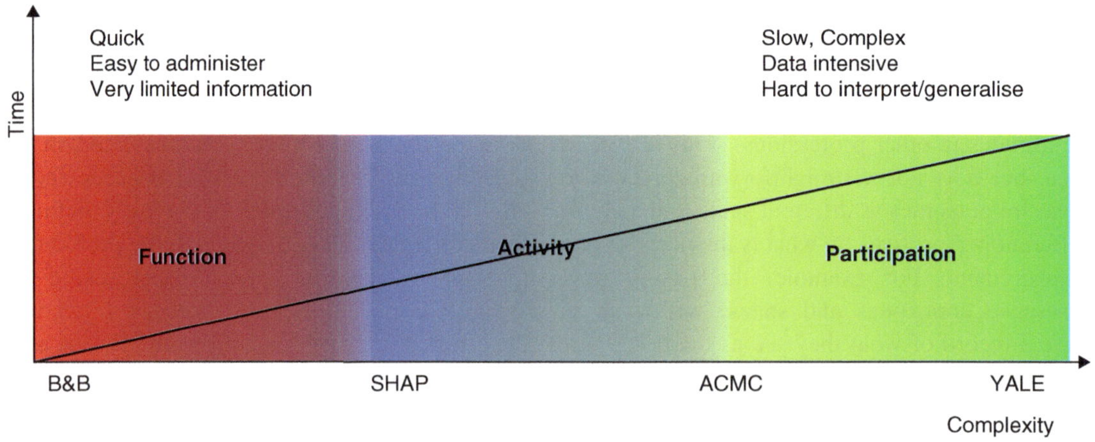

Fig. 7.1 The analysis paradox—There is a trade-off between complexity of the measurement and the level of information gained and the time taken to gather the information. It also maps well onto the WHO ICF domains. The vertical scale is the time taken for the assessment to

analyse. In contrast, a ruler makes a simple measurement of length. It could detect the widest gape of a hand and this *could* be related to the largest object the hand might be able to hold, but this would have to be demonstrated to be a valid measure of function. Otherwise it is merely implied and at the level of anecdote. The ruler is clearly quick and easy to administer, but the result only tells us one tiny thing about what the hand can do. Similarly, a stopwatch alone could be used to measure how fast the hand can open, but it tells us nothing about how useful that speed is when used in the field; this would require context, i.e. performing specific tasks (see SHAP).

The purpose of any test is to gain an idea of the level or quality of whatever is being measured. It is similar to written examinations, at the end of a period of study; the content of the exam should measure some underlying knowledge or ability. The content and form of the exam is aimed to get an indication of the person's level of knowledge. Performance at an exam should reflect performance in the subject, even if taking the exam is different to the real-world activity that is being measured. So a written exam for an engineer aims to test their ability to practice as an engineer. It is hoped that performance in the exam is an indication of performance at engineering. If it is possible to pass the exam when you are not good at the target subject, then the design of the exam is poor. So in the same way, any of the assessment tools need to be designed to measure something that reflects the aspect of interest and not something totally different.

Since measurements on humans can be different on different days due to matters that have nothing to do with the things being measured (the person's functional ability might be affected by being unwell, not because their skills have changed). The simpler the test, the more likely it will be the same for the same person under similar conditions. Unfortunately, if it is simple, it tells us less about a more general question. However, complex tests may not measure what we are interested and are more likely to be sensitive to other unrelated factors. Hence the design of the test has to be precisely configured to measure what we want to know, nothing more. A hand function test where the hand is used to perform activities clearly can tell us something about the use the hand is put to. The drawback is if the task is complex and can be done multiple ways, it cannot easily be employed to compare one user to another, one hand to another, and, even perhaps, one day to another. So the alternative is to control the test more rigidly and so more precisely and get a more limited picture.

As an example, if a test involves feeding oneself, a person can pick up a fork in at least one of two distinct ways: between the tips of the fingers or in the fist. What is needed is to know *why* they used one form over another. If that difference is because the prosthesis has limited range of motion and can perform only one grasp and a different more sophisticated hand design used the more common grasp, then this change in action from one hand is a subtle indicator of the performance of the prosthesis. It will tell us much about the capabilities of the two hands. However, if the difference in the way the fork is used is cultural, then the result will tell nothing about the prosthesis and much about the background of the subject. It is useless in this context.

If the task is made much simpler (such as picking up blocks and moving them over a barrier and dropping them in a container), then its relation to what happens in the real world is less clear, but it is much more controlled and so easier to be able to compare hands, users and days. This is why the tool design comes down to what one wants out of a measure.

Another aspect of the design is the level of difficulty. A test that is too easy to do will give everyone a high score and not differentiate the best from the merely good (this is the *ceiling effect*). The reverse, a hard to achieve test that results in many participants not scoring at all, has a *floor*.

Recently, the ideas about what needs to be measured and the technology to measure it have become more sophisticated. Earlier designs of tests used the basic tools of measurement, such as timers and structured activities with form boards and apparatus [6]. More recent innovations work in more unstructured environments using newer technology for monitoring movements (such as body motion recording [7], activity monitoring [8] or eye motion tracking [9]).

The design of the measurement tool does impact upon its use and utility. The aim for all observing participants would be a test that is easy and quick to administer and that also produced an unambiguous result quickly. For the technical and administrative observer, this would ideally be a single, easy-to-interpret number ('*he is a five with this hand and a twelve with that one, which makes him more than twice as functional*'). The therapist would more likely wish for an activity that enabled them to interpret easily the capabilities of the subject, ('*they can do all they need to look after themselves*'). Even then it would still need metrics that were easily shared between independent observers.

A good test relies on its psychometric properties, which need to be properly demonstrated. These properties are those needed of any measurement: that it measures what you think it does, that the same result means the same level of quality and that the same person measured by the same person on different days would have the same score. Similarly, if measured by *different* persons or on different *days*, they must get a very similar score. If not, then the test will not tell us anything useful. So if we consider the use of a ruler to measure length, the test is easy to use and gets the same value for a set length irrespective of the day or observer. When the property is poorly defined and hard to detect (such as intelligence, beauty, or grip function), then establishing the psychometric properties requires considerable work to establish if the properties are as expected. The way to do this is to perform many repeated measurements on large groups of people, over a long period of time. This statistically tests the findings [10]. Virginia Wright estimated this process is involved and can take as long as a decade [1]. There are some shortcuts (such as spreading the measurements across a wide range of test centres, so getting access to a wide range of persons and circumstances), but there is still a great deal of work involved in the process.

Standardising Tests

The Upper Limb Prosthetics Outcome Measures group was formed by professionals across the disciplines, and they considered all upper limb assessments they could find in the literature and tested them against a set of criteria on validity and usability. At the time a very few tests were already usable as fully validated, without more work to complete the process [2, 6, 11]. Since that time a range of new tests have been published. Some have clearly taken the findings of the group and built upon them [12]. Others have chosen to create their own tests [13–15]. Table 7.1 is the summary of tools that were reported to be used in upper limb prosthetics. It is modified and brought up to date from the original list in Hill [11]; it includes (were available) publications where the tool was *used with prostheses*, rather than the original publication or when used with a different patient group.

Since the ultimate aim of a prosthetic hand is that it will be used to perform tasks, this suggests that a test of functional capabilities should be based on its ability to perform representative tasks; however, these tasks need to be reproducible. The problem with the design of any reproducible task is that it could contain very many factors, i.e. bimanual/monomanual, short/long time, simple/complex actions and also how the object is held, if the grip changes, if operator stands in one place or moves about and so on. These can vary. The more controlled and defined a task becomes, the easier it is to get consistent results (the measurement/analysis paradox). However, the further the test is taken from a real situation, the less credible some clinicians find the results of the test.

An example of a simple test is time picking up a single object and moving it to a different place. The act itself requires a wide range of skills and abilities, from seeing the object, understanding the task and deciding how to tackle the problem, to being able to put the prosthesis in the right position to be able to achieve the task efficiently. During the task, the operator then needs to control the prosthesis to open at the right time and release the object at the appropriate moment,

Table 7.1 Tools that have been used for upper limb prosthesis assessment; references are to studies where they have been used rather than the publications that outline the tool, unless no such publication is available

ICF domain	Function	Box and Blocks [16] SHAP [17] Sollerman [18]
	Activity	ACMC [19] ARAT [20] (B)AM-ULA [21] Box and Blocks [16] CAPP-PSI [22] Jebsen [23] (M)CRT [24] OPUS [19] PUFI [25] SHAP [19] UFES [26] UNB Test [26] TAPES [27]
	Participation	CAPP-PSI CHQ [28] COPM [29] OPUS [19] PUFI [25] TAPES [27]

Modified and updated from [11]
ACMC assessment of capacity for myoelectric control, *ARAT* Action Research Arm Test, *(B)AM-ULA* (brief) activities measure for upper limb amputees, *CAPP-PSI* Child Amputee Prosthetics Project-Prosthesis Satisfaction Inventory, *COPM* Canadian occupational performance measure, *DASH* disability of the arm, shoulder, and hand measure, *Jebsen* Jebsen Standardized Test of Hard Function, *(M)RTC* (modified) clothespin relocation task, *OPUS* Orthotics and Prosthetics User Survey, *PUFI* Prosthetic Upper Extremity Functional Index, *SHAP* Southampton Hand Assessment Procedure, *Sollerman* Sollerman Hand Function Test, *TAPES* Trinity Amputation and Prosthetic Experience Scales, *UFES* upper extremity functional status, *UNB Test* University of New Brunswick Test of Prosthetic Function

without dropping it in mid transfer. This task is far more easily specified than the complex task of making a meal, but the activities within preparing the meal do map to repeated pick and place actions. What is important is if the skill to move the object and the skill to make the meal are similar, then the abstract object movement test will measure the subject's ability to also make a meal, so that if the subject is poorer at the simpler task, then the chances are that the more complex task will be achieved similarly poorly. For example, if the speed the person moves object is half as fast with a different prosthesis, then it is likely that the meal preparation will also be much more slowly executed. The speed of action of the simple task is very easy to measure. Some have raised concerns that timed tests do not measure anything meaningful. However, timed tests have been routinely used in many areas of assessment for a long time [30], and studies show that speed of action is related to ability, so that timing a hand task will give the observer some measure of performance [31].

Even so, the more complex the measured task, the more likely that the result may depend on something other than what is being measured. It may be related to a third factor, such as ability to switch through EMG commands. This might depend on the fit of the prosthesis, rather than the design of the switching method [32]. On the other hand, it may depend on something entirely unrelated, such as the ability to comprehend a complex prosthesis control format. This is not likely to simply reflect the intelligence of the user, but rather depend on the education or socio-economic background of the subject. It brings in a very wide range of influences that cannot easily be constrained or accounted for. Thus there are many coherent reasons to create simple tasks to allow measurement to be robust and straightforward.

One barrier to universal use is the different perspectives of the different disciplines of prosthetic provision (prosthetist, occupational therapist, rehab engineer). ULPOM took the field towards a consensus and recommended some of the simpler tests that were validated and did so for the particular domain they had validity for. This reflected the level of technology of the tools and the prostheses at the time; in recent times other tools have been developed in response to more complex designs of prosthesis (AM-ULA) [12] and advances in measurement tools [33] (see below).

Test Development

An indication of the problems associated with the design and testing of measures can be given by the study of some of the more popular tests. The first two are simple *Activity* tests, and they were developed from training equipment. Users were asked to perform the task once, or more often, and they were advised how to improve their approach or technique. The activity's potential for assessment was recognised, and the devices were modified to create standardised tools. These tests had merit, but also flaws in their original forms; examples of the modifications illustrate the process of creating a reliable tool.

One additional aspect of all of the tools is that they have users operate their prosthesis to perform them which gives the tester the chance to observe the operator under controlled conditions, so that even if the tool does not capture the totality of the prosthetic operation, an experienced observer can still see changes in the operation or compensation, which can feed back into the subjective assessment.

Box and Blocks: Small cubical wooden blocks (25 mm a side) are held in one of two shallow boxes (see Fig. 7.2), and the operator selects a block, picks it up, moves the block over a barrier along its midline and drops it into the second box. The number of blocks transferred in a set time is the metric [16, 34].

Flaws: Loose blocks are easy to pick up; if all of the blocks supplied are in the box during the measurement, then most blocks are very closely packed and so far harder to pick up. It becomes a challenge even to an unimpaired subject.

Modifications: Reducing the number of blocks to allow them to be easily selected. An alternative solution placed the blocks in standardised grid positions, and the subject picks them up in a fixed order [35]. This allows the motion of different subjects to be compared as they follow similar paths. That there are two different approaches shows that there is yet to be a standardised method.

Clothespin Relocation Task: The operator moves pegs/pins of a known resistance from a horizontal bar to a vertical one (or vice versa) in a set time (Fig. 7.3). This test requires flexion and extension as well as pro/supination of the hand/wrist, and so compensations are very obvious to both the operator and the observer. This was a training tool adapted by Stubblefield and colleagues [36].

Flaws: The rig has asymmetric bars so the vertical bar is only on one side; a left-handed operator has to reach across their body, while a right-handed one does not. The order and number of the pegs are uncontrolled as are the starting and ending locations of the pegs. There are colour-coded pegs of different spring resistances; different testers used different colours.

Fig. 7.2 Box and Blocks test set-up with small wooden blocks in one compartment that need to be moved by the prosthetic user to the other compartment within a given time (usually 1 minute). © Anna Bösendorfer and Agnes Sturma

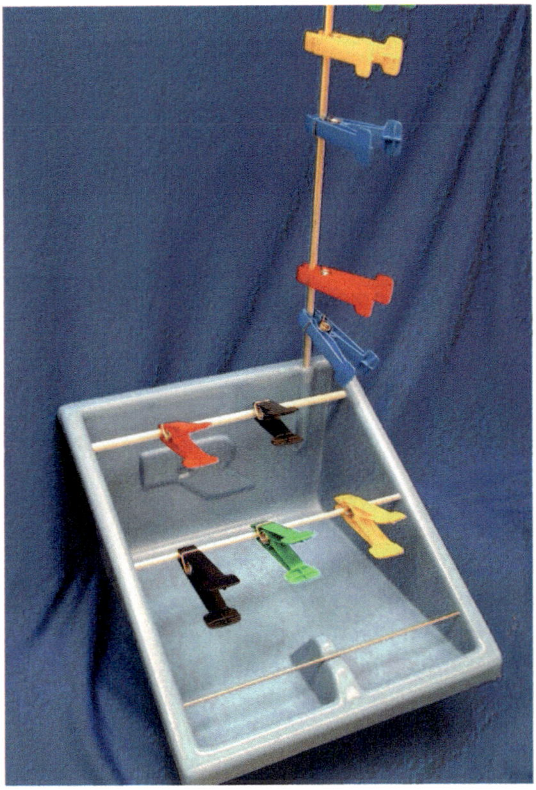

Fig. 7.3 The clothespin relocation task—This training tool has been turned into an effective assessment device. It makes the compensations that the user makes clear to both subject and assessor. (c) Peter Kyberd

Modifications: Standardised through modification to two vertical bars and peg order, plus the use of the same resistance pegs [24, 37].

The process for both these tools to be developed was that occupational therapists chose a tool they knew and trusted and created a useful measure. However, they both required modifications to further standardise them. Once the variability of both tasks was reduced, the validity increased.

Southampton Hand Assessment Procedure (SHAP)

This was developed at a time when there were no validated standardised tests for adult upper limb prosthetics users. One of the few other tests for prostheses that existed was one developed in Canada (the UNB test [38]). The UNB test was not a useful test for assessing adults, as it had been carefully designed to be used by children of different age groups. The UNB test had separate tests for each age group as their cognitive and physical development would influence their performance. The UNB is an observation test; a level of familiarity and practice with an activity is necessary; otherwise, the tool will test the subject's ability to learn, not their ability to perform the task. The UNB subtests were aimed to use recognisable tasks that the target group could easily understand and perform at their best. These were specific to North American youth in the 1980s, and so they included activities such as baseball, which could be assumed to be universal among some socio-economic and cultural groups not for European children.

This example showed that the design of SHAP had to be undertaken with care. The tasks were chosen to be usable by adults with a general background in the sort of devices and activities that could be assumed to be universal in Europe. It used a combination of standardised abstract objects to encourage the use of standard grip forms, plus a set of simulated activities of daily living (ADLs) that the literature suggested were representative and repeatable (see Fig. 7.4) [39]. The score was related to the time to execute the task but was weighted to reflect the frequency that the grip appropriate to the task was used in daily life. Timing of task completion was chosen as it was easily achieved, and other work suggested the score would reflect the capabilities of the subject. The subject initiated and finished the timing, as self-timing was also seen to be the most reliable way to time activities [40].

The area for the activities was limited to a form board and a central timer. The overall score was out of 100. There were secondary scores out of 100 related to the different grip patterns. The tool was validated for various psychometric properties, and its appeal was the simplicity of its execution and scoring. It has been tested on different conditions and age groups [17, 41]. However, it has not enjoyed universal acceptance. The aim of the measurement was to assess the person (or hand) under investigation. Thus most

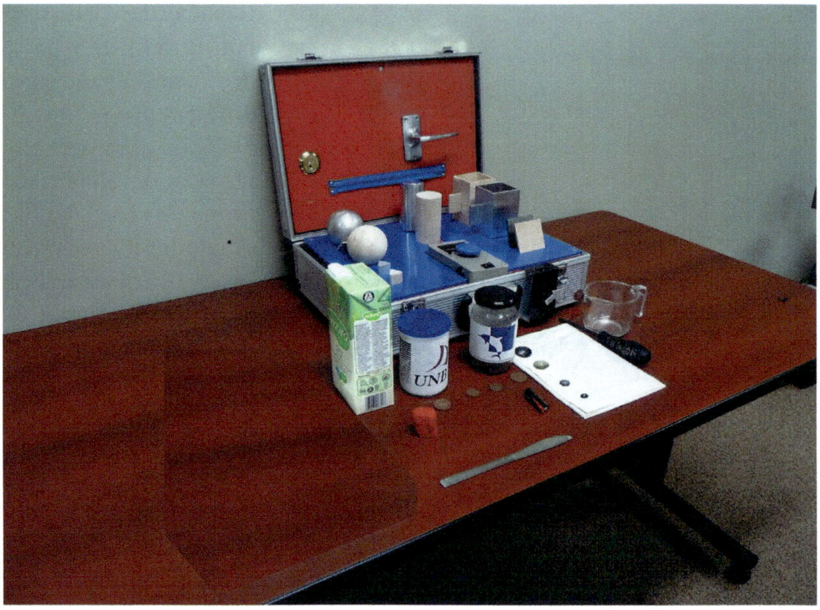

Fig. 7.4 The Southampton Hand Assessment Procedure (SHAP)—The complete set of tests based on activities of daily living are supplied in a single case. The test takes about 20 min to complete and provides a single number that measures the functional range for the hand. (c) Peter Kyberd

of the tasks were monomanual; so with a prosthesis under test, the device was used as the *dominant* hand (when more commonly, the prosthesis is used as a helper). Critics suggested that it is not measuring the real circumstance, forcing the subject to perform a task in a way they do not do habitually. This is a misinterpretation of SHAP's role. SHAP was aimed to work in the Function/Activity domains. Thus it is used to measure the performance of the user with the prosthesis. It does not measure what the user *does* with their hand; this is outside its scope in the *Participation* domain.

AM-ULA

For comparison is the more recently developed Activities Measure for Upper Limb Amputees (AM-ULA) [12]. This was also designed to capture performance in the Activity domain. It too cannot capture all aspects of living with a prosthesis, but the tasks chosen were bimanual and the subjects are allowed to use the hand of choice. It was specifically designed following the development of multifunction hands and arms, when other tools were either aimed at natural hands and were too complex for prosthetic assessment, or there were only simpler tests for simpler prostheses (such as Box and Blocks). AM-ULA followed on from the work of the ULPOM carefully working within the philosophy of the group to produce a valid test with easily understood results.

The scoring runs from 0 to 40 with the higher value relating to higher performance. The score considers the extent of completion of subtasks, the speed of completion, movement quality, skill of prosthetic use and independence. Scoring is based on a combination of observation and timing. The 18 tasks chosen for the test were based on those that were shown to have some validity (interrater and test-rest ICC > 0.5). Its validity was compared with other existing tests establishing the convergent validity as well. Having been developed it has been used in a number of tests particularly with the US Army's DEKA arm program [12, 26]. More recently the team has worked on making it quicker to deliver while retaining its psychometric properties; hence, there is now the *Brief* AM-ULA (BAM-ULA) [21].

Both of these last two tools have been used in programs to study prosthesis use where they are not the single measurement, but one of a set of tests to gather data on all three domains [26, 42].

Observational Analysis

Underlying much of subjective assessment of users is the clinical judgement. Based, at least in part, on the observation of the user in action. Thus the desire to quantify this is a reasonable response. However this is a complex task; it requires careful design to create a test with good psychometric properties. Liselotte Hermansson's *Assessment of Capacity for Myoelectric Control* (ACMC) is the most well controlled of all such tools [43].

Assessment of Capacity for Myoelectric Control (ACMC)

The aim of the ACMA was to capture the very wide range of skills that could be shown by users of myoelectric hands. The basic idea of any such test is to pick task or tasks that the subject is familiar with and ask them to perform the task while the clinician observes. The flaw with other tests is that if the test was based on an activity that was too hard, few would be able to complete it. So few would get any credit for their skills. If it was set too easy, then all bar a few would get the maximum rating (*ceiling* and *floor* effects). The ACMC was based on many years of experience of observing users and is a controlled form of observation. Familiarity with a task means the person is as good and fluid in the execution. So an assessment based on that task is likely to be the best and fairest test of the subject. An ACMC test is performed by a trained user observing the subject undertaking a task they are used to performing, and it is chosen in consultation with the subject. The test is conducted in circumstances that will trigger as many behaviours that are routine to the user as practical. So someone used to making a cake can do so easily, but doing so in a strange kitchen will force the subject to search for the right tools, rather than the job being performed entirely automatically. The result is that the subjects will require creative and imaginative solutions to solve the problems set to them. This will show off the capabilities and limitations of the subject.

For the assessment, the observer then looks for the recognisable actions that reveal how confident and capable the subject is at using their myoelectric command channel. The observer will watch out for bimanual use of the hand. They would also observe confident use of the EMG input. Thus if the user reaches with the prosthesis and grasps the object directly, they have great capacity. If they pick the object up with their sound hand and place it in the prosthesis' grasp, they are less confident in their ability to open and close their hand. Similarly, do they carry objects across the room with the prosthesis, or just their sound hand? Do they only use the prosthesis for pushing or clamping? The observer is able to combine the observations in such a way that the result can give a meaningful rating of the ability of the person to use their myoelectric hand. This tool has been carefully designed and rigorously tested so that variability between trained raters and measurements is small [44]. The test's biggest disadvantage is that the raters need to be experienced therapists who are trained to undertake the assessment, and not every team has the benefit of such a person.

ACMC is designed to encourage careful observation of the users. Its philosophy is precisely targeted at this group of users; what it is *not* designed for is to make any statements about users of any other sort of prosthetic limb, be it body powered or externally powered, and switch controlled, and the ACMC should *not* be used to measure these devices. However, its observational techniques are useful and would be beneficial training for anyone involved in prosthetic research, development or application. Training to conduct the ACMA can simply make an observer's ability to look at prosthetic use a little closer to that of a person of considerable experience (such as Dr. Hermansson). It is a gap in the range of tools that the therapist has to use that there is not yet a body-powered equivalent of the ACMC, although plans are afoot to develop this too.

Motion Tracking

Conventional motion tracking is based on recording the three-dimensional motion of retroreflective markers placed on the landmarks of the body. This data is then combined to create a three-dimensional model of the objects being tracked. This was originally developed for walking gait [45]. Its application in this context was easier to achieve, as walking is a stereotypical cyclic action that is repeated and varies only little between people and between different walking speeds. The traces of a person's motions or forces can be scaled and temporally adjusted to allow the comparison between subjects, and so deviations from population means can be easily discerned. In addition, it is possible to infer the forces generated within the musculoskeletal system from the forces at the foot, as measured using force plates in the ground. Forces at the foot translate to those at the ankle and then up to the knee, hip and so on.

Once the technology had developed sufficiently to detect the smaller motions associated with arm and hand motion, the idea of measuring upper limb motion became practical. The problem for achieving something similar with upper limb motion tracking was that the arms do not have a single *functional* stereotypic motion, and it is much tougher to register forces onto the body that can be used to infer internal muscular forces.

The measurement of motion can be analysed for clinical insight in a number of different ways. The common approach for walking is that the analysis attempts to see how close the motion is to that of the unimpaired. In addition to this, they also look to see how symmetrical the motion is. The assumption is that this is the optimal way to approach the problem: *Someone who walks symmetrically and like an unimpaired walker will use less energy and have fewer side effects than a person with differences and asymmetries.* This seems reasonable, but only the energy consumption can be measured entirely objectively. Attempts to design an experiment to *prospectively* measure asymmetry and the musculoskeletal damage that might result would be ethically impossible. Any retrospective study would require very large numbers to generate any significant conclusion. Moreover there is not even consensus within the profession as to what are the right measures of symmetry to base decisions on. See Pryor [46] for a thorough and entertaining review of the literature.

Upper limb motion is far less constrained than gait. However some rules can be applied to analysis of upper limb motion. Overuse injuries result from the use of increased ranges of motion, increased forces or increased repetitions of a specific motion of a joint [47]. In a person with a limb impairment, these criteria apply to the joints of the rest of the body as much as they do to the shortened limb. Any joint can be used to compensate for the limited abilities of the limb, or the prosthesis. Studies of the motions of subjects performing unimanual and bimanual tasks show that these compensations occur throughout the body [48]. The subject will shift their position so that the limited reach of the effected limb and prosthesis can get then hand to the target object and place it in the correct orientation to grasp the target. For example, a person with a loss below mid humerus might use scapular abduction to raise the elbow sideways and substitute for lack of pro/supination of the wrist. The particular task chosen for the assessment can make this more or less apparent. The clothes peg/pin relocation task requires the peg to be rotated in the hand from horizontal to vertical and back. With this motion any compensations become very apparent to operator *and* tester alike.

The choice of which method a user employs to put the hand in the correct position can be a very telling measure of the effectiveness of the control strategy. So even if the user has been given an electric wrist rotator, they might still choose to use humeral abduction of their remnant limb as it is easier and quicker to use, over mode switching in a myoelectric controller [17]. Potentially, it could be that the change in biomechanics reveals the change in the ease of control. So if a new pattern recognition (PR) system was easier to use, then the wearer might stop abducting the elbow and electronically pro/supinate with the PR system, which would be very clear.

As the tools have become more sophisticated, it is now possible to measure a much wider set of variables associated with the use of a prosthesis. The problem now becomes to select the right parameter, namely, selecting the parameter that has real bearing on the question one is attempting to answer. One factor not previously studied is the spread in the variation of the speeds of action of a prosthetic hand, i.e. how often the hand is opened quickly or slowly, or anywhere or in between. Studies by the author show that when performing a range of different manual tasks, the spread of speeds the joints move at is wide, from slow to fast depending on the action. By contrast, the original myoelectric hands operated at a single speed; thus, there was no spread of velocities; it was either moving at that speed or it was stopped. It might have been thought that this result might have changed with the proportional myoelectric control of modern hands, but with the current controllers there is still little spread, the hands being operated mostly as fast as they can, or stopped. There is some evidence that this might be different for pattern recognition-based control.

If a task is very well controlled, then people will tend to move their hands in the same ways as in walking, and so analysis of upper limb motion similar to lower limb motion is possible. Light et al. used the form board-controlled motions of SHAP to create similar motion patterns in subjects [7, 49, 50] and produced 'upper limb gait analysis'. Similarly Schmidt also proposed constraining the activity to create easier to interpret cyclic motions [51]. Other groups have chosen not to attempt to control the motions of the arms to facilitate easy comparison of upper limb motion, although the technology exists there are limited studies that use gait-like motion tracking to analyse upper limb motion.

The major factor that mitigates against any such studies is that it requires a considerable investment in time to set up a motion tracking system and even greater amount of time to process and analyse the data, so that upper limb motion tracking for clinical diagnosis remains more than 20 years behind lower limb motion analysis. An alternative approach was taken by Hussaini et al. who used baseline data from motion tracking of users of the clothes peg test to create an observational test aimed at the clinical setting [24]. The judgements are based on underlying experience from the motion tracking, but the scoring is only from a simple unmarked test, making the process of measurement much cheaper and quicker.

Recent Trends in Outcome Assessment

Pattern Recognition: With the rise of complex multifunction prosthetic hands and the reduction in the size and power consumption of computers, the long awaited clinical use of Pattern Recognition (PR) of EMG signals has begun. The earliest assessments of the technique came from the theoretical fields and used recognition rates and error rates to measure performance when selecting a mode (such as a grip) [52]. A rate of 95% would have been considered exceptional in this theoretical circumstance. Researchers in any other area where PR was studied would have considered the work a complete success and moved on to new areas of study. However, as it represents a 1 in 20 failure, it was clearly not nearly good enough for *practical* prosthetic applications.

At this time theoretical trials based on the idea of achieving a target were employed. Fitts' law seeks to measure the way a person achieves a target and has the potential to be a more practical approach [53]. It studies the time a subject takes to get to a target, or the way an operator might fail to get to the target. This is a measure based on an idea of usable performance. For example, it might be possible to get to a target every time, but if it always takes a long time to do so, then the controller is not going to be used in real prosthetics. If when one tries to get to a target quickly one overshoots and so has to repeatedly come back to the target in the opposite direction, the person might instead choose to move more slowly and so not to overshoot. Fitts' law testing not only measures the time but the quality of the movement to the target, so the number of overshoots will be factored in as well as the speed to the target. Fitts' law has been used in many fields to measure target acquisition, from computer mouse use to real weapons training.

Ultimately for a practical subject such as prosthetics, the actual use of the device is more relevant, so that Pattern Recognition tests are now used in conjunction with manual tasks. One chosen was the clothes peg relocation test (see above). This reflects its movement from a laboratory-based idea to a clinical solution. When used with PR, it originally was a computer-based simulation of the test. In this form it was used simply to measure how fast a user can achieve different control modes. When moved to the real world, it revealed when the user cheated and applied humeral abduction and adduction [24] rather than prosthetic joint movement, but it can also show if the controller was disturbed by shifts in the socket location. When the arm is raised to place the peg at the top of the bar, or to remove the peg to bring it down, the prosthesis may shift on the arm/shoulder of the operator and if so the EMG signals *will* change. The question is: Will the performance of the controller be affected? If the controller program is not robust enough to be useful in the real world, the settings will change and the control of the arm will be affected detrimentally; perhaps they will not be able to let go of the peg. If the controller *can* compensate for these changes, then it might be ready for genuine clinical application.

Cognitive Load: One area that promises to reveal considerable information about the way people use their limbs is through the measurement of the cognitive load. Cognitive load is how much concentration an operator needs to apply to control their limb. We achieve the level of dexterity we have with our bodies, in part at least, from the amount of practice we devote to learning to use our bodies as we grow. It is for this reason that we can conduct most routine operations without seeming to think about what we are doing. It is only when we try and learn a new skill, or are injured, we discover quite how hard many actions are to do and how long it takes to adapt. Routine practiced actions with our dominant hand take very little effort to perform, but the same task with the other hand requires effort and concentration. When we perform activities, we use a range of our senses without realising it; using a prosthesis generally employs only some of the senses and other senses in different ways. If it is always hard to control a prosthetic limb, it will always need a greater amount of cognitive effort to use. One consequence of the greater effort is that it will be less likely that the arm will be used routinely. So measuring the cognitive load needed when a user controls their prosthetic limb is something that may tell us about how good a particular control strategy or limb design is, compared with the conventional systems. It has potential to unlock the question of how effective a new prosthesis is.

Early attempts to measure load use the *dual task paradigm* [54, 55]. These attempts were unsuccessful because the tools were either inappropriate or unrepeatable. The tests employed tools that were based on memory or specific cognitive abilities, which were not universal across a population. Any system that variable cannot easily produce results standard enough to be useful in comparisons across a population [3].

For example, the basic concept for a dual task test is that a person only has a limited ability to process all the information needed to perform a task. If they perform two tasks separately, they will be able to achieve a certain level of competence in each. If they then perform the two tasks together and the amount of combined mental effort is more than they are capable of expending, the combined performance is less than the sum of both tasks alone, as both will be completed more slowly or less efficiently as the subjects divide their attention. This means that the impact of one task on the other becomes measureable, by observing a drop in performance. This paradigm fundamentally depends on the assumption that there are finite metal resources to process the information and act on it and the resources are similar across the population.

A simple example of its use is if the task is walking in different footwear. The first shoes are flat and the second are high heels. Without practice walking in heels is harder, so one would expect that the habitual users of heels would be less affected by using them than strangers to the task.

An experiment to measure this would be to have the subjects walk a fixed length and time. Then a second task is chosen to occupy their minds (counting backwards from 100 in sevens). This second task is performed for a fixed length of time, and the number they get down to is recorded. They are then asked to walk and count *at the same time*, and both the time to walk and the number they get to are recorded. Both should be lower than the solo runs. However, some subjects could concentrate on the counting and get to the same number and just walk more slowly. So the eventual score has to be derived from both numbers, i.e. the impact of the primary on the secondary and vice versa. If now the flat shoes are swapped with heels, and the solo footwear test and the combined runs are performed again, then a new score will be derived, and an idea of how much harder it is for the two groups, wearers and non-wearers, will be apparent. The impact on the scores of the habitual wearers will be small, while that of the non-wearers would be substantial. Two things will be measured; how much harder it is to walk in heels and how effective practice makes on wearing heels.

For a prosthetic task: One task is using a prosthetic hand (primary task) to perform some actions, and the secondary task is chosen that occupies the subject. Then repeating the test using a different (easier) prosthetic controller design would result in less impact on the performance. So if it was half as easy to do the task with the new controller, then the scores would go up as the subjects would more easily do each task. The difference measured in the second circumstance would be lower, and this would indicate this reduced impact with a better score. A lower score indicates an easier more subconscious controller.

Standard psychology literature uses visual displays for distracting tasks, such as spotting words in a list that do not match. Clearly this is not practical if the task being tested is employing a hand to perform grasping tasks. Alternatively, psychologists commonly use a mental arithmetic test as suggested above [56, 57]. However, the level of difficulty of these tasks is important. If it is too easy, it will not affect the score and will be of no use. The alternative of it being too hard to complete also makes it useless. The level of difficulty depends on the individual subject, and for a universal test, this is a drawback. Of even greater concern is that for psychology experiments the population of subjects are often undergraduate students and so are quite uniform in their abilities and backgrounds. They are far more likely to be comfortable with simple arithmetic than the general population. When the subjects are prosthesis users, this is a much broader population; some users might even be intimidated by mental arithmetic. Such subjects will either be unable to complete the task or they will withdraw in fear of being embarrassed by their inability. So despite these methods being well used and understood in theoretical fields, they cannot be readily transferred to prosthetics.

An alternative way to measure the cognitive load is the use changes in the brainwave signals when presented with something that does not make sense. This is the timing of *evoked potentials* during the task execution change when presented with phrases that make sense (*the boy runs the block*) compared with a phrase that makes grammatical sense but is nonsense (*the boy runs the apple*). The time the brain needs to process the information increases if it is nonsense. This difference can be seen in the recorded brainwave signals. An example of its use in prosthetics is in Deeny [58]. The disadvantage of this method is that it is far more invasive; it uses electrodes on the skull of the subject to record the signals. This limits its use to the more engaged volunteer with more time to give to the experiment.

A third approach is to use visual attention as a surrogate for cognitive load. The gaze of the subjects is monitored while they perform tasks with the prosthesis. The essential idea is that if they look in the same way at the same time as the unimpaired population, then the hand is easy to control. If they are distracted and they need to look elsewhere when they are performing manual tasks, then the prosthesis is less easy to control. This test has the advantage that the person's visual attention is very much part of the real-world operation of a prosthesis, so the results are more likely to reflect what happens outside the lab.

Results suggest that prostheses are not used in any way that is similar to the natural limb. Unimpaired users tend not to look at the task, but ahead of the hand, where it will be next in the task sequence. Such as the object that will be picked up as they move the hand to the object, and as soon as the hand makes contact, they gaze jumps to where the object is going to be moved to. What they don't do is look at the hand all the time the hand is moving to the object. Studies show that the prosthesis users look more at the hand than at the target [8, 59, 60]; they follow the hand and do not look at where it is going.

When the task needs a focus, such as pouring from a jug, then attention for the subject needs to be on the water being poured; the unimpaired person would look at the spout and the water. The prosthesis user needs to switch attention between the place where they are holding the jug to make sure the grip is secure and the stream of water. These two behaviours are clearly seen. In the future the question this tool might be able to answer is the impact of feedback (either extra artificial or through a body-powered harnessing) on how and where users look at a task.

Eye tracking is commonly used in a range of fields from advertising to driver distraction; the technology has advanced so that the cameras that record the activity are very small and the resulting system resembles cool-looking spectacles. One camera (the scene camera) records the view the subject sees, and a second camera looks at the eye of the subject and a calibrated computer infers where the person is looking. Studies from Sobuh [60] use only a two-dimensional set-up where the point of regard (where the person is looking) is mapped onto the two-dimensional image from the camera. Analysis of the data is slow, as the standard way to analyse such data is for an experienced observer to segment the data, by stepping through every frame of the data determining where the person is looking and recording it manually. Progressing beyond this will require the computer to recognise what the person is looking *at*, a difficult task that has occupied many computer engineers for decades. Popa [61] adopted an intermediate stage by using a recognisable set-up (a SHAP task) to limit what was being analysed. Only the simplest analysis was required (the colour of the object and its position within the frame) to identify what was being observed. This represents only a stage in the evolution of the technology. Herbert et al. have a binocular system that allows them to reproduce the focus of attention in three dimensions, so in a calibrated workspace they can more simply determine what the subject is looking at automatically [62, 63].

Future Trends: The Participation Domain

As the measurement systems became more compact and ubiquitous, it is becoming easier to objectively measure matters that fall within the Participation domain so we do not have to rely on questionnaire alone. The production of compact activity monitors for the personal fitness industry has created an opportunity to directly measure the activity of the limb a prosthesis is on, or contralateral limb [8]. Results are providing insights into the way people use their limbs.

The activity monitors show tendencies that might not have been expected. A simple prediction might be that the prosthesis would be hardly used at all, being only employed as the non-dominant hand. In bilateral measurement of activity, the arm carrying the prosthesis had levels of activity closer to the natural than might have been expected [8]. What was not clear from this set-up was if the results were due to 'balanced' activities, such as swinging one's arms while walking, rather than more active bimanual tasks, where the prosthesis was used in a dominant manner. This information can only be achieved by monitoring the use of the prosthesis very closely, using other tools.

Chadwell also undertook other measurements of Activity and Function with one of the largest cohorts of users to be included in a study such as this [8, 32]. What was perhaps the most significant result from a professional standpoint is that there was little support that the lab-based functional scores relate to prosthesis use in the field. This would suggest that measurements in the Participation domain need to be made and these measures need to capture more detailed information about real prosthesis use.

One promising development in this direction is that it is possible to wear a camera that is small enough to not be a burden and has the storage capacity to make the study potentially representative of true Participation. However, one factor that will continue to raise concerns is subject confidentiality. A decade or more ago, the idea of carrying a camera to record all one's actions would have been considered impractical, fantastical and an intolerable invasion of one's privacy. With the rise of cameras for self-filming sports, activities or one's driving, this means that people are happy to wear cameras and are very ready to share their results. Even so the cameras have the potential to invade subjects' privacy; as a profession we will need to be sure to find suitable solutions to this problem.

Aaron Dollar and colleagues at Yale have pioneered this research for prosthetics. The early work has come from studying the way that intact subjects perform complex tasks [64]. For this reason they have classified each action and the subtlety of every grasp. This has required a complex taxonomy of grasps, with passive use, single-digit and multi-digit grasps [65]. The analysis of which is a very long process. It is very similar to eye tracking analysis in that it needs an operator to segment the data up into individual actions and grasps. The disadvantage of this is that they will need a considerable amount of data to be recorded and analysed in order to be able to say anything definitive about routine grasping or how a particular group of users or prostheses designs impact on the way functional everyday handling is performed. The questions they will be able to answer are matters such as how much the prostheses are used at all and how much of this is passive pushing and pulling, even with allegedly 'active' mechanisms. At the time of writing, the results of the analysis are not yet available.

A trend in the last few years to start to try to understand the underlying processes of cognitive function and cognitive ability needed to use a prosthesis. This welcome trend stops treating the users as all the same and brings research to the sort of individualised treatment of the user that the clinic has always sought to achieve.

Users come from a wide range of social groups, but one group that is over-represented within the acquired loss population are young men. These are the individuals who are more likely (but not exclusively) to be undertaking more risky and physical jobs or activities (such as soldier, labourer or motorbike rider). Many will be suffering from multiple traumas that might impair their cognitive function in a way that impacts on their use of prosthetic limbs. This is one area that is beginning to get greater interest from the profession, and we can expect better information on it in the coming years. This, of course, requires a totally different set of measurement tools. The questions concerning psychology and the testing of this, area are beyond the scope of this chapter.

Conclusion

The choice of the measurement tool and the items to be measured is a matter that needs to be carefully considered. The expectation that one tool will cover all aspects of a prosthesis fitting must be resisted. The choice of tool should be made with care and attention to the needs of the user and the assessment. There are three domains of information, and each area requires different sorts of tools designed to measure this information. Simple tools which produce simple answers are easier to conduct and interpret, but they do not measure the totality of a prosthesis fitting. Recent years have seen a trend to increasingly sophisticated measurements being designed and used, and the new personal technologies of activity monitoring and self-filming are opening doors to a new realm of measurement which has the potential to transform our understanding of prosthesis use in the next few years.

References

1. Wright V. Measurement of functional outcome with individuals who use upper extremity prosthetic devices: current and future directions. Journal of Prosthetics and Orthotics. 2006;18(2):46–56.
2. Hill W, Stavdahl Ø, Hermansson LN, Kyberd P, Swanson S, Hubbard S. Upper limb prosthetic outcome measures (ULPOM): a working group and their findings. J Prosthet Orthot. 2009;21(4S):P69–82.
3. Kyberd PJ, Hill W, Hermansson LN, Hubbard S, Zinck A, Jones B, Murgia. A introduction to assessment. In: Parenti-Castelli, Troncossi M, editors. Grasping the future: advances in powered upper limb prosthetics V. Sharjah: Bentham Science Publishers; 2012.
4. Metcalf C, Adams J, Burridge J, Chappell P. A review of clinical upper limb assessments within the framework of the WHO ICF. Musculoskeletal Care. 2007;5:160–73.
5. World Health Organization. Towards a common language for functionary, disability and health: ICF Beginner's guide, (WHO/EIP/GPE/CAS/01.3). Geneva: WHO; 2002.
6. Wright V. Prosthetic outcome measures for use with upper limb amputees: a systematic review of the peer-reviewed literature, 1970 to 2009. J Prosthet Orthot. 2009;21:4S.
7. Murgia A, Kyberd PJ, Barnhill T. The use of kinematic and parametric information to highlight lack of movement and compensation in the upper extremities during activities of daily living. Gait Posture. 2010;31(3):300–6.
8. Chadwell A. The reality of myoelectric prostheses: how do EMG skill, unpredictability of prosthesis response, and delays impact on user functionality and everyday prosthesis use? Ph.D. thesis. Salford: University of Salford; 2018.
9. Sobuh M. Visuomotor behaviours during functional task performance with a myoelectric prosthesis. Ph.D. thesis. School of Health Sciences, University of Salford: Salford; 2012.
10. Streinier DL, Health NGR. Measurement scales (A practical guide to their development and use). 3rd ed. Oxford: Oxford University Press; 2003.
11. Hill W, Stavdahl Ø, Hermansson LN, Kyberd P, Swanson S, Hubbard S. Functional outcomes in the WHO-ICF model: establishment of the upper limb prosthetic outcome measures group. J Prosthet Orthot. 2009;21(2):115–9.
12. Resnik L, Adams L, Borgia M, Delikat J, Disla R, Ebner C, Smurr Walters L. Development and evaluation of the activities measure for upper limb amputees. Arch Phys Med Rehabil. 2013;94:488–94.
13. Carey SL, Highsmith MJ, Maitland ME, Dubey RV. Compensatory movements of transradial prosthesis users during common tasks. Clin Biomech. 2008;23:1128–35.
14. Thies SB, Tresadern PA, Kenney LP, Smith J, Howard D, Goulermas JY, Smith C, Rigby J. Movement variability in stroke patients and controls performing two upper limb functional tasks: a new assessment methodology. J Neuroeng Rehabil. 2009;6:2. https://doi.org/10.1186/1743-0003-6-2.
15. Valevicius AM, Boser QA, Lavoie EB, Murgatroyd GS, Pilarski PM, Chapman CS, Vette AH, Hebert JS. Characterization of normative hand movements during two functional upper limb tasks. PLoS One. 2018;13(6):e0199549. https://doi.org/10.1371/journal.pone.0199549.
16. Desrosiers J, Bravo G, Hebert R, Dutil E, Mercier L. Validation of the Box and Block Test as a measure of dexterity of elderly people: reliability, validity, and norms studies. Arch Phys Med Rehabil. 1994;75(7):751–5.
17. Kyberd PJ, Murgia A, Gasson M, Tjerks T, Metcalf C, Chappell PH, Warwick K, Lawson SEM, Barnhill T. Case studies to demonstrate the range of application of the southampton hand assessment procedure. Br J Occup Ther. 2009;72(5):212–8.
18. Sollerman C, Ejeskar A. Sollerman hand function test – a standardised method and its use in tetraplegic patients. Scand J Plast Reconstr Surg. 1995;29(2):167–76.
19. Bouwsema H, van der Sluis CK, Bongers RM. Guideline for training with a myoelectric prosthesis-based on scientific research. Groningen: Centre for Human Movement Sciences/Department of Rehabilitation Medicine, University Medical Centre Groningen, University of Groningen; 2013.
20. Lum PS, Black L, Holley RJ, Barth J, Dromerick AW. Internal models of upper limb prosthesis users when grasping and lifting a fragile object with their prosthetic limb. Exp Brain Res. 2014;232:3785–95.
21. Resnik L, Borgia M, Acluche F. Brief activity performance measure for upper limb amputees: BAM-ULA. Prosthet Orthot Int. 2018;42(1):75–83.
22. Pruitt SD, Varni JW, Setoguchi Y. Functional status in children with limb deficiency: development and initial validation of an outcome measure. Arch Phys Med Rehabil. 1996;77:1233–8.
23. Dromerick AW, Schabowsky CN, Holley RJ, Monroe BV, Markotic A, Lum PS. Effect of training on upper-extremity prosthetic performance and motor learning: a single-case study. Arch Phys Med Rehabil. 2008;89:1199–204.
24. Hussaini A, Kyberd P. Refined clothespin relocation test and assessment of motion. Prosthet Orthot Int. 2017;41(3):294–302. https://doi.org/10.1177/0309364616660248.
25. Wright FV, Hubbard S, Jutai J, Naumann S. The Prosthetic Upper Extremity Functional Index: development and reliability testing of a new functional status questionnaire for children who use upper extremity prostheses. J Hand Ther. 2001;14:91–104.
26. Resnik LJ, Borgia ML, Latlief G, Sasson N, Smurr-Walters L. Self-reported and performance-based

outcomes using DEKA arm. J Rehabil Res. 2014;51(3):351–62.
27. Desmond DM, MacLachlan L. Factor structure of the trinity amputation and prosthesis experience scales (TAPES) with individuals with acquired upper limb amputations. Am J Phys Med Rehabil. 2005;84:506–13.
28. Mary L, Carolyn B, Winnie D. Measuring occupational performance in basic activities of daily living. In: Measuring occupational performance: supporting best practice in occupational therapy. Thorofare, NJ: SLACK Inc; 2001. p. 134–5.
29. Law M, Polatajko H, Carswell A, McColl M. The Canadian occupational performance measure. 3rd ed. Ottawa, ON: CAOT Publications ACE; 1998.
30. Clawson DK, Souter WA, Carthum CJ, Hymen ML. Functional assessment of the rheumatoid hand. Clin Orthop Relat Res. 1971;77:203–10.
31. Bouwsema H, Kyberd PJ, Hill W, van der Sluis CK, Bongers RM. Determining skill level in myoelectric prosthesis use with multiple outcome measures. J Rehabil Res Dev. 2012;49(9):1331–48.
32. Chadwell A, Kenney L, Thies S, Galpin A, Head J. The reality of myoelectric prostheses: understanding what makes these devices difficult for some users to control. Front Neurorobot. 2016;10:7. https://doi.org/10.3389/fnbot.2016.00007.
33. Spiers A, Resnik L, Dollar A. Classifying and quantifying unilateral prosthesis use in home environments to inform device and treatment design. In: MEC17: a sense of what's to come. Fredericton, NB; 2017. ISBN: 978-1-55131-190-6.4.
34. Mathiowetz V, Federman S, Wiemer DM. Box and Block Test of manual dexterity: norms for 6–19 year olds. Can J Occup Ther. 1985;52(5):241–5.
35. Hebert JS, Lewicke J. Case report of modified Box and Blocks test with motion capture to measure prosthetic function. J Rehabil Res. 2012;49(8):1163–74.
36. Kuiken TA, Miller LA, Lipschutz RD, Stubblefield KA, Dumanian GA. Prosthetic command signals following targeted hyper-reinnervation nerve transfer surgery. In: Proceedings of the 2005 IEEE, engineering in medicine and biology 27th annual conference, Shanghai, China, 1–4 Sep. Shanghai: IEEE; 2005. p. 7652–5.
37. Kyberd PJ, Maillet G, Hussaini A. Characterisation of the clothespin relocation task as a functional assessment tool. J Rehabil Assist Technol Eng. 2018;5:1–7. https://doi.org/10.1177/2055668317750810.
38. Sanderson ER, Scott RN. UNB test of prosthetics function: a test for unilateral upper extremity amputees, ages 2–13. Fredericton, NB: University of New Brunswick; 1985.
39. Sollerman C. Sperling L evaluation of activities of daily living – especially hand function. Scand J Rehabil Med. 1978;6:139–43.
40. Light CM, Chappell PH, Kyberd PJ. Establishing a standardized clinical assessment tool of pathologic and prosthetic hand function. Arch Phys Med Rehabil. 2002;83:776–83.
41. Metcalf CD, Woodward H, Wright V, Chappell PH, Burridge JH, Yule T. Changes in hand function with age and normative unimpaired scores when measured with the Southampton Hand Assessment Procedure. Br J Hand Ther. 2008;13(3):79–83.
42. Deijs M, Bongers RM, van Leusen NDMR, van der Sluis CK. Flexible and static wrist units in upper limb prosthesis users: Functionality scores, user satisfaction and compensatory movements. J Neuroeng Rehabil. 2016;13:26. https://doi.org/10.1186/s12984-016-0130-0.
43. Hermansson LM, Fisher AG, Bernspang B, Eliasson A-C. Assessment of capacity for myoelectric control: a new Rasch-built measure of prosthetic hand control. J Rehabil Med. 2005;37:166–71.
44. Hermansson LM, Bodin L, Eliasson AC. Intra- and inter-rater reliability of the assessment of capacity for myoelectric control. J Rehabil Med. 2006;38:118–23.
45. Kirtley C. Clinical gait analysis, theory and practice. Amsterdam: Elsevier; 2006.
46. Pryor WB. Symmetry and attention in amputee walking. Ph.D. thesis. Melbourne, VIC: Department of Human Biosciences, Faculty of Health Sciences, La Trobe University, Australia; 2014.
47. Kidd PS, McCoy C, Steenbergen L. Repetitive strain injuries in youths. J Am Acad Nurse Pract. 2000;12(10):413–26. https://doi.org/10.1111/j.1745-7599.2000.tb00147.x.
48. Zinck A. Investigation of compensatory movements in prosthesis users and the design of a novel wrist. M.Sc., thesis. Fredericton, NB: University of New Brunswick; 2008.
49. Light CM. An intelligent hand prosthesis and evaluation of pathological and prosthetic hand function. Ph.D. thesis. Southampton: Electrical Engineering Department, University of Southampton; 2000.
50. Murgia A, Kyberd PJ, Chappell PH, Light CM. The use of gait analysis techniques in the measure of hand function. J Hand Surg. 2005;30B(Suppl 1):83–4.
51. Schmidt R, Disselhorst-Klug C, Silny J, Rau G. A marker-based measurement procedure for unconstrained wrist and elbow motions. J Biomech. 1999;32:615–21.
52. Hargrove LJ, Englehart K, Hudgins B. A comparison of surface and intramuscular myoelectric signal classification. IEEE Trans Biomed Eng. 2007;54(5):847–53, May.
53. Smith LH, Kuiken TA, Hargrove LJ. Real-time simultaneous and proportional myoelectric control using intramuscular EMG. J Neural Eng. 2014;11(6):066013. https://doi.org/10.1088/1741-2560/11/6/066013.
54. Abernathy B. Dual-task methodology and motor skills research: some applications and methodological constraints. J Hum Mov Stud. 1988;14:101–32.
55. Jacob RJK. The use of eye movements in human-computer interaction techniques: what you look at is what you get. ACM Trans Inform Syst. 1991;9(3):152–69.
56. Beauchet O, Dubost V, Herrmann FR, Kressig RW. Stride-to-stride variability while backward

counting among healthy young adults. J Neuroeng Rehabil. 2005;2:26. https://doi.org/10.1186/1743-0003-2-26.
57. Haggard P, Cockburn J, Cock J, Fordham C, Wade D. Interference between gait and cognitive tasks in a rehabilitating neurological population. J Neurol Neurosurg Psychiatry. 2000;69:479–86.
58. Deeny S, Chicoine C, Hargrove L, Parrish T, Jayaraman AA Simple ERP. Method for quantitative analysis of cognitive workload in myoelectric prosthesis control and human-machine interaction. PLoS One. 2014;9(11):e112091. https://doi.org/10.1371/journal.pone.0112091.
59. Popa F, Kyberd P, Hussani A, Gosine P, Hill W. Assessment of prosthesis use by visual attention analysis. In: Proceedings of TIPS'12 Trent international prosthetics symposium Loughborough, Leicestershire, 21–24 May; 2012.
60. Sobuh MM, Kenney LP, Galpin AJ, Thies SB, McLaughlin J, Jai Kulkarni J, Kyberd PJ. Visuomotor behaviours when using a myoelectric prosthesis. J Neuroeng Rehabil. 2014;11:72.
61. Popa FA, Kyberd PJ. Identification of patterns in upper limb prosthetic usage by analysis of visual attention to areas of interest. In: Proceedings of MEC'11 symposium – raising the standard Fredericton, NB, 14–19 Aug 2011. ISBN:978-1-55131-160-9; 2011. pp. 226–9.
62. Hebert J, Lavoie E, Boser Q, Valevicius A, Vette A, Pilarski PM, Chapman C. 3D-gaze and movement: a novel metric of visual attention to measure upper limb prosthetic function, In: Proceedings of MEC'17 symposium – a sense of what's to come, Fredericton, NB, Canada, 15–18 Aug 2017. ISBN 978-1-55131-190-6; 2017. p. 37.
63. Valevicius A, Boser Q, Lavoie E, Chapman C, Pilarski PM, Vette A, Hebert J. Kinematic insights from a novel gaze and movement metric for upper limb function: normative and prosthetic comparison. In: Proceedings of MEC'17 symposium – a sense of what's to come, Fredericton, NB, 15–18 Aug 2017. ISBN 978-1-55131-190-6; 2017. p. 43.
64. Bullock IM, Zheng JZ, De La Rosa S, Guertler C, Dollar AM. Grasp frequency and usage in daily household and machine shop tasks. IEEE Trans Haptics. 2013;6(3):296–308.
65. Feix T, Pawlik R, Schmiedmayer HB, Romero J, Kragic D. A comprehensive grasp taxonomy. In: Robotics, science and systems: workshop on understanding the human hand for advancing robotic manipulation, vol. 2(3). Stuttgart: Max Planck Institute for Intelligent Systems; 2009. p. 2–3.

8 Biologic Alternatives to Prosthetic Hand Replacement

Stefan Salminger, Johannes Mayer, Simo Vilkki, and Oskar C. Aszmann

The population who sustains severe upper extremity injuries is predominantly young, male, and employed in heavy labor occupations. Limb salvage should always be attempted if there is a reasonable potential to restore useful hand function, but may not be appropriate in complex injuries. Most traumatic amputations involve only parts of the hand, with some remaining digits [8]. Improvements in acute trauma care and microsurgical techniques enable limb salvage in most of severe hand trauma cases. Even if multiple digit amputations are necessary, a sensate partial hand is superior to all prosthetic options [19]. However, in some cases due to the severity of the injury, the entire hand must be amputated. Still, in distal amputations, there are some options for functional biological reconstruction [21]. The decision on the type of reconstruction should be made very individually fitting the patient's needs.

S. Salminger (✉) · O. C. Aszmann
Division of Plastic and Reconstructive Surgery, Department of Surgery, Medical University of Vienna, Vienna, Austria

Clinical Laboratory for Bionic Extremity Reconstruction, Department of Surgery, Medical University of Vienna, Vienna, Austria
e-mail: stefan.salminger@meduniwien.ac.at

J. Mayer
Clinical Laboratory for Bionic Extremity Reconstruction, Department of Surgery, Medical University of Vienna, Vienna, Austria

S. Vilkki
Department of Hand and Microsurgery, Tampere University Hospital, Tampere, Finland

Krukenberg Procedure

Before the microsurgical era, the Krukenberg procedure was a sophisticated method for the rehabilitation of below-elbow amputees with long residual limbs as well as in cases of congenital absence of the hand [7]. The Krukenberg procedure involves surgical separation of the ulna and radius to provide a pincerlike grasp that is motored by the pronator teres and biceps brachii muscles. Prerequisites for this operation are a stump length over 10 cm from the olecranon and good range of motion of the elbow joint. The success of the Krukenberg procedure depends directly on the strength and muscle excursion of the pronator teres and biceps muscles, the quality and sensibility of the skin surrounding ulna and radius, and the mobility of the ulna and radius at the proximal radio-ulnar joint. Therefore, especially in patients with bilateral amputations, the Krukenberg procedure offers the capability of a sensate pinching forceps. Nevertheless, Krukenberg procedure is rarely performed in the modern era due to its poor cosmesis. However, the motion of radius and ulna can also be used within a hybrid hand body-powered prosthetic fitting, where opening of the forceps is used to close the prosthetic hand. This offers a highly

O. C. Aszmann, D. Farina (eds.), *Bionic Limb Reconstruction*,
https://doi.org/10.1007/978-3-030-60746-3_8

precise control and may be an option especially in poor countries.

As microsurgical techniques were introduced to surgical practice, free vascularized tissue transfers added new options, and the Krukenberg procedure became less popular [21]. Nowadays it is possible to reconstruct some prehensile function using microsurgical techniques such as single or multiple toe transfer procedures, and in very selected cases, a hand transplantation might be an option [14, 20, 22].

Toe Transfer for Biologic Grasp Reconstruction

In 1983, Furnas and Achauer [4] published a simple method where a great toe was used on the side of the radius to create a pinch grip. Vilkki then modified this technique in 1985 using a three-jointed second toe with a hemipulp flap from the great toe after osteoplastic modeling of the distal radius [20].

The prerequisites for a reconstruction with a toe transfer depend on the length of the stump, the relationship between remaining lengths of radius and ulna, availability of adequate skin coverage, distance to sensory nerve stumps, and, most importantly, adequate motors to move the transferred toe. One single toe with three joints will need at least five motors to work properly. Two extensors, two flexors, and the intrinsics must be brought into balance in order to create the necessary power and mobility to stabilize and move the toe accurately. For adequate reconstruction of a useful sensibility, the best donor nerves are the median and radial nerve.

The single toe is positioned 7–8 cm proximally from the stump tip. This has several advantages. It will limit the need for grafts, better sensory function is achieved, and microsurgical anastomosis becomes easier. The second toe from ipsilateral foot is used within the Vilkki procedure. If the radial side of the distal stump is covered with scarred tissue of poor quality and sensibility, a glabrous hemipulp flap or in some cases a wrap-around flap from the great toe is performed to provide a sensory cover or "a mini-palm" for the opposing area of the toe. The cosmetic appearance and function at the foot are not disturbed if primary closure is achieved. Only a small skin graft is usually needed in case of performing a hemipulp flap.

This procedure is indicated in uni- or bilateral hand loss patients at wrist level or with a transradial stump of a minimum of 15–17 cm of ulna length. Especially in blind amputees, this method is able to reconstruct a sensate grasping organ, whereas rehabilitation with prosthetics in blind amputees will not be successful [21].

After a successful toe transfer, patients need a comprehensive rehabilitation of several weeks or months [20]. A protective sensation at the transferred toe will return at about 5–6 months, and functional rehabilitation will be completed about 1–1.5 years postoperatively [21]. The power of the reconstructed pinch grip is dependent on use and practice; however, a pinch gauge measurement over 10 kg is achievable in very successful cases [21]. The opening of the web space is limited to about 4 cm [21]. The timing of the procedure should be as early as possible after the amputation to achieve best results. In delayed cases, intensive pre-operative muscle training is needed to relearn activation of the distinct muscle and loosen possible contractions.

Toe Transfers for Congenital Finger and Hand Amputations

The rehabilitation of children or young adolescents with congenital amputations is an important issue. Prosthetic fitting often covers important sensory and functional surfaces of the remaining limb, and therefore abandonment of an expensive prosthetic device is very common. The microvascular transfer of one or more toes is a valuable option for biologic reconstruction of grasping function also in congenital cases.

Even in small children a congenital hand defect can be successfully reconstructed using microneurovascular toe-to-hand transplantation. This is a very effective procedure to improve grasping ability especially in the monodactylous symbrachydactyly type of congenital adactyly. Microvascular metatarsophalangeal joint transfer can also be used to stabilize a radial club hand after alignment correction. The recommended age in congenital cases is from 2 to 5 years. These complex operative procedures should be performed in specialized microsurgically oriented centers with dedicated teams in order to improve quality and ensure effective follow-up.

Human Hand Transplantation

Vascularized composite allotransplantation (VCA) is a rising field establishing new treatment strategies for patients missing various parts of their body. Although VCA necessitates lifelong immunosuppression with all known side effects, hand transplantation has the unique potential to fulfill Sir Harold Gillies principle of "replacing like with like" as such restoring a functional sensate hand [3, 13, 18]. Since 2013 hand and face allografts are recognized as organs and therefore listed in transplantation networks [17]. Still, hand transplantation does not represent a live-saving procedure; therefore, the risk-benefit ratio must be weighed carefully and patient selection is crucial [6, 14].

Since the first successful hand transplantation in Lyon in 1998, over 100 upper extremity transplantations have been performed in 26 centers worldwide listed in the international registry on hand and composite tissue transplantation [16]. However, there were 24 known re-amputations due to non-compliance, bacterial infection, or arterial ischemia, and almost every single patient experienced single or multiple episodes of rejection [5, 16]. Although homologous limb transplantation at the above-elbow level has been performed in a few cases, the functional outcome is questionable and not comparable with below-elbow hand transplants. Moreover, immunosuppression increases the risk of systemic infection, neoplasia, organ failure, or metabolic disorders [3]. Additionally, combined procedures of face and hand transplantation had to accept fatalities as a direct consequence of immunosuppressive medication [17].

In unilateral amputees, regardless of reconstructive technique, the remaining healthy hand will always have better functional capacities. Indeed, unilateral amputees with one remaining healthy hand, which becomes dominant, can usually perform up to 90% of the ADL [2]. The reconstructed hand by whatever means will always be a helping hand [1, 12, 15]. Thus, choosing the right treatment should be dependent on what is most beneficial for the patient with the least risk of harm. Apart from long-lasting rehabilitation and inpatient treatment, the greatest risk of allotransplantation is the immunosuppressive side effects [5, 9].

A multicenter study comparing hand function of below-elbow transplanted and prosthetic hands with objective functional outcome measurements showed, when only motor function is considered, that there is no significant difference between outcomes of these two groups [14]. Both provide reliable and sufficient hand function for the most relevant ADL. Still, hand transplantation represents a unique method of restoring a hand both from functional—motor and sensory—and psychosocial aspects such as the restoration of bodily integrity, strength, and even a sense of psychological closure related to the initial traumatic event. These factors need to be considered in the decision-making process leading to patient selection. If immunosuppression is tolerated by the patient along with post-operative rehabilitation, the overall goal of restoring like with like is best achieved with hand transplantation. Therefore, in bilateral below-elbow amputees, the benefits of motor and sensory restoration may outweigh the risks of lifelong immunosuppression [10, 11, 15]. In unilateral below-elbow amputees, a prosthesis represents a useful tool assisting the remaining limb [1, 2]. Unilateral amputees are able to compensate the majority of the functional deficit using their healthy hand and a prosthesis [5]. Studies have shown that, given the lower risks associated with prosthetic fitting, this represents the standard treatment for upper-limb amputees, especially in unilateral cases [14].

Conclusion

Procedures such as ray amputations, pollicizations, and an array of other microsurgical options will most probably lead to a superior hand function compared to any prostheses in selected cases. In young heavy working patients, return to work is the primary goal. The decision on the type of reconstruction should be made very individually fitting the patient's needs. Compared to biological reconstruction, prostheses provide a stronger and more robust grip with a shorter recovery time. Still, at this time, providing some sensation and tactile prosthetic feedback is in focus of current research, however, not in clinical use so far. Thus, especially bilateral amputees benefit of a sensate biological reconstruction.

References

1. Aszmann OC, Roche AD, Salminger S, Paternostro-Sluga T, Herceg M, Sturma A, Hofer C, Farina D. Bionic reconstruction to restore hand function after brachial plexus injury: a case series of three patients. Lancet. 2015; https://doi.org/10.1016/S0140-6736(14)61776-1.
2. Bhaskaranand K, Bhat AK, Acharya KN. Prosthetic rehabilitation in traumatic upper limb amputees (an Indian perspective). Arch Orthop Trauma Surg. 2003;123:363–6. https://doi.org/10.1007/s00402-003-0546-4.
3. Brenner MJ, Tung TH, Jensen JN, Mackinnon SE. The spectrum of complications of immunosuppression: is the time right for hand transplantation? J Bone Joint Surg Am. 2002;84-A:1861–70.
4. Furnas DW, Achauer BM. Microsurgical transfer of the great toe to the radius to provide prehension after partial avulsion of the hand. J Hand Surg Am. 1983;8:453–60.
5. Hautz T, Engelhardt TO, Weissenbacher A, Kumnig M, Zelger B, Rieger M, Rumpold G, Pierer G, Ninkovic M, Gabl M, Piza-Katzer H, Pratschke J, Margreiter R, Brandacher G, Schneeberger S. World experience after more than a decade of clinical hand transplantation: update on the Innsbruck program. Hand Clin. 2011;27:423–31, viii. https://doi.org/10.1016/j.hcl.2011.07.004.
6. Jowsey-Gregoire S, Kumnig M. Standardizing psychosocial assessment for vascularized composite allotransplantation. Curr Opin Organ Transplant. 2016;21:530–5. https://doi.org/10.1097/MOT.0000000000000351.
7. Kallio KE. Recent advance in Krukenberg's operation. Acta Chir Scand. 1948;97:165–88.
8. Luff R, Forrest J, Huntley J. The amputee statistical database for the United Kingdom; 2009.
9. Petruzzo P, Dubernard JM. The international registry on hand and composite tissue allotransplantation. Clin Transpl. 2011;2011:247–53.
10. Piza-Katzer H. My reflections and opinions on hand transplantation. J Hand Surg Br. 2001;26:518. https://doi.org/10.1054/jhsb.2001.0672.
11. Piza-Katzer H, Ninkovic M, Pechlaner S, Gabl M, Hussl H. Double hand transplantation: functional outcome after 18 months. J Hand Surg Br. 2002;27:385–90.
12. Piza-Katzer H, Wechselberger G, Estermann D, Gabl M, Arora R, Hussl H. [Ten years of hand transplantation experiment or routine?]. Handchirurgie, Mikrochirurgie, Plast Chir Organ der Deutschsprachigen Arbeitsgemeinschaft für Handchirurgie Organ der Deutschsprachigen Arbeitsgemeinschaft für Mikrochirurgie der Peripher Nerven und Gefässe Organ der Vereinigung der. Deut. 2009;41:210–6. https://doi.org/10.1055/s-0029-1225628.
13. Salminger S, Roche AD, Sturma A, Mayer JA, Aszmann OC. Hand transplantation versus hand prosthetics: pros and cons. Curr Surg Rep. 2016;4:8. https://doi.org/10.1007/s40137-016-0128-3.
14. Salminger S, Sturma A, Roche AD, Hruby LA, Paternostro-Sluga T, Kumnig M, Ninkovic M, Pierer G, Schneeberger S, Gabl M, Chelmonski A, Jablecki J, Aszmann OC. Functional and psychosocial outcomes of hand transplantation compared with prosthetic fitting in below-elbow amputees: a multicenter cohort study. PLoS One. 2016;11:e0162507. https://doi.org/10.1371/journal.pone.0162507.
15. Schuind F, Abramowicz D, Schneeberger S. Hand transplantation: the state-of-the-art. J Hand Surg Eur Vol. 2007;32:2–17. https://doi.org/10.1016/j.jhsb.2006.09.008.
16. Shores JT, Brandacher G, Lee WPA. Hand and upper extremity transplantation: an update of outcomes in the worldwide experience. Plast Reconstr Surg. 2015;135:351e–60e. https://doi.org/10.1097/PRS.0000000000000892.
17. Siemionow M, Gharb BB, Rampazzo A. Successes and lessons learned after more than a decade of upper extremity and face transplantation. Curr Opin Organ Transplant. 2013;18:633–9. https://doi.org/10.1097/MOT.0000000000000021.
18. Simmons PD. Ethical considerations in composite tissue allotransplantation. Microsurgery. 2000;20:458–65.
19. Tintle SM, Baechler MF, Nanos GP, Forsberg JA, Potter BK. Traumatic and trauma-related amputations: part II: upper extremity and future directions. J Bone Joint Surg Am. 2010;92:2934–45. https://doi.org/10.2106/JBJS.J.00258.

20. Vilkki SK. [Free toe transfer to the forearm stump following wrist amputation—a current alternative to the Krukenberg operation]. Handchirurgie, Mikrochirurgie, Plast Chir Organ der Deutschsprachigen Arbeitsgemeinschaft für Handchirurgie Organ der Deutschsprachigen Arbeitsgemeinschaft für Mikrochirurgie der Peripher Nerven und Gefässe Organ der Vereinigung der. Deut. 1985;17:92–7.
21. Vilkki SK, Kotkansalo T. Present technique and long-term results of toe-to-antebrachial stump transplantation. J Plast Reconstr Aesthet Surg. 2007;60:835–48. https://doi.org/10.1016/j.bjps.2007.02.018.
22. Yu ZJ, Ho HG, Chen TC. Microsurgical reconstruction of the amputated hand. J Reconstr Microsurg. 1984;1:161–5. https://doi.org/10.1055/s-2007-1007070.

Part III

Selective Nerve Transfers in Upper Limb Amputees

Motor Unit Characteristics After Selective Nerve Transfers

9

Konstantin Bergmeister, Martin Aman, Matthias Sporer, Anna Willensdorfer, Silvia Muceli, Ivan Vujaklija, Oskar C. Aszmann, and Dario Farina

Selective nerve transfers are used in biological and bionic extremity reconstruction to restore and improve extremity function. Here, peripheral nerves are rerouted to various target muscles, and thereby the structural composition of motor units is surgically altered. Previous studies have shown a high success rate of successful reinnervation of above 90% after these nerve transfers. In targeted muscle reinnervation, nerve transfers are applied to reroute amputated nerves to more proximal muscles in the stump and thereby increase the number of prosthetic control signals. Because donor nerves physiologically supply multiple muscles but are transferred to a single target muscle, the innervation ratio between donor and recipient is substantially altered. This changes the characteristics of the motor unit of the target muscles that we extensively investigated in a novel nerve transfer animal model. In this chapter, we illustrate this model, the effect of nerve transfers on motor unit physiology, as well as the implications on improving the interface between man and machine in prosthetic extremity reconstruction. In addition, first results on the effect of targeted muscle reinnervation on human motor unit physiology are described.

K. Bergmeister (✉)
Clinical Laboratory for Bionic Extremity Reconstruction, Department of Surgery, Medical University of Vienna, Vienna, Austria

Karl Landsteiner University of Health Sciences, Department of Plastic, Aesthetic and Reconstructive Surgery, University Hospital St. Pölten, Krems, Austria

Department of Plastic, Aesthetic and Reconstructive Surgery, University Hospital St. Poelten, St. Poelten, Austria

M. Aman · M. Sporer · A. Willensdorfer
Clinical Laboratory for Bionic Extremity Reconstruction, Department of Surgery, Medical University of Vienna, Vienna, Austria

S. Muceli
Division of Signal Processing and Biomedical Engineering, Department of Electrical Engineering, Chalmers University of Technology, Gothenburg, Sweden

I. Vujaklija
Department of Electrical Engineering and Automation, Aalto University, Espoo, Finland

O. C. Aszmann
Clinical Laboratory for Bionic Extremity Reconstruction, Department of Surgery, Medical University of Vienna, Vienna, Austria

Division of Plastic and Reconstructive Surgery, Department of Surgery, Medical University of Vienna, Vienna, Austria

D. Farina
Department of Bioengineering, Imperial College London, London, UK

O. C. Aszmann, D. Farina (eds.), *Bionic Limb Reconstruction*,
https://doi.org/10.1007/978-3-030-60746-3_9

Surgical Rewiring of the Motor Unit

The motor unit is the smallest controllable unit of force in the human body and consists of a motor neuron, whose axon connects to the neuromuscular junction(s) and thereby to multiple muscle fibers, termed the muscle unit [21]. These components of the motor unit are aligned in their physiological properties because of neonatal development and are termed as slow, intermediate, or fast motor units.

Slow (S) motor units comprise of muscle fibers expressing "slow oxidative" type I myosin heavy chain (MHC) protein, are resistant to fatigue, and can perform constant contractions. Contrary, intermediate, or fast fatigue-resistant (FR) motor units express fast oxidative glycolytic MHC IIA, are also responsible for continuous motions, but can perform faster movements yet fatigue moderately. Last, fast, or fast fatigable (FF) motor units express fast glycolytic MHC IIB and perform rapid contractions necessary for running and jumping. Depending on the function of the muscle, the total number of motor units and the ratio of their subpopulations (S, FR, and FF) varies. For example, muscles responsible for posture such as the soleus muscle or the autochthonous back muscles contain a high proportion of slow (S) motor units that are less prone to fatigue. The small muscles of the hand comprise a high degree of slow (S) and fatigue-resistant (FR) motor units for precise movements and a relatively higher number of motor units (in relation to their size) in comparison to proximal stump muscles as the biceps [21, 28]. The particular motor unit composition of the hand allows wide ranges of motion spanning from delicate to gross hand function [9, 21, 25, 27–29, 34, 39, 40].

Surgical nerve transfers are used to reinnervate target muscles in extremity reconstruction and to provide more prosthetic control signals for myoelectric prostheses in a procedure termed targeted muscle reinnervation (TMR). During these surgeries, the motor unit components are rewired at the axonal or peripheral nerve level, as the donor nerve's axons and their motor neurons are connected to a different target muscle. Consequently, the donor nerve reinnervates different neuromuscular junctions and muscle fibers, which may alter the motor unit's physiological properties [28]. Despite the routine clinical use of nerve transfers for the past decades [2, 4, 10–14], their effects on the motor unit have been relatively unknown.

Previous studies have used the hind limb of rats, cats, and rabbits as experimental model, to investigate the effects of cross-innervation using antagonist donor nerves [1, 38, 36]. These studies indicate that when one or multiple antagonistic donor nerves are redirected to a target muscle with less motor units, this can result in hyperinnervation of the target muscle by a higher number of motor units. However, this quantitative change in motor unit numbers may result in further currently unknown effects on the target muscle and motor unit's physiology and consecutively their functionality. Therefore, we decided to extend the investigation to all levels of the motor unit in a clinically relevant experimental model. As the majority of nerve transfers and TMR procedures are used in upper limb reconstruction [2, 3, 31, 32], it became necessary to develop a more realistic model in the smaller anatomy of the rat forelimb. This is especially relevant, as both antagonistic donor nerves and the lower extremity perform significantly different than the clinical standard of agonistic nerve transfers in the upper extremity. The lower extremity provides mostly antagonistic donor nerves and therefore requires more complex cognitive rewiring, which may also explain the clinically unfavorable results and minor use in the lower extremity [37].

In this chapter, we present investigations on the phenomenon of hyper-reinnervation, defining the reinnervation of a target muscle with a higher axonal load than its original motor branch, as previously described by Kuiken et al. in a hind limb model [26]. Further, we investigated the effects of nerve transfers on the motor unit levels using a novel experimental model and established structural and functional assessments. This chapter also explores the implications of these effects for optimizing the surgical technique for both biological reconstructions and improving the man-machine interface for bionic reconstruction (Fig. 9.1).

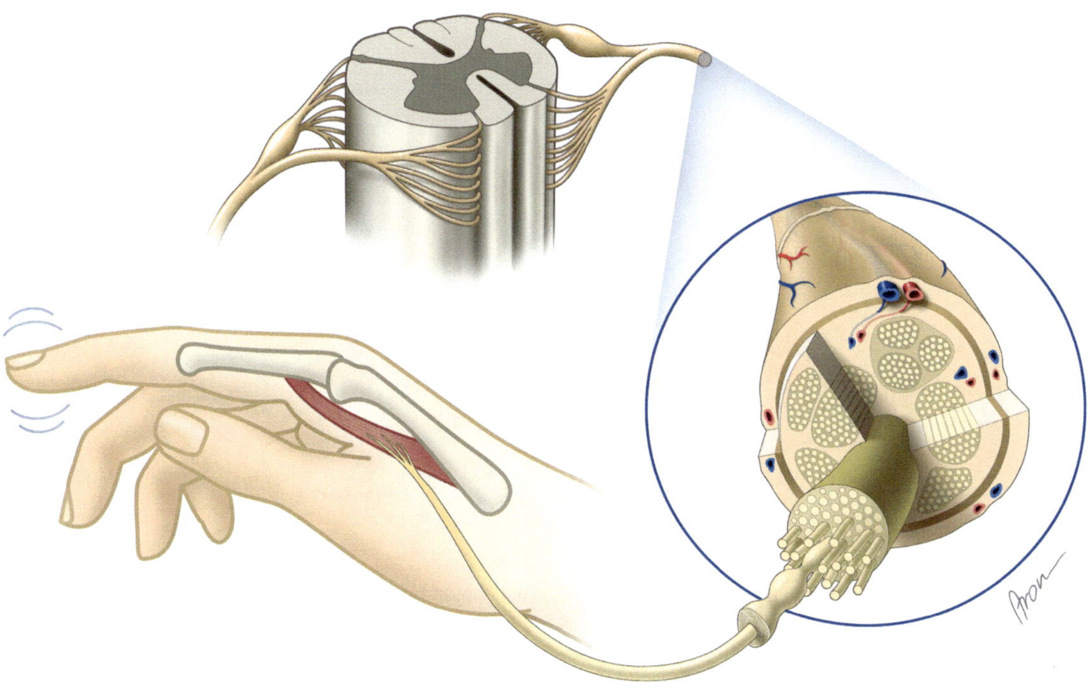

Fig. 9.1 Motor unit composition: Motor neurons are located in the anterior gray column of the spinal cord. Their axons descend grouped with other axons as peripheral nerves to the target muscles (Courtesy of Oskar Aszmann and Jeanette Schulz[†])

Experimental Nerve Transfer Model

Most experimental nerve transfer models rely on the hind limb due to its bigger anatomy [18, 26, 35]. The forelimb model has only been applied for the specific use in brachial plexus reconstruction [22, 30, 48]. Only one study has addressed the need to specifically investigate the effects of TMR experimentally, although with the limitations of using a rabbit model [24]. Based on the standard rat hind limb model for neuromuscular studies [15, 44, 47], we established a nerve transfer model in the rat forelimb to investigate the effects of nerve transfers on the motor unit [6]. Contrary to the standard hind limb model, the forelimb model provides a more realistic simulation of the clinical application of nerve transfers and especially TMR, regarding the ability to use an agonistic donor nerve while providing consistent anatomy [6]. Furthermore, significant physiological and anatomical differences exist between the fore- and hind limb in both humans and rats, regarding muscle fiber types, motor unit numbers, motor unit populations (slow, intermediate, fast), and cortical representation [8, 33, 45, 46]. Based on current nerve transfer matrices for TMR and the Oberlin/Mackinnon nerve transfer procedure [31, 32], in this model (see Fig. 9.2), the ulnar nerve is transferred to the lateral head of the biceps.

Due to the size of the donor nerve and its target, the surgery is performed using a surgery microscope, microsurgery instruments, and 11–0 sutures. For the nerve transfer procedure, the upper extremity is exposed using a Z-shaped skin incision from the pectoral muscle to the humerus' medial epicondyle. After blunt mobilization, the venous connection between the brachial and

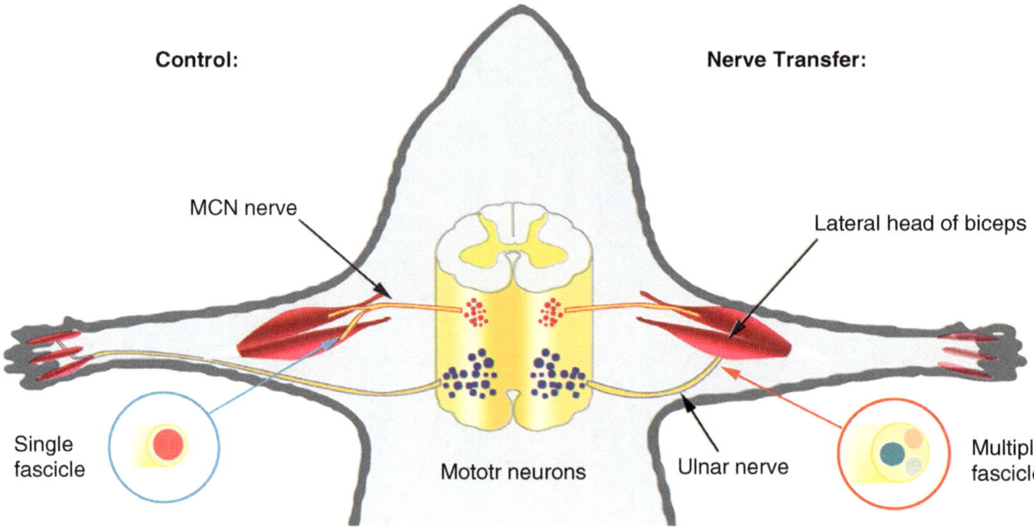

Fig. 9.2 Nerve transfer model: The nerve transfer procedure is performed on one side, whereas the other serves as internal control for structural and functional analyses. During the surgery, the ulnar nerve is dissected and transferred to the lateral head of the biceps instead of its original motor branch. This concept is similar to biological reconstruction such as the Oberlin transfer for elbow flexion or common targeted muscle reinnervation transfer matrices to improve prosthetic control

cephalic vein is electrocuted and removed to allow exploration of the biceps motor branch. The pectoral muscle is retracted for maximum exposure, and the ulnar nerve is dissected from its proximal origin to the medial epicondyle. As shown in Fig. 9.3, the ulnar nerve is now cut distally close to the medial epicondyle and mobilized to the origin from the brachial plexus' medial cord to gain sufficient length (approx. 1.5 cm) for the nerve transfer. In between the biceps' two heads, the motor branch of the biceps' long head is resected from the muscular insertion to its origin from the musculocutaneous nerve. The ulnar nerve is neurotized to the epimysium of the motor branch's insertion point via two epineural 11–0 sutures.

Postoperatively, animals are allowed to move freely, and no restrictions are imposed. Similar to other experimental nerve surgery models, analyses were performed 12 weeks after surgery to allow for complete reinnervation. From our experience, the nerve transfer surgery does not significantly affect the use of the limb in daily activities, and visual examination showed almost normal

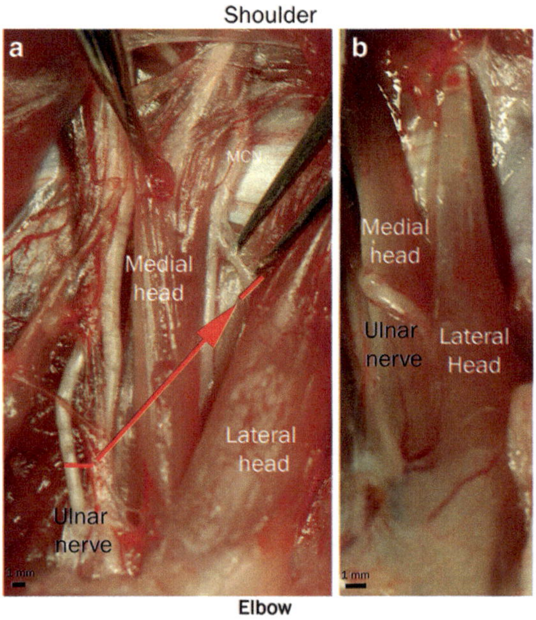

Fig. 9.3 Anatomical study of the nerve transfer procedure: (**a**) The ulnar nerve is harvested distally and transferred to the lateral head of the biceps instead of its original motor branch. (**b**) Intraoperative situs after 12 weeks when the nerve has successfully reinnervated the lateral head of the biceps (Image from Bergmeister KD et al., used with permission [6])

gait except for partially lost lumbrical function. In about 5% of animals, small ulcers occur on the forelimb's fifth digit as a result of sensory denervation of the ulnar nerve and self-mutilation. These typically heal within days after the initial surgery under conservative treatment.

This model can also be adapted to study nerve transfers in neonates as used, for example, in obstetric brachial plexus lesions, where nerve transfers are used to reroute intact nerve function to more important extremity function. Due to the high neonatal plasticity, the effects of these transfers may substantially vary compared to adults and are therefore of special interest. A model was created using the above nerve transfer procedure in neonates within 24 h after birth. Due to the delicate nerve structures in neonates, fibrin glue is used instead of sutures to neurotize the transfer. Importantly, a special dam preconditioning protocol and special considerations concerning anesthesia are required to ensure surviving of the neonate rat.

Experimental Effects of Nerve Transfers

In our studies, several effects occur on all motor unit levels following the neuroanatomical rewiring of nerve transfers. Most importantly, the hyper-reinnervative nerve transfers used in TMR and many biological reconstructions lead to excellent functional muscle reinnervation and change the motor unit composition both structurally and functionally on all motor unit levels [5, 6].

Based on the forelimb model described above, we performed analyses of the course of reinnervation after hyper-reinnervative nerve transfers and used functional muscle assessment with progressive electrical stimulation to estimate the number of reinnervating motor units.

This assessment showed that high axonal load nerve transfers lead to a high degree of muscle force regeneration that was not statistically different compared to not operated control animals at 12 weeks [5, 6]. Furthermore, smaller isolable motor units (average motor unit force) were identifiable, and additionally approximately 10–15% more motor units were estimated in the target muscle compared to the contralateral control side. Retrograde labeling of the target muscles confirmed hyper-reinnervation suggesting an increase of about 70% more motor units. Hence, these analyses illustrate a structural and functional hyper-reinnervation of the target muscle that resulted in the formation of smaller motor units. Interestingly, the donor nerve provided a total close to 300 motor neurons for possible reinnervation, but approximately only one fifth innervated the target muscles. The remaining motor units of the donor nerve's motor neurons contributed to the formation of a neuroma at the insertion of the nerve into the muscle, as was evident in multiphoton microscopy. These effects presumably resulted from axons not finding sufficient neurotrophic support for making sufficient contact with the muscle's neuromuscular junctions, representing an axonal surplus exceeding the muscle's maximum innervation capacity. This was also confirmed by labeling of motor end plates, where we did not find any denervated or polyinnervated motor end plates [5] (Fig. 9.4). In

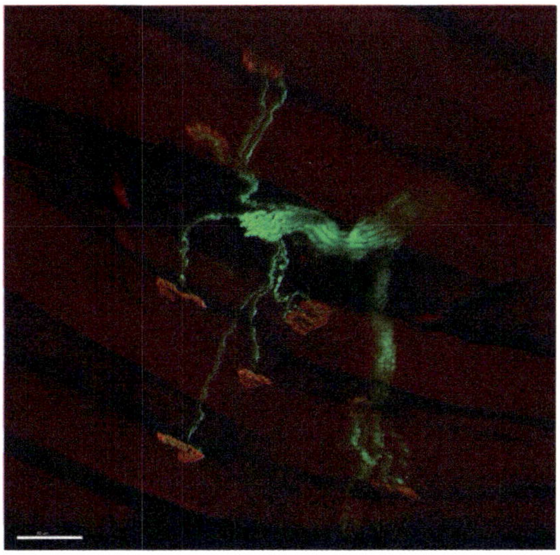

Fig. 9.4 Muscle fiber innervation after the nerve transfer: The donor nerve's axons (green) reinnervate the neuromuscular junctions (light red circular formations) and thereby innervate muscle fibers (red) [41] (Sample, 300-μm longitudinal cross-section of Thy1-GFP rat, stained with alpha-bungarotoxin for neuromuscular junction staining. Bar, 20 μm. KD Bergmeister). (© Konstantin Bergmeister)

this regard, it is important to note that previous studies not using hyper-reinnervative nerve transfers showed less success in muscle force regeneration compared to direct repair, highlighting the impact of hyper-reinnervation for successful reinnervation via nerve transfers [43].

Also, sensory axons that make up a high portion of the ulnar nerve contributed largely to the neuroma as they did not find a target organ, e.g., denervated skin to reinnervate [20]. However, prior studies have shown that nerve transfers after neuroma resection into intact muscle result in the formation of atypical intramuscular neuromas without the typical pain [16, 17]. This finding was also confirmed in an experimental TMR model in rabbits [24]. Therefore, we assume that the surplus of the nerve unable to innervate the muscle leads to the formation of neuromas but without the typical neuroma pain. This finding is in accordance with clinical observations that TMR does not result in the development of neuroma pain [42]. Likewise, an experimental study indicated that TMR favorably alters the histomorphometric characteristics of nerves after neuroma resection by decreasing myelinated fiber counts and increasing their fascicle diameter in the transferred nerves [24]. TMR may, therefore, prevent the formation of painful neuromas and may instead be a viable treatment option for neuromas [16, 42].

A further effect of nerve transfers resulted from rerouting motor neurons of different physiological properties to the target biceps muscle, which thereby changed muscle fiber types after reinnervation. Whereas control biceps samples showed a regular pattern of predominantly fast (fatigue-resistant) fibers, this pattern changed substantially after the nerve transfer with the main portion of fibers switching from fast to intermediate. The small portion of slow fibers remained constant even after the nerve transfer and visually in the same location. Although this might initially result from the fiber denervation during the nerve transfer procedure, the effect remained constant with a more vivid expression of the fiber types at 6 and 12 weeks. In comparison, this pattern is similar to muscles naturally innervated by the ulnar nerve, as, for example, the third lumbrical muscle at the rat's paw, which almost exclusively consists of slow and intermediate fibers as shown in Fig. 9.5.

In summary, the nerve transfers used in TMR lead to excellent functional muscle reinnervation/regeneration and change the motor unit composition. Following the nerve transfer, the muscle is hyper-reinnervated by the axons of the

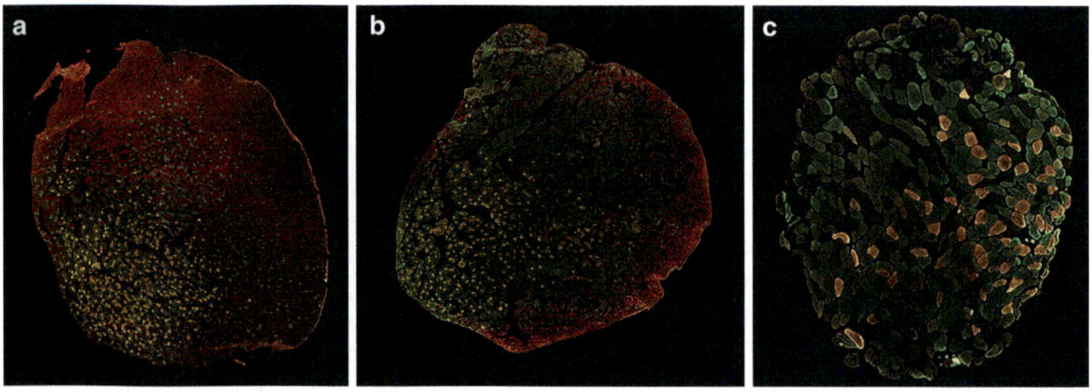

Fig. 9.5 Muscle fiber populations after nerve transfers: During the nerve transfer, axons of a different motor neuron pool are rerouted to different target muscles. Due to different physiological properties of the reinnervating motor neurons, the target muscle fibers change their protein expression to represent the physiological properties of the reinnervating motor neuron. (**a**) The control muscle is shown with predominantly red fast myosin heavy chain fibers. (**b**) After the nerve transfer, the proportion of green intermediate fibers is significantly increased. (**c**) This is similar to the original target muscles of the donor ulnar nerve as, for example, the lumbricals that consist mainly of intermediate fibers. (© Konstantin Bergmeister)

larger motor neuron pool of the donor nerve. Consequently, muscle fibers adapted to the physiological properties of these reinnervating motor neurons, and the physiological motor unit integrity was restored. These results illustrate that targeted muscle reinnervation can surgically alter the motor units of target muscles in terms of possibly finer controllability or restoring lost nerve function in anatomically different target muscles.

Clinical Impact Experimental studies have shown several neurophysiological effects that take place during nerve transfers or TMR and their subsequent reinnervation in animal models. These studies provide insights not only on neuromuscular physiology but also on potential refinements of the surgical techniques (e.g., optimal donor-to-recipient ratio) as well as on possible interface improvements for the control of prostheses. However, these experimental neurophysiological and anatomical analyses cannot be performed in humans. Instead, EMG technology, as used for prosthetic control, can be used to assess voluntary muscle function and thereby assess long-term neuromuscular effects [19]. In a clinical study performed in TMR patients, it was shown that the reinnervation by an increased motor neuron number results in a higher number of smaller functional motor units in humans, similar to the effects found in animal models [23, 26]. Due to this hyper-reinnervation, targeted muscles could potentially be controlled in a finer way than with their original innervation.

Furthermore, surface EMG analyses detected different characteristics of reinnervated motor units following TMR compared to the able-bodied controls. Here, overall smaller surface areas and shorter action potential durations were detected. This might result from the target muscle being reinnervated by physiologically different motor neurons of the donor nerve [23]. The change of muscle fiber MHC expression after TMR found in animal trials, likewise, indicates that the target muscles undergo profound neurophysiological changes due to the different reinnervating motor neurons (Fig. 9.5).

Currently, the full extent of hyper-reinnervation in humans and its potential benefit on the prosthetic interface is not yet completely explored. Further investigations are undergoing in several study groups regarding cortical reafferentiation, spino-cortical remodeling, and the effects of nerve transfers on motor unit in humans.

Conclusion and Perspective

In conclusion, this chapter describes the relevance of motor unit structure and function for surgical nerve transfers and their implications for biological and bionic extremity reconstruction. It summarizes the status quo of investigations regarding the effects of the nerve transfers on the motor unit and experimental models. We show that following high-capacity nerve transfers, hyper-reinnervation of the target muscle occurs and increases muscle functionality and controllability. In the future, it will be of great interest to identify the optimal donor to recipient axon ratios for nerve transfers and thereby refine this surgical technique and reroute surplus axons to other muscle targets. Using the high EMG sensitivity of modern implantable EMG systems (as described in Part V), these improved signals could be reliably recorded to provide a high number of intuitive muscle signals [7]. Therefore, understanding the effects of nerve transfers on the motor unit provides an essential foundation to improve and refine the function of prosthetic interfaces, which could eventually lead to improved nerve transfer procedures and thereby handlike prosthetic function in upper extremity amputees [7]. In addition, these investigations will evolve our understanding of motor unit plasticity, structure, and physiology [19].

References

1. Amphlett GW, Perry SV, Syska H, Brown MD, Vrbova G. Cross innervation and the regulatory protein system of rabbit soleus muscle. Nature. 1975;257(5527):602–4.

2. Aszmann OC, Dietl H, Frey M. Selective nerve transfers to improve the control of myoelectrical arm prostheses. Handchir Mikrochir Plast Chir. 2008;40(1):60–5. https://doi.org/10.1055/s-2007-989415.
3. Aszmann OC, Vujaklija I, Roche AD, Salminger S, Herceg M, Sturma A, Hruby LA, Pittermann A, Hofer C, Amsuess S, Farina D. Elective amputation and bionic substitution restore functional hand use after critical soft tissue injuries. Sci Rep. 2016;6:34960. https://doi.org/10.1038/srep34960.
4. Barbour J, Yee A, Kahn LC, Mackinnon SE. Supercharged end-to-side anterior interosseous to ulnar motor nerve transfer for intrinsic musculature reinnervation. J Hand Surg Am. 2012;37(10):2150–9. https://doi.org/10.1016/j.jhsa.2012.07.022.
5. Bergmeister KD, Aman M, Muceli S, Vujaklija I, Manzano-Szalai K, Unger E, Byrne RA, Scheinecker C, Riedl O, Salminger S, Frommlet F, Borschel GH, Farina D, Aszmann OC. Peripheral nerve transfers change target muscle structure and function. Sci Adv. 2019;5(1):eaau2956. https://doi.org/10.1126/sciadv.aau2956.
6. Bergmeister KD, Aman M, Riedl O, Manzano-Szalai K, Sporer ME, Salminger S, Aszmann OC. Experimental nerve transfer model in the rat forelimb. Eur Surg. 2016;48(6):334–41. https://doi.org/10.1007/s10353-016-0386-4.
7. Bergmeister KD, Vujaklija I, Muceli S, Sturma A, Hruby LA, Prahm C, Riedl O, Salminger S, Manzano-Szalai K, Aman M, Russold MF, Hofer C, Principe J, Farina D, Aszmann OC. Broadband prosthetic interfaces: combining nerve transfers and implantable multichannel EMG technology to decode spinal motor neuron activity. Front Neurosci. 2017;11:421. https://doi.org/10.3389/fnins.2017.00421.
8. Bertelli JA, Taleb M, Saadi A, Mira JC, Pecot-Dechavassine M. The rat brachial plexus and its terminal branches: an experimental model for the study of peripheral nerve regeneration. Microsurgery. 1995;16(2):77–85.
9. Bottinelli R, Betto R, Schiaffino S, Reggiani C. Unloaded shortening velocity and myosin heavy chain and alkali light chain isoform composition in rat skeletal muscle fibres. J Physiol. 1994;478(Pt 2):341–9.
10. Boutros S, Nath RK, Yuksel E, Weinfeld AB, Mackinnon SE. Transfer of flexor carpi ulnaris branch of the ulnar nerve to the pronator teres nerve: histomorphometric analysis. J Reconstr Microsurg. 1999;15(2):119–22. https://doi.org/10.1055/s-2007-1000081.
11. Boyd KU, Nimigan AS, Mackinnon SE. Nerve reconstruction in the hand and upper extremity. Clin Plast Surg. 2011;38(4):643–60. https://doi.org/10.1016/j.cps.2011.07.008.
12. Brandt KE, Mackinnon SE. A technique for maximizing biceps recovery in brachial plexus reconstruction. J Hand Surg Am. 1993;18(4):726–33. https://doi.org/10.1016/0363-5023(93)90328-z.
13. Brown JM, Mackinnon SE. Nerve transfers in the forearm and hand. Hand Clin. 2008;24(4):319–40. https://doi.org/10.1016/j.hcl.2008.08.002.
14. Brown JM, Shah MN, Mackinnon SE. Distal nerve transfers: a biology-based rationale. Neurosurg Focus. 2009;26(2):E12. https://doi.org/10.3171/foc.2009.26.2.e12.
15. Brushart TM. Nerve repair. New York, NY: Oxford University Press; 2011.
16. Dellon AL, Mackinnon SE. Treatment of the painful neuroma by neuroma resection and muscle implantation. Plast Reconstr Surg. 1986;77(3):427–38.
17. Dellon AL, Mackinnon SE, Pestronk A. Implantation of sensory nerve into muscle: preliminary clinical and experimental observations on neuroma formation. Ann Plast Surg. 1984;12(1):30–40.
18. Deslivia MF, Lee HJ, Zulkarnain RF, Zhu B, Adikrishna A, Jeon IH, Kim K. Reinnervated split-muscle technique for creating additional myoelectric sites in an animal model. Plast Reconstr Surg. 2016;138(6):997e–1010e. https://doi.org/10.1097/prs.0000000000002768.
19. Farina D, Negro F, Muceli S, Enoka RM. Principles of motor unit physiology evolve with advances in technology. Physiology (Bethesda). 2016;31(2):83–94. https://doi.org/10.1152/physiol.00040.2015.
20. Gesslbauer B, Hruby LA, Roche AD, Farina D, Blumer R, Aszmann OC. Axonal components of nerves innervating the human arm. Ann Neurol. 2017; https://doi.org/10.1002/ana.25018.
21. Heckman CJ, Enoka RM. Motor unit. Compr Physiol. 2012;2(4):2629–82. https://doi.org/10.1002/cphy.c100087.
22. Jia X, Yu C, Yang J. Outcome comparison of nerve transfer with different donor nerves in a rat model. Can J Neurol Sci. 2016;43(6):786–90. https://doi.org/10.1017/cjn.2016.302.
23. Kapelner T, Jiang N, Holobar A, Vujaklija I, Roche AD, Farina D, Aszmann OC. Motor unit characteristics after targeted muscle reinnervation. PLoS One. 2016;11(2):e0149772. https://doi.org/10.1371/journal.pone.0149772.
24. Kim PS, Ko JH, O'Shaughnessy KK, Kuiken TA, Pohlmeyer EA, Dumanian GA. The effects of targeted muscle reinnervation on neuromas in a rabbit rectus abdominis flap model. J Hand Surg Am. 2012;37(8):1609–16. https://doi.org/10.1016/j.jhsa.2012.04.044.
25. Kostrominova TY, Reiner DS, Haas RH, Ingermanson R, McDonough PM. Automated methods for the analysis of skeletal muscle fiber size and metabolic type. Int Rev Cell Mol Biol. 2013;306:275–332. https://doi.org/10.1016/b978-0-12-407694-5.00007-9.
26. Kuiken TA, Childress DS, Rymer WZ. The hyper-reinnervation of rat skeletal muscle. Brain Res. 1995;676(1):113–23.
27. Larsson L, Moss RL. Maximum velocity of shortening in relation to myosin isoform composition in single fibres from human skeletal muscles. J Physiol. 1993;472:595–614.

28. Lieber RL. Skeletal muscle structure, function, and plasticity. Baltimore, MD: Lippincott Williams & Wilkins; 2009.
29. Lieber RL, Fridén J. Clinical significance of skeletal muscle architecture. Clin Orthop Relat Res. 2001;383:140–51.
30. McGrath AM, Lu JC, Chang TN, Fang F, Chuang DC. Proximal versus distal nerve transfer for biceps reinnervation—a comparative study in a rat's brachial plexus injury model. Plast Reconstr Surg Glob Open. 2016;4(12):e1130. https://doi.org/10.1097/gox.0000000000001130.
31. Oberlin C, Ameur NE, Teboul F, Beaulieu JY, Vacher C. Restoration of elbow flexion in brachial plexus injury by transfer of ulnar nerve fascicles to the nerve to the biceps muscle. Tech Hand Up Extrem Surg. 2002;6(2):86–90.
32. Oberlin C, Béal D, Leechavengvongs S, Salon A, Dauge MC, Sarcy JJ. Nerve transfer to biceps muscle using a part of ulnar nerve for C5–C6 avulsion of the brachial plexus: anatomical study and report of four cases. J Hand Surg Am. 1994;19(2):232–7. https://doi.org/10.1016/0363-5023(94)90011-6.
33. Papalia I, Tos P, D'Alcontres FS, Battiston B, Geuna S. On the use of the grasping test in the rat median nerve model: a re-appraisal of its efficacy for quantitative assessment of motor function recovery. J Neurosci Methods. 2003;127(1):43–7. https://doi.org/10.1016/S0165-0270(03)00098-0.
34. Pette D, Staron RS. Myosin isoforms, muscle fiber types, and transitions. Microsc Res Tech. 2000;50(6):500–9. https://doi.org/10.1002/1097-0029(20000915)50:6<500::aid-jemt7>3.0.co;2-7.
35. Poppler LH, Schellhardt LM, Hunter DA, Yan Y, Mackinnon SE, Wood MD, Moore AM. Selective nerve root transection in the rat produces permanent, partial nerve injury models with variable levels of functional deficit. Plast Reconstr Surg. 2017;139(1):94–103. https://doi.org/10.1097/prs.0000000000002874.
36. Prewitt M, Salafsky B. Effect of cross innervation on biochemical characteristics of skeletal muscles. Am J Physiol. 1967;213(1):295–300.
37. Ray WZ, Chang J, Hawasli A, Wilson TJ, Yang L. Motor nerve transfers. A comprehensive review. Neurosurgery. 2016;78:1):1–26. https://doi.org/10.1227/NEU.0000000000001029.
38. Romanul FA, Van Der Meulen JP. Slow and fast muscles after cross innervation: enzymatic and physiological changes. Arch Neurol. 1967;17(4):387–402. https://doi.org/10.1001/archneur.1967.00470280053006.
39. Schiaffino S. Fibre types in skeletal muscle: a personal account. Acta Physiol (Oxf). 2010;199(4):451–63. https://doi.org/10.1111/j.1748-1716.2010.02130.x.
40. Schiaffino S, Reggiani C. Myosin isoforms in mammalian skeletal muscle. J Appl Physiol (1985). 1994;77(2):493–501.
41. Silbernagl S. Taschenatlas physiologie. New York, NY: Thieme; 2012.
42. Souza JM, Cheesborough JE, Ko JH, Cho MS, Kuiken TA, Dumanian GA. Targeted muscle reinnervation: a novel approach to postamputation neuroma pain. Clin Orthop Relat Res. 2014;472(10):2984–90. https://doi.org/10.1007/s11999-014-3528-7.
43. Swanson AN, Wolfe SW, Khazzam M, Feinberg J, Ehteshami J, Doty S. Comparison of neurotization versus nerve repair in an animal model of chronically denervated muscle. J Hand Surg Am. 2008;33(7):1093–9. https://doi.org/10.1016/j.jhsa.2008.02.025.
44. Tos P, Ronchi G, Papalia I, Sallen V, Legagneux J, Geuna S, Giacobini-Robecchi MG. Methods and protocols in peripheral nerve regeneration experimental research: part I—experimental models. Int Rev Neurobiol. 2009;87:47–79. https://doi.org/10.1016/s0074-7742(09)87004-9.
45. Tosolini AP, Morris R. Spatial characterization of the motor neuron columns supplying the rat forelimb. Neuroscience. 2012;200:19–30. https://doi.org/10.1016/j.neuroscience.2011.10.054.
46. Whishaw IQ, Pellis SM, Gorny BP. Skilled reaching in rats and humans: evidence for parallel development or homology. Behav Brain Res. 1992;47(1):59–70.
47. Wood MD, Kemp SW, Weber C, Borschel GH, Gordon T. Outcome measures of peripheral nerve regeneration. Ann Anat. 2011;193(4):321–33. https://doi.org/10.1016/j.aanat.2011.04.008.
48. Yang W, Yang J, Yu C, Gu Y. End-to-side neurotization with different donor nerves for treating brachial plexus injury: an experimental study in a rat model. Muscle Nerve. 2014;50(1):67–72. https://doi.org/10.1002/mus.24110.

Targeted Muscle Reinnervation in Upper Limb Amputees

10

Stefan Salminger, Agnes Sturma, Johannes A. Mayer, and Oskar C. Aszmann

Above-elbow or even shoulder-level amputations pose a major disability leading to a challenging prosthetic replacement [5]. Although homologous limb transplantation at the above-elbow level of amputation is possible, the functional outcome is questionable, and the risks of immunosuppression still remain high; thus prosthetic limb replacement represents the standard of care up to date [13, 14]. However, conventional myoelectric control limits speed and natural fluency of movements; thus selective nerve transfers are used to multiply myoelectric sites and enhance prosthetic control.

Targeted Muscle Reinnervation

Until the advent of targeted muscle reinnervation (TMR) technique, prosthetic function for upper arm amputations was poor [5, 7]. Conventional myoelectric upper arm prostheses are operated via two surface electrodes that are controlled by two separately innervated muscle groups. The various prosthetic joints are chosen by co-contractions and controlled linearly by these two muscles. To overcome these limitations, in TMR surgery, amputated nerves of the brachial plexus, median, ulnar, musculocutaneous, and radial nerve, respectively, are selectively transferred to the remaining stump muscles to create up to six myosignals for intuitive and simultaneous control of the different prosthetic joints [7]. In this way, an efficient and more harmonious control of the prosthetic device is possible without the need to change between the different prosthetic joints [1]. At the same time, possible neuromas are treated or prevented, and painless wearing of the prosthesis may be achieved [16].

Surgical Technique

The surgery is performed under general anesthesia without muscle relaxation (so motor nerves can be stimulated), in a supine position with moderate elevation of the shoulder. The stump is circumferentially prepared and the head slightly

S. Salminger (✉) · O. C. Aszmann
Division of Plastic and Reconstructive Surgery, Department of Surgery, Medical University of Vienna, Vienna, Austria

Clinical Laboratory for Bionic Extremity Reconstruction, Department of Surgery, Medical University of Vienna, Vienna, Austria
e-mail: stefan.salminger@meduniwien.ac.at

A. Sturma
Clinical Laboratory for Bionic Extremity Reconstruction, Department of Surgery, Medical University of Vienna, Vienna, Austria

Department of Bioengineering, Imperial College London, London, UK

J. A. Mayer
Clinical Laboratory for Bionic Extremity Reconstruction, Department of Surgery, Medical University of Vienna, Vienna, Austria

O. C. Aszmann, D. Farina (eds.), *Bionic Limb Reconstruction*,
https://doi.org/10.1007/978-3-030-60746-3_10

turned to the opposite side. Depending on the circumstances of the accident, the surgery begins with a supraclavicular exploration of the brachial plexus. Especially after high-velocity trauma and potential traction injuries of the brachial plexus, it is important to evaluate and also treat possible proximal nerve damage. If the nerve lesion does not affect the residual muscles of the stump, time is not an issue in contrast to conventional brachial plexus reconstructions, as nerve transfers reinnervate healthy muscles. In patients were the amputation stump is paralyzed, timely brachial plexus reconstruction is mandatory as can be seen in the case below.

In above-elbow amputees, through a medial approach reaching into the axillary fold, the median, ulnar, and musculocutaneous nerves are prepared and dissected. The motor branches to the medial and lateral head of the biceps and to the brachial muscle are identified. Dissection and separation of the different branches is important, and stimulation of the different branches should only provoke twitches in the targeted muscles selectively. Once the target muscles with their branches are identified, the short head of the biceps muscle is detached from its origin at the coracoid process to displace it to the medial distal aspect of the stump and separate it clearly from the long head of the biceps. To maximize the amplitude and separation of electromyographic signals, a subcutaneous fat flap is placed between the short and the long head of the biceps in above-elbow amputees [5]. Furthermore, subcutaneous fat over the targeted muscles is thinned to maximize electromyographic amplitude [8].

According to the anatomical prerequisites, the nerve transfers are performed after microsurgical preparation and exploration of all donor nerves and their new targets. The donor nerves have to be neurotomized at least to a level of palpable and healthy-appearing fascicles. With accurate and individualized planning, a primary tensionless coaptation can be achieved even in patients previously treated with neuroma resections. However, a standard nerve transfer scheme has been established which has been used in most of the patients (Table 10.1).

Through a second lateral incision, blunt dissection between the triceps heads is performed, to separate and displace the lateral head. The branch to the lateral head is further dissected, and the distal radial nerve is followed until the end-bulb neuroma has been identified. In patients with long stumps and the presence of the proximal brachioradialis muscle, the distal radial nerve can be split intraneurally into two parts—one to reinnervate the lateral head of the triceps and one to reinnervate the brachioradialis muscle. These separate signals will allow separate prosthetic hand opening and supination.

In shoulder disarticulation amputees, the existing amputation scar is opened to expose the pectoralis major muscle as well as the underlying plexus of pectoral nerves from laterally to medially. On the deep surface, the single muscle branches originating from the superior, middle, and inferior pectoral nerves to the clavicular, sternocostal, and abdominal part of the pectoralis muscle as well as the pectoralis minor muscle can be identified, prepared, and stimulated individually [2]. At this amputation level, the median, ulnar, musculocutaneous, and radial nerves are typically embedded in scarred tissue. Extensive neurolysis has to be performed to separate the individual nerves. The dissected nerves can be identified based on their relationship to the axillary artery. Within this preparation of the entire brachial plexus, the

Table 10.1 Nerve transfers at the above-elbow level of amputation

Targeted muscles	Nerves	Prosthetic function	Innervation
Biceps long head	Musculocutaneous	Elbow flexion	Original
Biceps short head	Ulnar	Hand close	Transferred
Brachialis	Median	Pronation	Transferred
Triceps long and medial head	Radial	Elbow extension	Original
Triceps lateral head	Split deep radial branch	Hand open	Transferred
Brachioradialis	Split deep radial branch	Supination	Transferred

thoracodorsal and suprascapular nerves are also identified. The latissimus dorsi and the infraspinatus muscle serve as additional targets to the pectoralis major and minor muscles. To achieve maximum distribution of the different muscle signals, the pectoralis minor muscle is detached from its origin and mobilized pedicled on its vascular supply toward the mid-axillary line. In most cases, a standard nerve transfer scheme can be used (Table 10.2). At this amputation level, electromyographic signals are well distributed on the chest and back, and no additional separation or interpositional adipofascial flaps are necessary (Fig. 10.1).

For all nerve transfers, the motor branches of the targeted muscles are transected close to the muscle to achieve a short regeneration time. The proximal part of the motor branches is transected a few centimeters back and buried deep to prevent it from reinnervating the targeted muscles. All nerve transfers are performed under loupe magnification in an end-to-end fashion using 8–0

Table 10.2 Nerve transfers at the shoulder disarticulation level of amputation

Targeted muscles	Nerves	Prosthetic function	Innervation
Pectoral major clavicular part	Musculocutaneous	Elbow flexion	Transferred
Pectoral minor	Ulnar	Hand close	Transferred
Pectoral major sternocostal part	Median	Hand close/wrist rotation	Transferred
Pectoral major abdominal part	Median	Wrist rotation	Transferred
Latissimus dorsi	Radial	Elbow extension	Transferred
Infraspinatus	Deep radial branch	Hand open	Transferred

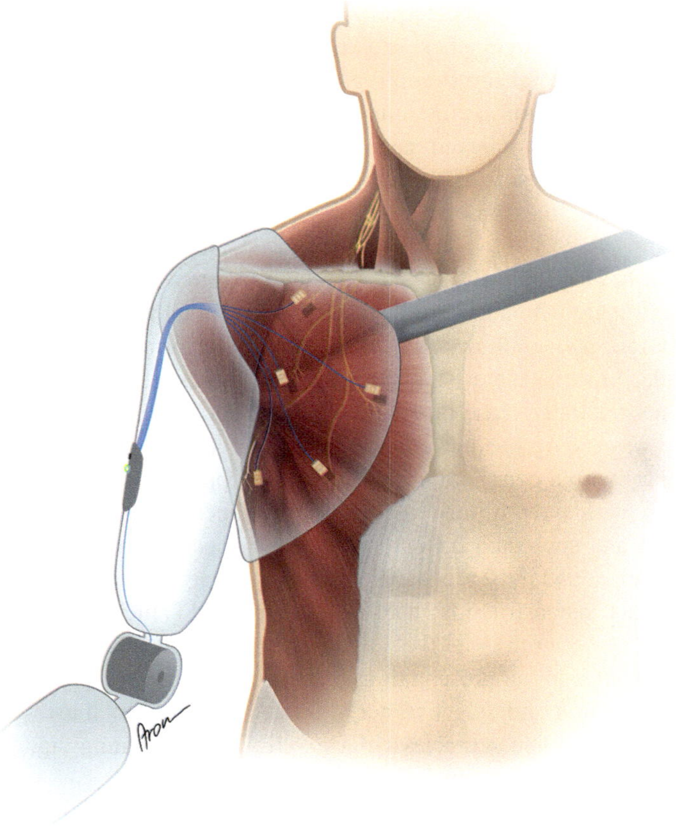

Fig. 10.1 Schematic illustration of the nerve transfers and the signal pickup in a shoulder-disarticulated patient. (© Aron Cserveny for Oskar Aszmann)

or 9–0 nylon sutures and fibrin glue. The distal neuromas are not excised, as this would necessitate additional and not beneficial further dissection.

Indications and Adapted Concepts

The ideal patient for TMR surgery has a history of a sharp amputation without traction or soft tissue injuries [11]. In these cases, the brachial plexus should be intact, the residual muscle healthy, and the nerves sufficient length to facilitate transfer. In above-elbow amputees, 50–70% of the normal length of the remaining humeral bone are necessary to achieve best prosthetic fitting and functional outcome [10, 12]. Preferably, the remaining joints should reveal full range of motion. Additionally, the patient should be motivated and cognitively capable for the time-consuming rehabilitation process.

There are, however, patients who do not fit these standard inclusion criteria who can still be candidates for TMR and myoelectric prosthetic usage. For patients with a short above-elbow stump, lack of muscles at the stump region, soft tissue deficits, or brachial plexus injuries, treatment concepts such as adapted nerve transfer schemes and additional soft tissue surgery allow improved prosthetic function.

Brachial Plexus Injury

A 24-year-old male who suffered a car accident resulting in an above-elbow amputation and infraclavicular brachial plexus injury of his left arm offers an illustrative case on adaptive nerve transfer. He was referred to our clinic 4 months after the injury with a complete palsy of his left residual limb with only limited motion in the shoulder joint and suspected damage of at least the axillary, radial, and musculocutaneous nerve. Brachial plexus reconstruction had not been attempted.

Four weeks after initial presentation, brachial plexus reconstruction and selective nerve transfers were performed. The supraclavicular exploration showed all roots in continuity, though only the pectoralis major and latissimus dorsi muscles showed response to electrical stimulation. The infraclavicular exploration revealed the axillary, radial, and musculocutaneous nerves embedded in massive scar tissue. There was no response to stimulation distal to the deltoid muscle. Through an additional incision on the medial aspect of the stump, the different nerves could be identified and dissected proximally into the scar tissue. Median and ulnar nerves, now without distal targets, seemed to be unaffected. Therefore, the ulnar nerve was used as a graft to reconstruct the musculocutaneous and the axillary nerves. After complete neurolysis of the radial nerve, the long head of the triceps showed slight response to stimulation and therefore was not grafted. The median nerve was transected proximal from its neuroma and coapted to the muscle branch of the short head of the biceps. Nine months after surgery, the patient was able to move his residual limb in three-dimensional space, and the triceps muscle showed two individual myosignals. Additionally, the reconstructed musculocutaneous nerve reinnervated the long head of the biceps providing a fourth individual myoelectric signal.

Very Short Transhumeral Amputation

A 29-year-old male presented after undergoing a very short above-elbow amputation after an agricultural accident (Fig. 10.2). Only the coracobrachialis muscle and the long head of the triceps remained. Since these two muscles did not offer enough myosignal options, nerve transfers were performed similar to shoulder disarticulation amputees [1]. The residual activity of the coracobrachialis muscle and the long head of the triceps, respectively, provided myosignals for prosthetic elbow flexion and extension. The ulnar nerve was transferred to the clavicular part of the pectoral major muscle, the median nerve to the sternocostal part of the pectoral major muscle, and some small fascicles of the median nerve to the pectoral minor muscle. Distal radial nerve fascicles (past the division to the triceps) were coapted to the thoracodorsal nerve to create a hand opening signal within the latissimus dorsi muscle.

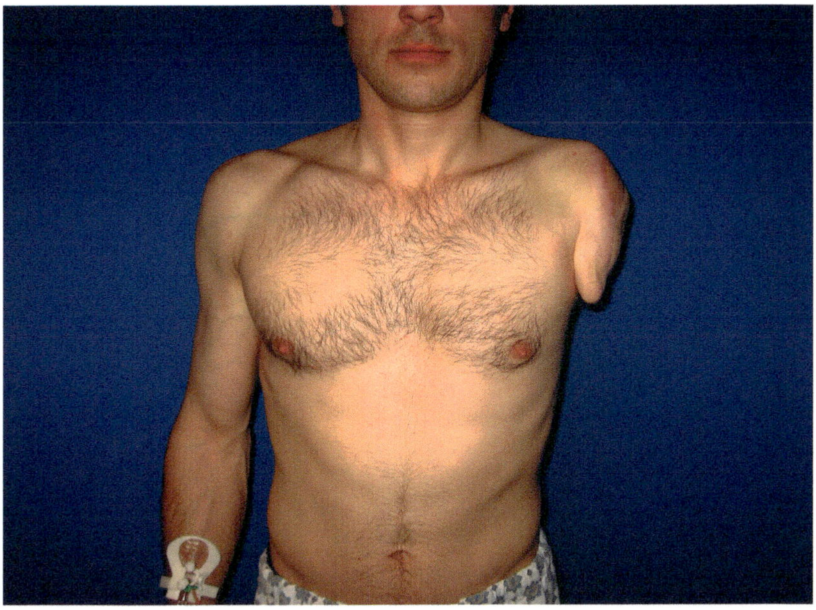

Fig. 10.2 Patient with a very short above-elbow amputation

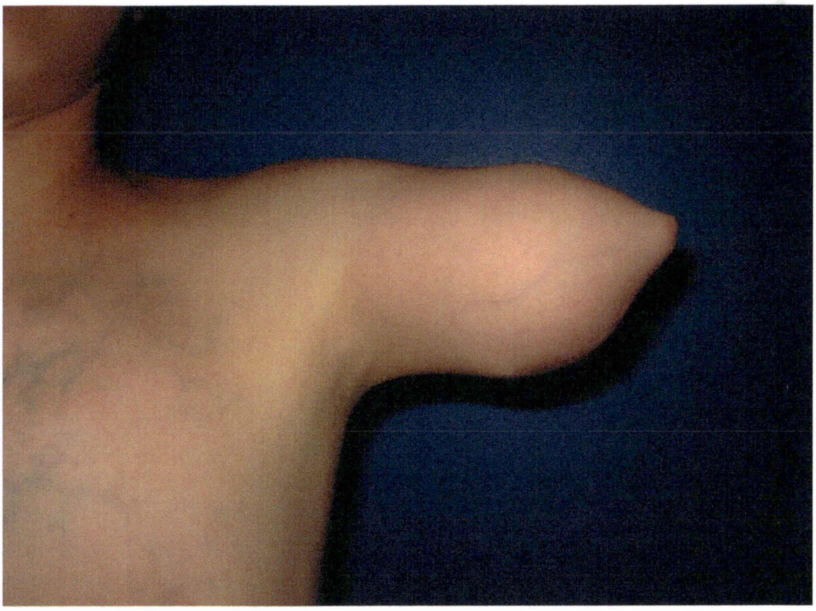

Fig. 10.3 Patient with above-elbow amputation and skin perforation

Insufficient Soft Tissue/Bony Overgrowth

A 14-year-old boy suffering progressive lymphaticovenous malformation of his left upper extremity underwent an above-elbow amputation after failing multiple sclerotherapy treatments and debulking operations. Due to appositional bone growth, the humerus perforated the skin 7 months after the amputation (Fig. 10.3). In consideration of the planned prosthetic reconstruction, we did not want to shorten the stump. However, as the skin was already perforated, for example, Marquardt's stump capping procedure for this patient was not applicable [9]. Therefore, we performed a pedicled myocutaneous latissimus dorsi flap to cover the bone and create a

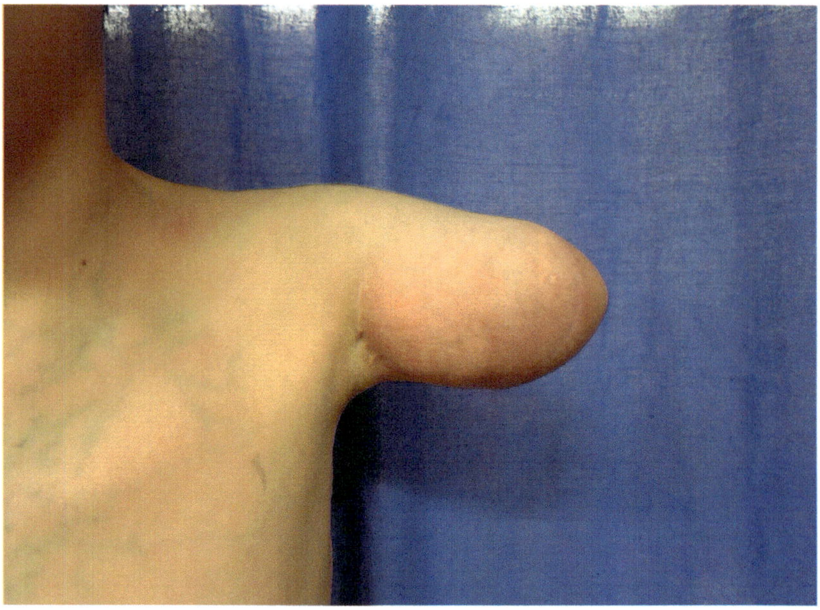

Fig. 10.4 Patient from Fig. 10.3 with tissue enlargement with pedicled musculocutaneous latissimus dorsi flap

proper stump. In this case, the new muscle provided an additional target, and selective nerve transfer of the distal radial nerve to the thoracodorsal nerve was performed in conjunction with the standard nerve transfers for above-elbow level amputations. Almost 5 years after this surgery, having reached maturity, the stump is of perfect shape, without any soft tissue problems (Fig. 10.4).

Rehabilitation

Since the long head of the biceps and the long and medial heads of the triceps maintain their original innervation, above-elbow amputees can use their conventional myoelectric prosthesis throughout the rehabilitation process. However, shoulder disarticulation amputees have to wait for reinnervation of target muscles before they are able to use a myoelectric device. As soon as the new myosignals are active, between 3 and 9 months postoperatively, a complex neurorehabilitation program is needed for the patient to learn how to activate and separate the different signals, especially in patients who have lost their arm many years ago. For this purpose, biofeedback systems are used to visualize the individual myosignals. This process needs professional guidance of a physiotherapist and/or an occupational therapist and is often underestimated [17]. The rehabilitation program in detail is presented in Part VII.

Conclusion

Function and comfort are the most important factors for successful prosthesis use, from both the amputees' and prosthetic experts' perspective [15]. The conventional two-signal control is limited in function, unnatural, and unintuitive [1, 6]. Amputees often do not experience sufficient improvement in their daily lives with these prosthetic devices resulting in up to 50% abandonment rates [4, 6, 18]. However, although the distal targets have been lost in high-level amputees, the neural signals for intuitive hand and arm function are still available and thus can be incorporated into more useful prosthetic function. Still, compliance and reliability of prostheses use after high-level amputations remains a significant concern [6].

In above-elbow amputees with a long stump and existing brachioradialis muscle, six individual and intuitive myosignals can be achieved with the above-described nerve transfers, thus generating one separate signal for each range of freedom of the prosthetic device [5]. In shoulder disarticulation amputees, a maximum of five myoelectric signals can be achieved [13]. However, in patients not fitting the standard inclusion criteria described above, the focus should lie on the creation of four individual myoelectric signals to control hand and elbow function independently. Two additional signals for wrist control are advantageous but not a priority. Hence, in challenging cases, the main goal is to perform one transfer of the median or ulnar nerve and one of the distal radial nerve.

In patients suffering additional nerve damage requiring repair, the nerve transfer scheme has to be adapted to minimize non-synergistic matching between cortical organization and target muscle function. Therefore, only cognitively "simple" nerve transfers should be performed. Transfers of both repaired median and ulnar nerves, for example, may be difficult to incorporate into prosthetic function since even amputees without nerve damage have difficulties separating these two signals.

The lack of sensory feedback represents one of the major ongoing obstacles to long-term prosthetic adoption [3]. Patients are currently forced to rely on visual feedback only. Despite this, targeted sensory reinnervation does not solve this problem as current prosthetic systems are not able to incorporate this interface and patients have reported about long-lasting dysesthesia and pain in the reinnervated areas.

References

1. Aszmann OC, Dietl H, Frey M. [Selective nerve transfers to improve the control of myoelectrical arm prostheses]. Handchirurgie, Mikrochirurgie, Plast Chir Organ der Deutschsprachigen Arbeitsgemeinschaft für Handchirurgie Organ der Deutschsprachigen Arbeitsgemeinschaft für Mikrochirurgie der Peripher Nerven und Gefässe Organ der Vereinigung der. Deut. 2008;40:60–5. https://doi.org/10.1055/s-2007-989415.
2. Aszmann OC, Rab M, Kamolz L, Frey M. The anatomy of the pectoral nerves and their significance in brachial plexus reconstruction. J Hand Surg Am. 2000;25:942–7. https://doi.org/10.1053/jhsu.2000.17818.
3. Biddiss E, Chau T. Upper-limb prosthetics: critical factors in device abandonment. Am J Phys Med Rehabil. 2007;86:977–87. https://doi.org/10.1097/PHM.0b013e3181587f6c.
4. Biddiss EA, Chau TT. Upper limb prosthesis use and abandonment: a survey of the last 25 years. Prosthetics Orthot Int. 2007;31:236–57. https://doi.org/10.1080/03093640600994581.
5. Dumanian GA, Ko JH, O'Shaughnessy KD, Kim PS, Wilson CJ, Kuiken TA. Targeted reinnervation for transhumeral amputees: current surgical technique and update on results. Plast Reconstr Surg. 2009;124:863–9. https://doi.org/10.1097/PRS.0b013e3181b038c9.
6. Farina D, Aszmann O. Bionic limbs: clinical reality and academic promises. Sci Transl Med. 2014;6:257ps12. https://doi.org/10.1126/scitranslmed.3010453.
7. Kuiken TA, Dumanian GA, Lipschutz RD, Miller LA, Stubblefield KA. The use of targeted muscle reinnervation for improved myoelectric prosthesis control in a bilateral shoulder disarticulation amputee. Prosthetics Orthot Int. 2004;28:245–53.
8. Kuiken TA, Lowery MM, Stoykov NS. The effect of subcutaneous fat on myoelectric signal amplitude and cross-talk. Prosthetics Orthot Int. 2003;27:48–54.
9. Marquardt E. Plastic surgery in imminent bone perforation at the infantile humerus stump. Preliminary report. Z Orthop Ihre Grenzgeb. 1976;114:711–4.
10. Mixter RC, Rao VK, De Angelis A, Donald CE. Salvage of proximal humeral amputations with a remnant forearm flap. Plast Reconstr Surg. 1991;87:965–8.
11. O'Shaughnessy KD, Dumanian GA, Lipschutz RD, Miller LA, Stubblefield K, Kuiken TA. Targeted reinnervation to improve prosthesis control in transhumeral amputees. A report of three cases. J Bone Joint Surg Am. 2008;90:393–400. https://doi.org/10.2106/JBJS.G.00268.
12. Salminger S, Gradischar A, Skiera R, Roche AD, Sturma A, Hofer C, Aszmann OC. Attachment of upper arm prostheses with a subcutaneous osseointegrated implant in transhumeral amputees. 2016; https://doi.org/10.1177/0309364616665732.
13. Salminger S, Sturma A, Herceg M, Riedl O, Bergmeister K, Aszmann OC. Prosthetic reconstruction in high amputations of the upper extremity. Orthopade. 2015; https://doi.org/10.1007/s00132-015-3113-0.
14. Salminger S, Sturma A, Roche AD, Hruby LA, Paternostro-Sluga T, Kumnig M, Ninkovic M, Pierer G, Schneeberger S, Gabl M, Chelmonski A, Jablecki J, Aszmann OC. Functional and psychosocial outcomes of hand transplantation compared with prosthetic fitting in below-elbow amputees: a multicenter cohort study. PLoS One. 2016;11:e0162507. https://doi.org/10.1371/journal.pone.0162507.
15. Schultz AE, Baade SP, Kuiken TA. Expert opinions on success factors for upper-limb prostheses. J Rehabil

Res Dev. 2007;44:483–9. https://doi.org/10.1682/JRRD.2006.08.0087.

16. Souza JM, Cheesborough JE, Ko JH, Cho MS, Kuiken TA, Dumanian GA. Targeted muscle reinnervation: a novel approach to postamputation neuroma pain. Clin Orthop Relat Res. 2014; https://doi.org/10.1007/s11999-014-3528-7.
17. Sturma A, Herceg M, Bischof B, Fialka-Moser V, Aszmann OC. Advanced rehabilitation for amputees after selective nerve transfers: EMG-guided training and testing. In: Replace, repair, restore, relieve – bridging clinical and and engineering solutions in neurorehabilitation. New York, NY: Springer; 2014. p. 169–77.
18. Wright TW, Hagen AD, Wood MB. Prosthetic usage in major upper extremity amputations. J Hand Surg Am. 1995;20:619–22. https://doi.org/10.1016/S0363-5023(05)80278-3.

Part IV

Phantom Pain in Limb Amputees

Epidemiology and Mechanisms of Phantom Limb Pain

11

Agnes Sturma, Laura Hruby, and Martin Diers

After amputation, most individuals describe that they can still feel their missing limb. This phantom limb sensation can range from the pure perception of the limb to non-painful cold and warm sensations and electrical feelings in the limb. At the same time, a majority of the amputees have a painful perception of the limb that has been amputated. This phenomenon is called "phantom limb pain (PLP)" and is estimated to occur in 50–80% of all amputees. PLP is usually classified as neuropathic pain, mediated by changes in the peripheral and central nervous system. Additionally, different factors such as cause of amputation, amputation level, demographic factors, or psychosocial factors all seem to contribute to the course and severity of PLP. Despite the fact that an extensive number of research projects have been focusing on PLP, it is still not fully understood and remains a challenging problem for both the patient and the clinician.

A. Sturma (✉)
Clinical Laboratory for Bionic Extremity Reconstruction, Department of Surgery, Medical University of Vienna, Vienna, Austria

Department of Bioengineering, Imperial College London, London, UK
e-mail: agnes.sturma@meduniwien.ac.at

L. Hruby
Clinical Laboratory for Bionic Extremity Reconstruction, Department of Surgery, Medical University of Vienna, Vienna, Austria

Department of Orthopedics and Trauma Surgery, Medical University of Vienna, Vienna, Austria

M. Diers
Department of Psychosomatic Medicine and Psychotherapy, LWL University Hospital, Ruhr University Bochum, Bochum, Germany

Background

In 1872 Silas Weir Mitchell was the first to coin the term "phantom limb" in relation to sensations felt in limbs no longer present after amputation [13, 21]. Still, the first reports of phantom phenomena date back to the mid-sixteenth century, when French military surgeon Ambrose Pare described that soldiers with amputation were still aware of the lost extremities [66].While phantom sensations and phantom pain encountered much skepticism and were rather seen as psychogenic in the past, these phenomena are widely accepted nowadays. We know that they not only are limited to extremity loss but can be seen with removal of nearly every part of the body [12, 26, 35, 45]. Characteristics of phantom phenomena can range from the non-painful perception that the amputated limb is still present to kinetic and kinesthetic sensations and both painful and debilitating conditions [21, 46].

O. C. Aszmann, D. Farina (eds.), *Bionic Limb Reconstruction*,
https://doi.org/10.1007/978-3-030-60746-3_11

Non-painful Phantom Phenomena

Nearly every amputee reports to have an awareness of the non-existing body part (phantom). This phantom limb awareness appears immediately in most cases, as soon as the anesthetic wears off and the patient is conscious [54]. While some amputees mention an embodiment of the missing limb without any further sensation [28, 58], others describe exteroceptive sensations. They are often referred to as non-painful tingling, electrical, or cold sensory feelings in the missing limb [27]. These sensations are rarely perceived as uncomfortable or limiting in daily life [46].Other sensations in the phantom can include vivid and not necessarily natural sensations of limb shape, length, posture, and movement [46].This means that the phantom limb can be perceived either in a position like normal limbs or in any position in space next to the body that normal anatomy would not allow. Similar to this, also the length of the limb can range from "normal" to shorter limbs. About one-third of all amputees report that their phantom limbs often retract inside the stump, a phenomenon referred to as "telescoping" [60]. In extreme cases of telescoping after upper limb amputations, the fingers are reported to be located directly on the end of or within the stump.

In some amputees, phantom sensations in the missing limb can be evoked by tactile non-painful stimulation of other body sites, e.g., the chest or parts of the face [55, 56]. These referred phantom sensations were described to be related to cortical re-mapping of the hand region in the somatosensory cortex of the brain, which is described in more detail later in this chapter. However, in contrast to the assumption of Ramachandran that touch-induced referred sensations are a perceptual correlate of cortical reorganization, more recent research could not find a significant correlation between cortical reorganization and referred sensations induced by stimulation [23].

Characteristics and Epidemiology of Phantom Limb Pain

PLP is usually characterized as painful sensations referred to the missing limb and can be throbbing, burning, stabbing, or cramping [13]. Other descriptions include squeezing, itching, tearing, and shooting sensations of electrical current flowing through the limb [44].The pain can be located in one region of the amputated body part or on the entire amputated limb [27].In many cases, amputees describe their pain as similar to the pain experienced before amputation [6]. Most patients with phantom pain have intermittent background pain, with intervals that range from one day to several weeks [68] and report of additional short exacerbations. These additional pain attacks usually only last for seconds or minutes and can have a different pain modality than the background pain.

Typically, PLP starts within the first months after amputation. Older studies report a short-term incidence of phantom pain of 72% one week after amputation and 67% half a year later [30]. Elsewhere, a short-term incidence of 54% for lower limb amputees and 82% in upper limb amputees is described [62].

Possible changes in pain modalities over time have been discussed (from sharp or sticking to burning or squeezing) [30] but still need to be confirmed by larger controlled studies. In a similar way, estimations of the exact long-term prevalence of phantom pain and phantom sensations vary considerably. These variations can be explained by different investigated samples, different definitions of phantom limb pain, and different methods used to evaluate phantom sensations. Especially earlier studies might not discuss different sensations, which can lead to an over- or underestimation of prevalence rates [44]. Furthermore, patients themselves often have difficulties distinguishing between phantom limb pain, phantom sensations, and residual limb pain [27]. Studies conducted within the last 20 years use a clearer definition of phantom limb pain and might therefore allow a better estimation of its real prevalence. Kooijman et al. report of a relatively low rate of phantom limb pain of 48.6% in their sample of 99 upper limb amputees [34]. In contrast to this, larger studies with 914 and 255 amputees found prevalence rates of 72 and 79.9% [10, 11]. Similar numbers were found by Dijkstra et al. for 536 subjects (upper and lower limb) and by Richardson et al. for 59 lower limb amputations caused by vascular diseases [8, 58]. Therefore, the prevalence of PLP can be estimated to be between 50 and 80%.

Mechanisms for PLP

Regarding the definition of mechanisms contributing to the development and long-term course of PLP, a widely agreed and accepted theory does not exist. Nevertheless, recent literature suggests an interaction of peripheral as well as central pathophysiological mechanisms [13, 14, 18, 64]. Additionally, sensory-motor incongruence through the loss of proprioceptive feedback after amputation is discussed as a factor contributing to PLP [5, 24]. Furthermore, emotional stress and psychiatric disorders are believed to trigger and exacerbate PLP in some patients [13].

Peripheral Mechanisms of PLP

Peripheral tissue damage or nerve injury often leads to pathological pain processes, such as spontaneous pain, hyperalgesia, and allodynia, that persist for years or decades after all possible tissue healing has occurred [4]. One of the causes of phantom limb pain is the elimination or interruption of sensory nerve impulses by destroying or injuring the sensory nerve fibers after amputation or deafferentation [68]. Amputation results in massive tissue and neuronal injury with the proximal portion of the severed nerve sprouting to form neuromas [63]. These neuromas are characterized by disorganized proliferation of nerve fascicles [37] and display spontaneous and abnormal evoked activity to mechanical and chemical stimuli. As the nerves which form neuromas originally innervated areas of the now missing limb, the central nervous system interprets impulses from them as coming from the phantom limb [66]. These ectopic discharges from stump neuromas (mainly enlarged and disorganized endings of C fibers and demyelinated A fibers [13]) represent a source of abnormal afferent input to the spinal cord and a potential mechanism for spontaneous pain and abnormally evoked pain [18]. As Flor and colleagues (2006) propose, the increased excitability of injured nerves resulting in ectopic firing seems to be due to alterations in the electrical properties of cellular membranes. This includes the upregulation or novel expression of voltage-sensitive sodium channels and decreased levels of potassium channel expression, as well as altered transduction molecules for mechano-, heat, and cold sensitivity in the neuroma itself. Additionally, they report on expression of novel receptors in the neuroma that are sensitive to cytokines, amines, and other mediators of inflammation and nociception in experimental injury. The development of new nonfunctional connections between axons (ephapses) also seems to enhance spontaneous afferent input to the spinal cord [27]. Since phantom limb pain often occurs right after amputation before neuromas could have formed in the stump, another source of ectopic activity has been searched for and identified as the dorsal root ganglion (DRG), which summates with ectopic discharge of neuromas in the stump [18]. Ectopia in the DRG can lead to cross-excitation and instigate the depolarization of neighboring neurons [13]. Nystrom [50] showed that the anesthetic blockage of a neuroma eliminated nerve activity related to the stimulation of the stump but not spontaneous activity, which may be originating in the DRG. Ectopic discharges generated in neuromas as well as in the DRG substantially increase the overall barrage of abnormal afferent input to the spinal cord and upper centers, such as the brainstem, thalamus, or cortex [33].

Central Mechanisms of PLP

While evidence suggests that the peripheral nervous system contributes to PLP, the mechanisms described above seem to be further complicated by central mechanisms, which maintain this chronic pain syndrome. As peripheral changes inevitably occur after amputation, this cannot explain why some amputees experience severe phantom limb pain, while others are completely pain-free [44].Therefore, an involvement of the central nervous system has been suggested. Central mechanisms contributing to the etiology of PLP can be broadly divided into spinal and supraspinal mechanisms [27, 68].

In the spinal cord, PLP is associated with changes in the dorsal horn. After deafferentation the loss of peripheral input leads to decreased impulses from reticular areas of the brainstem. As they normally inhibit transmission of sensory information, this leads to increased autonomous activity of the dorsal horn in the form of "sensory epileptic discharges" [31]. This ectopic activity in the dorsal horn may be related to sudden pain bursts as patients report of shooting pain attacks that last several seconds to minutes.

Supraspinal mechanisms describe neural changes of anatomical structures within the brain as changes in the cortical representation of areas adjacent to the amputated limb, with the intensity of PLP being related to these changes. It has been suggested that "cortical re-mapping" might be related to PLP and that its restitution might relieve it [7]. Furthermore, more severe PLP is associated with a greater shift of activation in both the primary somatosensory and motor cortices [17, 32], as indicated in Fig. 11.1. In people with arm or hand amputations, a shift of the mouth into the hand representation area in the primary somatosensory cortex has been proven in several studies, with a larger shift of the mouth representation area into the zone that formerly represented the amputated hand and arm in patients with greater phantom pain [17, 39, 68]. A schematic illustration of this process is displayed in Fig. 11.2. Pons and colleagues [52] observed massive cortical reorganization of about 1–2 cm in adult macaque monkeys that had been subjected to deafferentation of one upper extremity 12 years earlier. Tactile stimulation of the face was found to activate an area of the somatosensory cortex which previously represented the finger, the ventral hand area, the upper arm, and the neck [22]. The functional significance of these reorganizational changes was reported by Birbaumer and colleagues [1], who showed that local anesthesia eliminating PLP was associated with a reversal of this phenomenon: the cortical representation of the lip shifted into a more caudal position, which approximated the actual lip location in the somatosensory cortex as compared to the intact side. Interestingly, behavioral interventions such as imaginative resonance training (IRT) [43], mental imagery [40], prosthesis use [38,

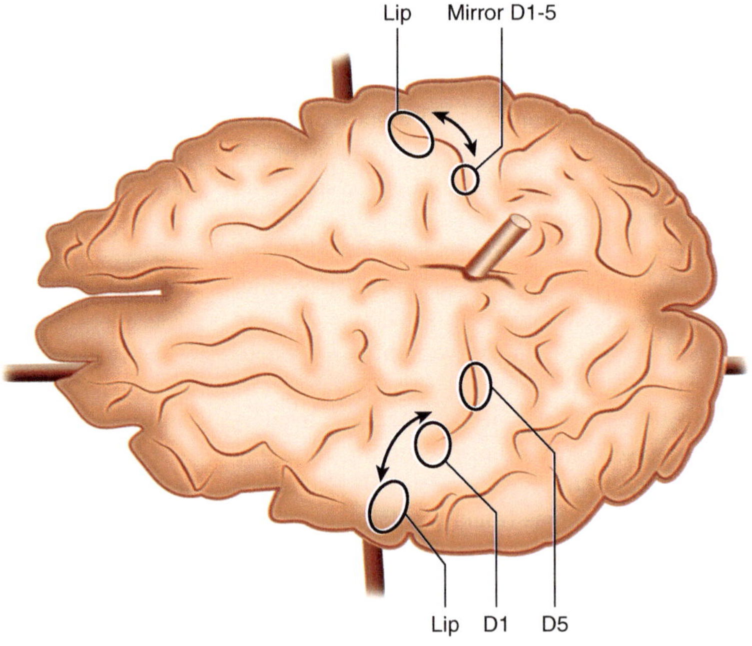

Fig. 11.1 As demonstrated by Flor et al. 1995 [17], with methods of magnetic source imaging, there is a strong relationship between the amount of PLP and the amount of cortical reorganization. This could be observed by investigating the focus of cortical activation during facial stimulation in participants with and without PLP. In participants with high levels of PLP, facial stimulation elicited activation in areas that were expected to serve the amputated hand

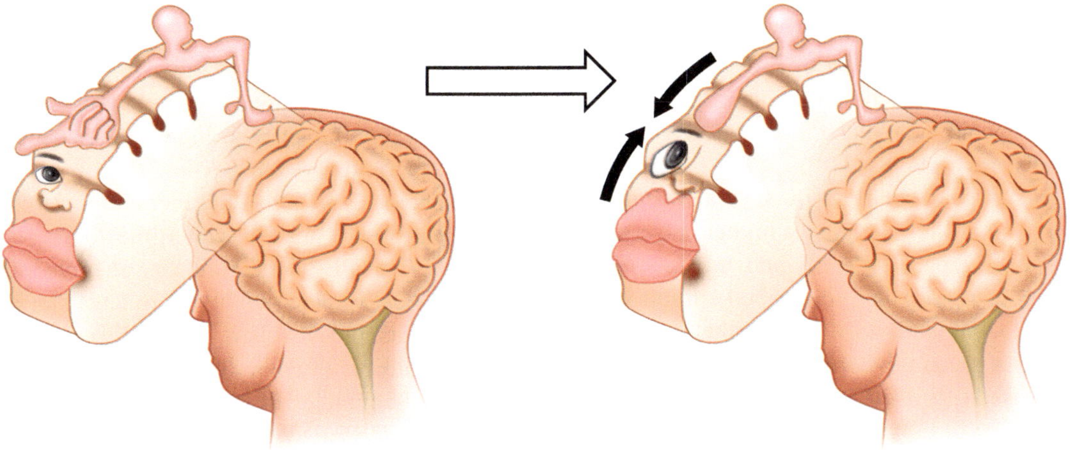

Fig. 11.2 In amputees, phantom limb pain is associated with functional cortical changes. While sensory-motor areas of the amputated hand decrease, a shift of the representation of the mouth in these areas can be seen

67], sensory discrimination training [15], virtual reality training with sensory feedback [5], or mirror training [19] have shown that a reduction in PLP intensity levels was accompanied by reversed or rather eliminated cortical reorganization processes post-interventionally.

Besides the functional plasticity in the primary somatosensory and motor cortex, also other structures were identified to adapt as a consequence of amputation [9, 53].

Interestingly, research by Makin et al. suggests that PLP might rather be driven by the disrupting of interregional functional connectivity than by changes in the local cortical representation [42]. They could show that multiple factors contribute to PLP, including a preserved structural representations of the area of the amputated hand [41]. However, maladaptive reorganization and persistent representation of the limb are not necessarily mutually exclusive and may depend on the task used to measure cortical changes. For example, in a computational model of phantom limb pain, Boström et al. [3] showed that both the amount of reorganization during tactile stimulation (used by Flor 1995) and the level of cortical activity during phantom movements (used by Makin 2013) were enhanced in a scenario with strong phantom pain as compared to a scenario with weak phantom pain (for a further discussion, see [16]). Thus, depending on the experimental context or method chosen, one might find evidence for either cortical reorganization or preservation of the amputated limb representation. Both cortical reorganization and preservation might not be contradictory but rather complementary, which should be considered in future PLP models.

Factors Associated with PLP

Different factors contributing to PLP can be classified in amputation-related factors (e.g., cause and level of amputation, stump pain), demographic factors (age, gender), psychosocial factors (employment status), and environmental factors (weather).

Amputation- and Stump-Related Factors

There is a high agreement in literature that the prevalence of stump pain is associated with PLP. Kooijman et al. found that the relative risk of having PLP is nearly twice as high for amputees with stump pain compared to those without [34]. Also, Richardson et al. explored a strong link between PLP and stump pain [58]. Dijkstra et al. had similar findings in 536 subjects, indicating a strong correlation of PLP and stump pain. Additionally, they found a higher risk for PLP if the amputation is bilateral or caused by a vascular disease or diabetes [8].

It has been reported by several authors that the risk for phantom pain is higher in the case of preexistent pain [34, 48] and the modality of phantom limb pain is described similar to pain prior to amputation [31]. Nevertheless, the relevance of preemptive treatment is controversially discussed [36, 44, 47, 49].

Demographic Factors

Phantom limb pain is more frequent when the amputation occurs in adulthood, less frequent in child amputees, and virtually non-existent in congenital amputees [8, 18]. There is conflicting evidence whether there are sex differences in the development of PLP. Kooijman and colleagues, as well as Dijkstra and colleagues, could not find any [8, 34]. In their longitudinal study including 33 females and 52 males, Bosmans et al. found that PLP was more frequent in women than in men [2].

Psychosocial Factors

Psychological variables such as depression do not seem to contribute to the emergence but may instead affect the course and the severity of the pain [20, 25, 61]. Richardson et al. [57] found that a passive coping style (especially catastrophizing) prior to amputation was associated with PLP. In a study by Jensen et al. [29], it has been shown that cognition, coping style, and social environmental variables predict 43% of the variance of PLP intensity. Furthermore, personality factors such as rigid and compulsive self-reliant personality assessed directly after the amputation were significantly correlated with a higher PLP intensity one year after the amputation [51]. Nevertheless, not all studies could find such a relationship [61].

Environmental Factors

Many clinicians working with amputees have experienced their patients complaining about more phantom limb pain during weather changes or cold and rainy weather. A few of our middle European patients have even reported to move to warmer countries during wintertime to prevent high levels of PLP. Although this phenomenon has not been explained so far, Sherman et al. [61] describe this phenomenon. In their systematic questionnaire evaluation of 128 war veterans, they asked amputees to identify everything that induced their PLP. With 48% of all amputees with PLP, "weather" was the most prominent answer. As other causes amputees named prosthetic problems (8%), mental stress (6%), fatigue (4%), and gut and back problems (2%), while 26% could not name a trigger [61].

One major difficulty with most of the studies trying to identify factors that play a role in development and maintenance of PLP is that most of them include subjects years after amputation. Especially for the amputees with a long history of devastating PLP, it is hard to determine if these factors originally caused the pain or are themselves caused by the experience of pain. This consideration is especially important for the interpretation of psychosocial factors, as it is known that pain itself can cause behavioral changes.

Differences in PLP Between Upper and Lower Extremity Amputees

According to data from the 2005 to 2006 National Amputee Statistical Database Group (NASDAB 2005), 91% of all major limb amputees in the UK have a lower limb amputation, 4% have a congenital malformation of their extremities, and 5% are upper limb amputees [59]. It can be assumed that there are similar numbers for all first-world countries. The main cause of lower limb amputations in those countries is dysvascularity related to peripheral vascular disease and diabetes [59]. In contrast to that, the main cause for upper limb amputation is trauma [63]. While cancer is a cause of amputations of both upper and lower limb, there is a higher percentage of all upper limb amputations (13% vs 3% in the lower extremity) caused by neoplasia [59]. While patients with lower limb amputation caused by dysvascularity can be considered as inactive individuals with a mean age of 65-69 years (according to the NASDAB 2005 [59]) and a long history of pain, upper limb amputees are usually young, active, and otherwise healthy. Additionally, as phantom limb pain is associated with cortical changes, the fact that the upper limb has a bigger representation in the primary somatosensory and motor cortices [65] might influence PLP modalities. Although most of the underlying mechanisms seem to be similar for all extremity losses, findings and treatment strategies from lower limb amputees do not necessarily apply for upper limb amputees and vice versa. Nevertheless, most studies on phantom limb pain have not focused on possible differences in these groups. While some authors suggest that PLP occurs more often in lower limb amputees [8], others found a higher incidence in upper limb amputees [2, 62].These different numbers might be caused by differences (e.g., age, cause of amputation) between the groups investigated or unclear definitions of stump pain and PLP.

References

1. Birbaumer N, et al. Effects of regional anesthesia on phantom limb pain are mirrored in changes in cortical reorganization. J Neurosci. 1997;17(14):5503–8.
2. Bosmans JC, et al. Factors associated with phantom limb pain: a 31/2-year prospective study. Clin Rehabil. 2010;24(5):444–53.
3. Bostrom KJ, et al. A computational model unifies apparently contradictory findings concerning phantom pain. Sci Rep. 2014;4:5298.
4. Coderre TJ, et al. Contribution of central neuroplasticity to pathological pain: review of clinical and experimental evidence. Pain. 1993;52(3):259–85.
5. De Nunzio AM, et al. Relieving phantom limb pain with multimodal sensory-motor training. J Neural Eng. 2018;15(6):066022.
6. Desmond DM, Maclachlan M. Prevalence and characteristics of phantom limb pain and residual limb pain in the long term after upper limb amputation. Int J Rehabil Res. 2010;33(3):279–82.
7. Diers M, et al. Mirrored, imagined and executed movements differentially activate sensorimotor cortex in amputees with and without phantom limb pain. Pain. 2010;149(2):296–304.
8. Dijkstra PU, et al. Phantom pain and risk factors: a multivariate analysis. J Pain Symptom Manage. 2002;24(6):578–85.
9. Draganski B, et al. Decrease of thalamic gray matter following limb amputation. NeuroImage. 2006;31(3):951–7.
10. Ehde DM, et al. Chronic phantom sensations, phantom pain, residual limb pain, and other regional pain after lower limb amputation. Arch Phys Med Rehabil. 2000;81(8):1039–44.
11. Ephraim PL, et al. Phantom pain, residual limb pain, and back pain in amputees: results of a national survey. Arch Phys Med Rehabil. 2005;86(10):1910–9.
12. Fainsinger RL, de Gara C, Perez GA. Amputation and the prevention of phantom pain. J Pain Symptom Manag. 2000;20(4):308–12.
13. Flor H. Phantom-limb pain: characteristics, causes, and treatment. Lancet Neurol. 2002;1(3):182–9.
14. Flor H. Maladaptive plasticity, memory for pain and phantom limb pain: review and suggestions for new therapies. Expert Rev Neurother. 2008;8(5):809–18.
15. Flor H, et al. Effect of sensory discrimination training on cortical reorganisation and phantom limb pain. Lancet. 2001;357(9270):1763–4.
16. Flor H, Diers M, Andoh J. The neural basis of phantom limb pain. Trends Cogn Sci. 2013;17(7):307–8.
17. Flor H, et al. Phantom-limb pain as a perceptual correlate of cortical reorganization following arm amputation. Nature. 1995;375(6531):482–4.
18. Flor H, Nikolajsen L, Staehelin J. Phantom limb pain: a case of maladaptive CNS plasticity? Nat Rev Neurosci. 2006;7(11):873–81.
19. Foell J, et al. Mirror therapy for phantom limb pain: brain changes and the role of body representation. Eur J Pain. 2014;18(5):729–39.
20. Fuchs X, Bekrater-Bodmann R, Flor H. Phantom pain: the role of maladaptive plasticity and emotional and cognitive variables. In: Pickering G, Gibson S, editors. Pain, emotion and cognition: a complex nexus. Cham: Springer; 2015. p. 189–207.
21. Grusser SM, Diers M, Flor H. [Phantom limb pain: aspects of neuroplasticity and intervention]. Anasthesiol Intensivmed Notfallmed Schmerzther. 2003;38(12):762–6.
22. Grusser SM, et al. Remote activation of referred phantom sensation and cortical reorganization in human upper extremity amputees. Exp Brain Res. 2004;154(1):97–102.
23. Grusser SM, et al. The relationship of perceptual phenomena and cortical reorganization in upper extremity amputees. Neuroscience. 2001;102(2):263–72.
24. Harris AJ. Cortical origin of pathological pain. Lancet. 1999;354(9188):1464–6.
25. Hill A. Phantom limb pain: a review of the literature on attributes and potential mechanisms. J Pain Symptom Manage. 1999;17(2):125–42.

26. Hope-Stone L, et al. Phantom eye syndrome: patient experiences after enucleation for uveal melanoma. Ophthalmology. 2015;122(8):1585–90.
27. Hsu E, Cohen SP. Postamputation pain: epidemiology, mechanisms, and treatment. J Pain Res. 2013;6:121–36.
28. Hunter JP, Katz J, Davis KD. The effect of tactile and visual sensory inputs on phantom limb awareness. Brain. 2003;126(Pt 3):579–89.
29. Jensen MP, et al. Cognitions, coping and social environment predict adjustment to phantom limb pain. Pain. 2002;95(1–2):133–42.
30. Jensen TS, et al. Phantom limb, phantom pain and stump pain in amputees during the first 6 months following limb amputation. Pain. 1983;17(3):243–56.
31. Jensen TS, et al. Immediate and long-term phantom limb pain in amputees: incidence, clinical characteristics and relationship to pre-amputation limb pain. Pain. 1985;21(3):267–78.
32. Karl A, et al. Reorganization of motor and somatosensory cortex in upper extremity amputees with phantom limb pain. J Neurosci. 2001;21(10):3609–18.
33. Knotkova H, et al. Current and future options for the management of phantom-limb pain. J Pain Res. 2012;5:39–49.
34. Kooijman CM, et al. Phantom pain and phantom sensations in upper limb amputees: an epidemiological study. Pain. 2000;87(1):33–41.
35. Kroner K, et al. Immediate and long-term phantom breast syndrome after mastectomy: incidence, clinical characteristics and relationship to pre-mastectomy breast pain. Pain. 1989;36(3):327–34.
36. Lambert A, et al. Randomized prospective study comparing preoperative epidural and intraoperative perineural analgesia for the prevention of postoperative stump and phantom limb pain following major amputation. Reg Anesth Pain Med. 2001;26(4):316–21.
37. Lim KB, Kim YS, Kim JA. Sonographically guided alcohol injection in painful stump neuroma. Ann Rehabil Med. 2012;36(3):404–8.
38. Lotze M, et al. Does use of a myoelectric prosthesis prevent cortical reorganization and phantom limb pain? Nat Neurosci. 1999;2(6):501–2.
39. Lotze M, et al. Phantom movements and pain. An fMRI study in upper limb amputees. Brain. 2001;124(Pt 11):2268–77.
40. MacIver K, et al. Phantom limb pain, cortical reorganization and the therapeutic effect of mental imagery. Brain. 2008;131(Pt 8):2181–91.
41. Makin TR, Bensmaia SJ. Stability of sensory topographies in adult cortex. Trends Cogn Sci. 2017;21(3):195–204.
42. Makin TR, et al. Phantom pain is associated with preserved structure and function in the former hand area. Nat Commun. 2013;4:1570.
43. Meyer P, et al. Imaginative resonance training (IRT) achieves elimination of amputees' phantom pain (PLP) coupled with a spontaneous in-depth proprioception of a restored limb as a marker for permanence and supported by pre-post functional magnetic resonance imaging (fMRI). Psychiatry Res. 2012;202(2):175–9.
44. Murray C, editor. Amputation, prosthesis use, and phantom limb pain. New York: Springer; 2010. p. 203.
45. Namba Y, et al. Phantom erectile penis after sex reassignment surgery. Acta Med Okayama. 2008;62(3):213–6.
46. Nikolajsen L. Postamputation pain: studies on mechanisms. Dan Med J. 2012;59(10):B4527.
47. Nikolajsen L, et al. Randomised trial of epidural bupivacaine and morphine in prevention of stump and phantom pain in lower-limb amputation. Lancet. 1997;350(9088):1353–7.
48. Nikolajsen L, et al. The influence of preamputation pain on postamputation stump and phantom pain. Pain. 1997;72(3):393–405.
49. Nikolajsen L, Lindvig M. [Phantom pain after amputation of extremities]. Ugeskr Laeger. 2001; 163(24):3338–41.
50. Nystrom B, Hagbarth KE. Microelectrode recordings from transected nerves in amputees with phantom limb pain. Neurosci Lett. 1981;27(2):211–6.
51. Parkes CM. Factors determining the persistence of phantom pain in the amputee. J Psychosom Res. 1973;17(2):97–108.
52. Pons TP, et al. Massive cortical reorganization after sensory deafferentations in adult macaques. Science. 1991;252(5014):1857–60.
53. Preissler S, et al. Gray matter changes following limb amputation with high and low intensities of phantom limb pain. Cereb Cortex. 2013;23(5):1038–48.
54. Ramachandran VS, Hirstein W. The perception of phantom limbs. The D. O. Hebb lecture. Brain. 1998;121(Pt 9):1603–30.
55. Ramachandran VS, Rogers-Ramachandran D, Stewart M. Perceptual correlates of massive cortical reorganization. Science. 1992;258(5085):1159–60.
56. Ramachandran VS, Stewart M, Rogers-Ramachandran DC. Perceptual correlates of massive cortical reorganization. Neuroreport. 1992;3(7):583–6.
57. Richardson C, et al. A prospective study of factors associated with the presence of phantom limb pain six months after major lower limb amputation in patients with peripheral vascular disease. J Pain. 2007;8(10):793–801.
58. Richardson C, et al. Incidence of phantom phenomena including phantom limb pain 6 months after major lower limb amputation in patients with peripheral vascular disease. Clin J Pain. 2006;22(4):353–8.
59. Robin L. The amputee statistical database for the United Kingdom 2004/05. Edinburgh: Common Services Agency; 2005.
60. Schmalzl L, et al. "Pulling telescoped phantoms out of the stump": manipulating the perceived position of phantom limbs using a full-body illusion. Front Hum Neurosci. 2011;5:121.
61. Sherman RA, Sherman CJ, Bruno GM. Psychological factors influencing chronic phantom limb pain: an analysis of the literature. Pain. 1987;28(3):285–95.

62. Shukla GD, et al. Phantom limb: a phenomenological study. Br J Psychiatry. 1982;141:54–8.
63. Subedi B, Grossberg GT. Phantom limb pain: mechanisms and treatment approaches. Pain Res Treat. 2011;2011:864605.
64. Vase L, et al. Cognitive-emotional sensitization contributes to wind-up-like pain in phantom limb pain patients. Pain. 2011;152(1):157–62.
65. Vincent DJ, et al. The range of motor activation in the normal human cortex using bold FMRI. Brain Topogr. 2006;18(4):273–80.
66. Weinstein SM. Phantom limb pain and related disorders. Neurol Clin. 1998;16(4):919–36.
67. Weiss T, et al. Decrease in phantom limb pain associated with prosthesis-induced increased use of an amputation stump in humans. Neurosci Lett. 1999;272(2):131–4.
68. Wolff A, et al. 21. Phantom pain. Pain Pract. 2011;11(4):403–13.

Treatment Strategies for Phantom Limb Pain

12

Agnes Sturma, Laura Hruby, Ivan Vujaklija,
Kristin Østlie, and Dario Farina

Although 50–80% of amputees are confronted with the phenomenon of phantom limb pain, treatment strategies with high enduring positive effects are yet to be developed. Most conventional drug therapies provide pain reduction to a very limited extent and are ineffective in a significant number of patients. Psychological interventions, which are becoming more prominent, currently do not result in complete pain reduction and increase in quality of life on a long-term basis. However, during the past two decades, growing interest in research of underlying central mechanisms for phantom limb pain has promoted substantial progress in behavioral treatment strategies. New emerging treatment methods include biofeedback training, mirror therapy, graded motor imagery, rubber hand illusion training, and virtual reality immersion.

A. Sturma (✉)
Clinical Laboratory for Bionic Extremity Reconstruction, Department of Surgery, Medical University of Vienna, Vienna, Austria

Department of Bioengineering, Imperial College London, London, UK
e-mail: agnes.sturma@meduniwien.ac.at

L. Hruby
Clinical Laboratory for Bionic Extremity Reconstruction, Department of Surgery, Medical University of Vienna, Vienna, Austria

Department of Orthopedics and Trauma Surgery, Medical University of Vienna, Vienna, Austria

I. Vujaklija
Department of Electrical Engineering and Automation, Aalto University, Espoo, Finland

K. Østlie
Department of Physical Medicine and Rehabilitation, Innlandet Hospital Trust, Ottestad, Norway

D. Farina
Department of Bioengineering, Imperial College London, London, UK

General Considerations

Phantom limb pain (PLP), presented after an amputation, is caused by peripheral, spinal, and supraspinal processes, which are further complicated by patient-related factors and should all be considered within treatment [37]. Common approaches for pain relief include drug therapy and invasive as well as behavioral interventions. In recent years, behavioral treatment strategies, which influence body ownership, body illusion, and embodiment of artificial limbs upon tactile and visual stimuli (e.g., mirror therapy or virtual reality training), have been shown to effectively decrease PLP in affected patients [16, 19, 24, 38, 45, 73]. Despite the vast amount of research and the encouraging findings, the clinical reality of PLP treatment is less encouraging. 70% of phantoms remain painful even 25 years after limb loss [84], between 50 and 80% of amputees are confronted with the phenomenon of PLP [19, 23, 29, 32], and clinical treatment is

O. C. Aszmann, D. Farina (eds.), *Bionic Limb Reconstruction*,
https://doi.org/10.1007/978-3-030-60746-3_12

commonly deemed unsatisfactory for both the physician and the patient [27].

Evaluation of Phantom Limb Pain

Although PLP is recognized as an important factor that may limit the amputee's quality of life, many patients fail to receive any pain treatment. This issue has been already identified in 1984 by Sherman et al. [92]. In their cohort study, they found that while 78% of amputees complained about PLP, only 19% had received any treatment, out of which only 1% had benefited from it. Later, Kooijman et al. reported that only a minority of amputees included in their study had received any treatment [47].

Determining comprehensive current state of each patient is the first step for any successful treatment; therefore it is necessary to assess PLP as an integral part of the overall clinical examination of each amputee. Hence, it is recommended to include screening tools for pain to a full diagnostic algorithm, which should also contain medical history, physical examination, and further confirmatory tests [27]. While PLP itself can hardly be assessed by physical examination, trigger points that may provoke stump pain, or PLP may be localized upon examination and palpation of the stump. This assessment should be performed in conjunction with screening for other stump problems, including scar deformity and extensive soft tissue [104].

To further understand the severity of PLP, a series of standardized questionnaires have been developed. The most widely used scales for the evaluation of pain intensity are the visual analogue scale (VAS), the numerical rating scale (NRS), and the McGill pain questionnaire (MPQ) [27]. However, many more questionnaires for the evaluation of PLP, non-painful phantom sensations, and prosthetic use are available, some of them providing detailed information on the pain's impact on activities of daily life or quality of life. For the latter purpose a generic scale, such as the Short Form 36 (SF-36), may be beneficial as well [27].

Pharmacologic Interventions

PLP is often considered neuropathic because of the changes that involve the central and peripheral nervous systems. Hence, pharmacologic interventions have traditionally been based on guidelines for neuropathic pain. The effect on PLP, however, is not well-documented for most drugs, and for some agents the effect on PLP differs from the effect on other types of neuropathic pain [2]. Specific evidence-based recommendations for treatment of PLP were published as part of a Norwegian national guideline for rehabilitation after upper limb loss [75, 76]. These updated recommendations for pharmacologic interventions, based on the AGREE-II [10] and GRADE [34, 97] methodology, are mainly in accordance with those of the Italian Consensus Conference on Pain in Neurorehabilitation [27].

First-Line Medication

Currently available evidence suggests that the effectiveness of paracetamol and Non-steroidal anti-inflammatory drugs (NSAIDs) for a specific patient with PLP should be initially explored before other pharmacologic agents are prescribed. These agents are often the first choice when it comes to a range of painful conditions; however, only limited evidence of their effect on PLP is available. Still, these are cheap, easily accessible medications with few side effects [2, 36, 46, 63, 75, 86, 100]. If paracetamol and NSAIDs do not provide sufficient pain relief, the selective noradrenalin reuptake inhibitor (SNRI) duloxetine (Duloxetine, Cymbalta) should be explored before introduction of gabapentin (Neurontin) or finally pregabalin (Lyrica). The evidence for the effect of duloxetine on phantom pain is of the same quality as for gabapentin and pregabalin (moderate), and these are all suggested as first-line agents for other neuropathic pain syndromes [59, 60, 64, 71, 75, 86, 102]. However, duloxetine has the advantage of not requiring a gradual dose increase over time. Moreover, the side effects of duloxetine for the most common dose (60 mg) are at the placebo level [71]. For gabapentin and pregabalin, side effects are relatively common [2, 72, 88], and in our clinical experience, these may cause the patient to reject the medication before reaching the therapeutic dose.

Topical lidocaine and topical capsaicin can also be used for treatment of PLP. For topical lidocaine, there is a very limited evidence for its effect on PLP. However, this is an adjunctive first-line agent for other neuropathic pain syndromes with few side effects and low cost that is easily accessible and easy to apply. It may therefore be explored if other better documented agents have not given sufficient pain relief [18, 75, 87]. Similarly, topical capsaicin (Qutenza, Capsina) is also an adjunctive first-line agent for other neuropathic pain with low (direct) to moderate (indirect) quality of evidence for PLP. However, common and severe side effects, high cost, and the need for a trained personnel to apply the Qutenza patch make this the last choice among the first-line agents [43, 56, 57, 59, 68, 75, 86].

Combination therapy among first-line drugs for neuropathic pain may be more efficient and/or associated with less side effects than single drugs and can be considered also in patients with PLP [27].

Second-Line Medication

The recommended second-line medication in general has less proved evidence than the first-line agents. Moreover, side effects are common. Second-line medications for PLP include oral morphine, transdermal buprenorphine, tramadol, and the tricyclic antidepressants (TCAs).

Oral morphine is a second-line drug for other types of neuropathic pain [86] with low to moderate quality of evidence for short-term effect on PLP (4–8 weeks). Evidence for long-term effect is lacking [2, 39, 59, 105]. Side effects are common [2, 105], and, as for all opioids, there is a risk of tolerance and habituation. Transdermal buprenorphine is another opioid recommended as a second-line drug for other types of neuropathic pain [86], with only very low quality of evidence for effect on PLP [35, 51]. However, the transdermal dosage yields more stable plasma levels and less risk of side effects, tolerance, and habituation than oral morphine [35, 52]. The choice between these two should therefore be based on individual assessments and patient preferences [75]. Tramadol is a third opioid recommended as a second-line drug for other types of neuropathic pain [86] that has low quality of evidence for effect on PLP [2, 59, 103] and risk of tolerance and habituation. It may be explored if other, better documented medications, fail to provide sufficient pain relief [75].

The TCA amitriptyline is well-documented and a first-line agent for other types of neuropathic pain [25, 86, 87, 89], but high-quality evidence suggests that it is not better than placebo for PLP treatment [2] with commonly exhibited side effects. This class of drugs should therefore only be tried for PLP if all other treatment options have failed [75].

Not Recommended Medication

Several other agents are mentioned in the literature for treatment of neuropathic pain but currently lack evidence for effect on PLP [2, 26, 27, 59, 75, 87]. We therefore recommend against the use of intravenous morphine, botulinum toxin type A injections, Selective serotonin reuptake inhibitors (SSRIs), beta blockers, memantine, ketamine, muscle relaxants (such as baclofen), calcitonin, tapentadol, transdermal fentanyl, fentanyl nasal spray, and local injections of corticosteroids for the treatment of long-term PLP. In addition to the uncertain evidence of efficacy, several of these interventions are invasive, and for all, there is a risk of side effects. For corticosteroid injections, this includes the risk of subcutaneous atrophy at the injection site, which in turn may interfere with prosthesis use. For calcitonin, there is a small increased risk of cancer associated with long-term use [27]. Larger and more rigorous randomized controlled trials are needed to clarify which of these medications would be useful for clinical practice in the future [2, 75].

Non-pharmacologic Interventions and Alternative Therapies

The fact that PLP can be influenced by cortical reorganization processes has led to the development of treatments that are based on active stimulation of the corresponding areas in the brain [104]. One of the benefits of these treatment strategies is that harmful side effects are virtually nonexistent and they can easily be combined with pharmacologic or surgical interventions.

Mirror Therapy

Mirror therapy was first proposed by Ramachandran and Rogers-Ramachandran in the mid-1990s [85]. They postulated that after amputation the brain still transmits motor commands to the missing limb. As there are no adequate afferent sensory feedback to the brain or control by vision, the brain cannot confirm that the limb has been moved, causing the illusion of paralysis. This deafferentation and the associated sensory-motor incongruence were identified as the main reason for pain. This led to the idea of a mirror box that allowed unilateral amputees to see the reflection of their intact hand where they would expect the phantom, causing a visual illusion that both hands were moving (see Fig. 12.1). With this illusionary movement, they postulated that the brain would receive visual information consistent with that of the movement of the amputated limb. This approach has been shown to successfully reduce pain [85]. Since then, several studies were conducted to further investigate the influence of this therapy on PLP. These included several case studies and longitudinal studies without a control group but also a number of controlled trials. A sham-controlled crossover trial with an intervention period of 4 weeks showed that mirror therapy has a greater effect than mental visualization or covered mirror therapy [11]. Tilak et al. [95] compared transcutaneous electrical nerve stimulation (TENS) applied to the contralateral extremity and mirror therapy. Twenty-six randomly assigned subjects had 4 days of treatment with either one of these interventions. Both groups showed a significant

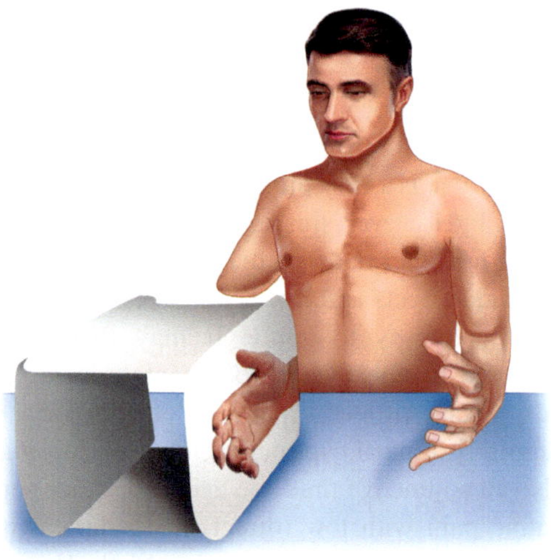

Fig. 12.1 Mirror therapy allows the patient to see a reflection of the sound side at the position where the amputated limb is expected

decrease in pain, but there was no significant difference between the groups. Another randomized controlled study on mirror therapy for upper limb amputees [28] showed that the participants in the control group ($n = 6$) that trained with a covered mirror or worked with mental imagery had no significant improvement of their pain levels in the VAS, while the treatment group ($n = 9$), which performed mirror therapy for 15 min 5 days/week for 4 weeks has done so in a significant manner. This supports the hypothesis that the use of mirror therapy may reduce PLP [28]. Although previous studies investigating the effects of mirror therapy on PLP have limitations concerning the sample size and setup, mirror therapy seems to be beneficial in a subset of patients. Since no harmful side effects were reported, the use of mirror therapy can be recommended to treat PLP [27].

Virtual Reality and Virtual Mirror Therapy

The use of augmented or virtual reality to treat PLP is another approach aiming to control negative changes within the central nervous system after amputation [73]. These systems can either directly mirror movements of the healthy limb in unilateral amputees [62] or use motion tracking

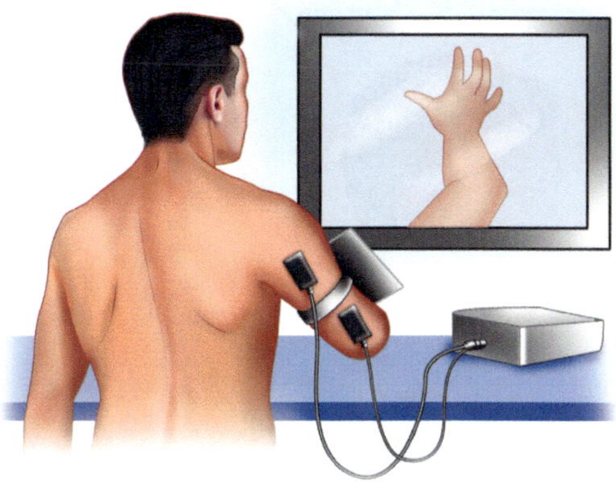

Fig. 12.2 Motion and EMG sensors can be used to enable movement of a virtual limb

systems [13] and/or biological signal acquisition on the affected side to predict indented movements [74]. The provided visualization can include a simple representation of the missing hand or can be combined with virtual reality tasks such as picking up and moving objects, as shown in Fig. 12.2. Additionally, EMG signals can be used to play serious games, e.g. allowing the patient to control a car in a racing game by using muscular activity [82]. More advanced setups of may also include multimodal feedback [16]. Some of these strategies are also used for signal training in prosthetic rehabilitation as described in Chap. 21.

The evidence for this treatment approach is limited to some case studies [70, 74] and case series [13, 16, 41, 62, 73, 77, 91] that indicate positive effects on PLP. Therefore, virtual and augmented reality still needs to be seen as experimental [22]. However, since virtual reality training presents a noninvasive option with no apparent side effects [73] and is known to increase patient's motivation during training [80, 81], it may be considered as a supplementary treatment strategy.

Mental Imagery and Graded Motor Imagery

Graded motor imagery (GMI) and its components build upon mirror therapy and similarly aim to enhance cortical activation of painful (phantom) limbs and therefore reduce pain. The concept of GMI consists of three stages which gradually engage cortical motor networks in order to avoid triggering pain as a protective response by an overly sensitive nociception system [9, 66]. The first step includes left/right discrimination from photographs of healthy body parts as depicted in Fig. 12.3. Here, the affected areas are chosen, i.e. a patient with upper limb amputation is asked to discriminate between left and right hands and arms. The second stage is motor imagery where patient is asked to imagine movements of the painful (phantom) limb. Finally, mirror therapy is introduced, and the patient is prompted to think about bilateral movements while watching the reflection of the non-affected limb [67].

Evaluating the effects of graded motor imagery and its components on PLP in a systematic way is rather complicated due to the heterogeneity of the interventions, intervention groups, and follow-up periods [9]. Studies investigating the effect and applicability of GMI in the clinical practice came to contrasting conclusions [53]. A recent systematic literature review concluded that GMI is widely used to treat PLP, yet there is limited evidence supporting this approach on the short- to medium-term basis [5].

Prosthesis Use

There is some limited evidence indicating possible positive effects of prosthesis use on PLP as shown in Fig. 12.4. However, most of the relevant studies originate in the late 1990s, when Lotze and Weiss investigated this topic.

Weiss fitted one group of patients with a cosmetic prosthesis, while the other one was fitted with a Sauerbruch prosthesis. This mechanical device is surgically connected to one of the muscles of the arm through a perforated tunnel enabling direct control over the gripper. While there was no change in PLP in the cosmetic prosthesis group, he found a significant decrease in PLP in those using the active Sauerbruch prosthesis [101].

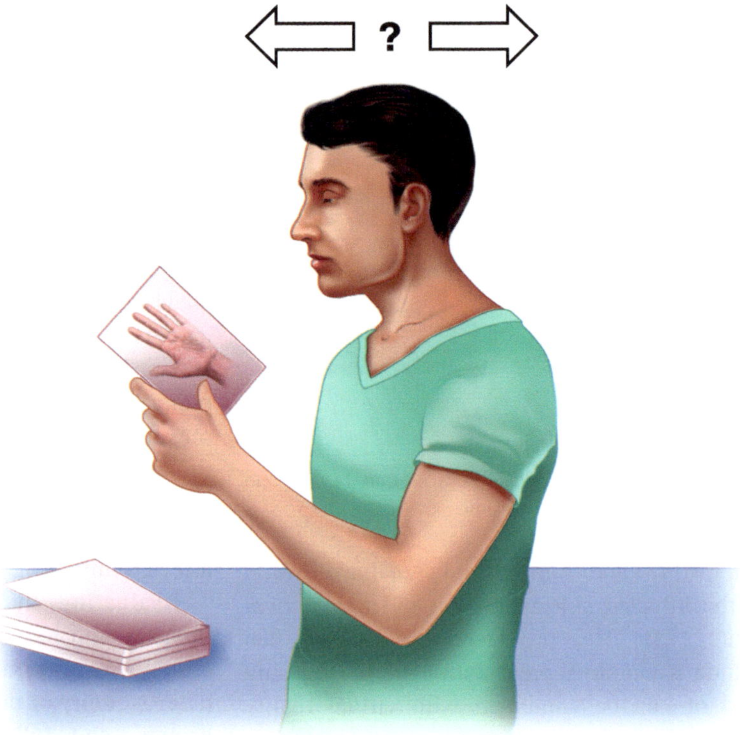

Fig. 12.3 Cards with left and right hands or arms can be used for lateralization training to enhance cortical activation of the upper extremity

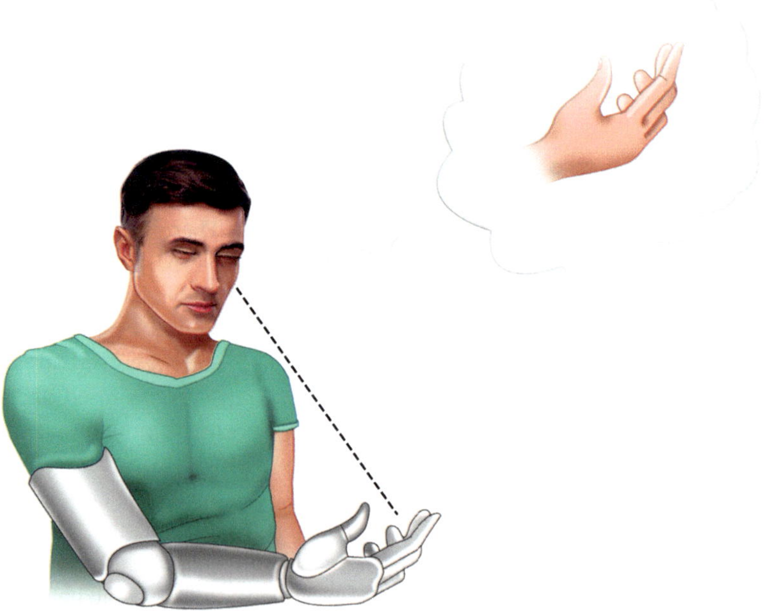

Fig. 12.4 An active prosthesis might trigger cortical reafferentiation as the patient sees a hand and can voluntarily control its movements

At the same time, Lotze et al. investigated cortical reorganization and PLP in a cohort of 14 unilateral upper limb amputees. They found a negative correlation between extensive use of a myoelectric prosthesis and cortical reorganization, as well as PLP [54].

A recent study included ten upper limb amputees with PLP in a 2-week training protocol with a myoelectric prosthesis with somatosensory feedback. Patients reported a reduction in PLP. Furthermore, the authors described a gain in prosthesis function and cortical changes in the brain areas involved in vision and pain processing [83].

While this may be considered as weak positive evidence for the use of active prostheses to prevent or relieve PLP, prosthesis use is also associated with more self-confidence and better functioning in daily life [15, 98]. Additionally, prosthesis use does not seem to have harmful side effects, and there are no contraindications.

Transcutaneous Electrical Nerve Stimulation (TENS)

Transcutaneous electrical nerve stimulation (TENS) is a form of nerve stimulation, with electrodes placed onto the surface skin that is widely used to treat chronic pain conditions [3]. It is usually administrated via portable battery-powered devices that can also be used in the absence of a clinician [42] as shown in Fig. 12.5. The mechanism of pain relief is explained through the gate control theory. This includes the assumption that peripheral stimulation activates large (Aβ) nerve fibers, which activate spinal inhibitory neurons. They inhibit the activity of small nociceptive (C) fibers, which leads to reduced pain [61].

TENS seems to be better than placebo in controlled trials investigating its effect on general neuropathic pain [14]. Also, studies with small sample sizes were conducted to specifically investigate its effects on PLP. In a case study, Giuffrida et al. showed a positive effect of contralateral TENS application on PLP [33]. A pilot study with ten transtibial amputees has also found it to be beneficial for pain treatment [69]. In a controlled trial, Tilak et al. could show that it is as effective as mirror therapy to treat PLP [95].

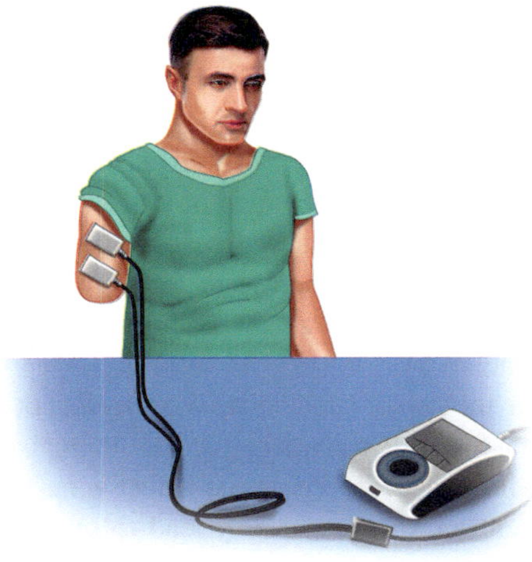

Fig. 12.5 A typical setup for electrical stimulation of the stump intended to mediate PLP

However, like in most interventions for PLP treatment, literature on TENS lacks the methodological rigor and robust reporting needed to come to firm conclusion on its effectiveness [27, 42]. Nevertheless, it presents a cheap and noninvasive treatment that is relatively easy to apply and can be combined with other approaches. Therefore, TENS might also be considered as a supplementary intervention for treating PLP.

Other Strategies

Sensory discrimination training focuses on tactile input to the stump area. While different stimuli are applied to the stump, the amputee is asked to identify the correct location (and frequency) of the applied stimulus [30, 40, 65]. A positive effect on PLP as well as a reversal of cortical reorganization was reported in a small sample of amputees [30]. Similarly, a recent study showed a relief in pain using tactile discrimination training [99]. Comparable positive results could be demonstrated in a cohort study combining visual and sensory feedback [16]. As no harmful side effects were reported so far and studies with larger sample sizes confirm its efficacy in patients with complex regional pain syndrome (CRPS), we think that this treatment can be offered in patients who show interest in behavioral treatment approaches.

The positive effect of *hypnosis* is well-documented for other types of pain. However, a recent systematic review [5] found limited evidence for use of hypnosis in PLP. Therefore, it may be tried only as an additional treatment approach [75].

On the other hand, *acupuncture* is not suggested for PLP treatment [75] as there is only very limited evidence on its effectiveness [27], while it is invasive and may lead to serious side effects such as infections or pneumothorax.

Surgical Therapies

Usually, clinicians only fall back to invasive interventions when other treatment methods have been proven ineffective [94] or if there are additional indications for surgery.

Deep Brain Stimulation

Several approaches that directly stimulate cortical neural activity have been shown to be useful in the management of PLP [50]. Deep brain stimulation to contralateral regions (such as the thalamus or medial regions of periventricular and periaqueductal gray regions) has been found to effectively reduce phantom pain [6]. Thalamic deep brain stimulation is thought to normalize ectopic thalamocortical activation and abnormal projections to the cortex [78]. Similarly, chronic epidural motor cortex stimulation (MCS) has been found to transiently relieve phantom pain. A recent study in rats has shed light on the underlying mechanisms of MCS and how they affect pain-signaling pathways: apart from a modulation of c-fos and serotonin expression, MCS suppresses neuronal activity in the ventral posterolateral nucleus of the thalamus [44], which is known to maintain neuropathic pain as part of an ascending pathway. Additionally MCS regulates neuropathic pain by affecting descending pathways, in particular the striatum, periaqueductal gray, cerebellum, and other thalamic areas [44].

Less invasively, repetitive transcranial magnetic stimulation (externally applied stimulation that modulates cortical excitability) [50] has been proven successful in the treatment of phantom pain [7, 8, 96].

Spinal Cord Stimulation

Direct stimulation of the spinal cord has been investigated in the management of neuropathic pain, including PLP. A recent systematic review (which also highlights neurophysiological data on the mechanism of SCS) found that almost half of the included studies had not achieved pain relief following the procedure [1].

The invasive nature of spinal cord stimulation and its associated complications [58] in combination with the lack of evidence preclude its routine application for patients with phantom pain.

Targeted Muscle Reinnervation

As described in previous chapters, targeted muscle reinnervation (TMR) is a surgical intervention to improve the neuromuscular interface for prosthesis control [48, 49]. By re-routing nerves that originally served the now missing limb, additional myosignals can be created [90]. This method was originally developed to allow more intuitive prosthesis control for above-elbow amputees as well as after shoulder disarticulation [4, 20, 48, 90].

As shown by Chen et al., TMR has the potential to restore cortical reorganization occurring after amputation. By using high-density electroencephalography (EEG), they assessed cortical activity during motor tasks for the intact and missing limb [12]. After amputation, but before TMR, a shift of the representation of the missing limb to other cortical areas was observed. In a second assessment after undergoing the nerve transfer surgery, they found that the missing limb representation had shifted back closer to the normal representation areas [12]. As cortical reorganization after amputation is correlated with PLP [31, 32, 55], reversing this phenomenon might also cause pain resolution.

Apart from central mechanisms, TMR also has the potential to change mechanisms in the peripheral nervous system which are known to contribute to postamputation pain. Selective nerve transfers have been used in reconstructive surgery to prevent or treat pain caused by neuromas for decades [17]. Case studies as well as a recent randomized controlled trial on amputees undergoing TMR surgery have confirmed its applicability for treatment and prevention of neuroma formation and PLP after upper and lower limb amputation [21, 79, 93]. Additionally, this effect might be even stronger when also fitting patients undergoing TMR surgery with an active prosthesis which they learn to intuitively control. As described above, the use of an active device is also correlated with less PLP [54, 101].

Conclusion

With a PLP incidence between 50 and 80% [19, 23, 29, 32] and most people still suffering from pain years after amputation [84], it is fair to say that current treatment approaches in clinical practice are mostly unsatisfactory [27]. However, with ongoing research efforts in the field, more and more evidence has become available in the recent years to help clinicians make informed decisions on patient treatment. Clinical interventions should start with an evaluation of the type of pain and by communicating current knowledge about the origins of PLP and available treatment strategies to the patient. As psychological factors might influence pain, they should be considered within treatment as well [37]. Current guidelines [27, 75, 76] suggest the use of paracetamol and NSAIDs as first-line medication for PLP. As a next step the SNRI duloxetine (Duloxetine, Cymbalta) should be explored before introduction of gabapentin (Neurontin) or finally pregabalin (Lyrica). Combination therapy among these first-line drugs for neuropathic pain can be considered in patients with PLP [27]. Second-line medication including oral morphine, transdermal buprenorphine, tramadol, and the tricyclic antidepressants (TCAs) usually show more side effects, and there is relatively little evidence for their effects on PLP [75]. Other treatments already available in the clinical setting include mirror therapy, graded motor imagery, prosthesis use, electrical stimulation, and TMR that all seem to positively influence pain levels [5, 95, 101]. Research approaches include the use of virtual reality and sensory feedback with promising results [16, 73, 99]. Combining these methods (as TMR used to control a prosthesis with sensory feedback) might lead the path towards more successful PLP treatment in the future.

References

1. Aiyer R, et al. A systematic review on the treatment of phantom limb pain with spinal cord stimulation. Pain Manag. 2017;7(1):59–69.
2. Alviar MJM, Hale t, Dungca M. Pharmacologic interventions for treating phantom limb pain (Cochrane review). Cochrane Database Syst Rev. 2012;2012(2):56.
3. American Society of Anesthesiologists. Practice guidelines for chronic pain management: an updated report by the American Society of Anesthesiologists Task Force on Chronic Pain Management and the American Society of Regional Anesthesia and Pain Medicine. Anesthesiology. 2010;112(4):810–33.
4. Aszmann OC, Dietl H, Frey M. Selective nerve transfers to improve the control of myoelectrical arm prostheses. Handchir Mikrochir Plast Chir. 2008;40(1):60–5.
5. Batsford S, Ryan CG, Martin DJ. Non-pharmacological conservative therapy for phantom limb pain: a systematic review of randomized controlled trials. Physiother Theory Pract. 2017;33(3):173–83.
6. Boccard SG, et al. Long-term outcomes of deep brain stimulation for neuropathic pain. Neurosurgery. 2013;72(2):221–30; discussion 231.
7. Bolognini N, et al. Motor and parietal cortex stimulation for phantom limb pain and sensations. Pain. 2013;154(8):1274–80.
8. Bolognini N, et al. Long-term analgesic effects of transcranial direct current stimulation of the motor cortex on phantom limb and stump pain: a case report. J Pain Symptom Manag. 2013;46(4):e1–4.
9. Bowering KJ, O'Connell NE, Tabor A, Catley MJ, Leake HB, Moseley GL, Stanton TR. The effects of graded motor imagery and its components on

chronic pain: a systematic review and meta-analysis. J Pain. 2012;14(1):3–13.
10. Brouwers M, et al. AGREE II: advancing guideline development, reporting and evaluation in health care. Can Med Assoc J. 2010;182(18):E839–42.
11. Chan BL, et al. Mirror therapy for phantom limb pain. N Engl J Med. 2007;357(21):2206–7.
12. Chen A, Yao J, Kuiken TA, Dewald JPA. Cortical motor activity and reorganization following upper-limb amputation and subsequent targeted reinnervation. Neuroimage Clin. 2013;3:498–506.
13. Cole J, et al. Exploratory findings with virtual reality for phantom limb pain; from stump motion to agency and analgesia. Disabil Rehabil. 2009;31(10):846–54.
14. Cruccu G, et al. EFNS guidelines on neurostimulation therapy for neuropathic pain. Eur J Neurol. 2007;14(9):952–70.
15. Davidson J. A comparison of upper limb amputees and patients with upper limb injuries using the Disability of the Arm, Shoulder and Hand (DASH). Disabil Rehabil. 2004;26(14–15):917–23.
16. De Nunzio AM, et al. Relieving phantom limb pain with multimodal sensory-motor training. J Neural Eng. 2018;15(6):066022.
17. Dellon AL, Mackinnon SE. Treatment of the painful neuroma by neuroma resection and muscle implantation. Plast Reconstr Surg. 1986;77(3):427–38.
18. Derry S, et al. Topical lidocaine for neuropathic pain in adults. Cochrane Database Syst Rev. 2014;2014(7). Art.No.: CD010958. https://doi.org/10.1002/14651858.CD010958.pub2.
19. Diers M, Flor H. Phantomschmerz. Schmerz. 2013;27(2):205–13.
20. Dumanian GA, et al. Targeted reinnervation for transhumeral amputees: current surgical technique and update on results. Plast Reconstr Surg. 2009;124(3):863–9.
21. Dumanian GA, et al. Targeted muscle reinnervation treats neuroma and phantom pain in major limb amputees: a randomized clinical trial. Ann Surg. 2019;270(2):238–46.
22. Dunn J, et al. Virtual and augmented reality in the treatment of phantom limb pain: a literature review. NeuroRehabilitation. 2017;40(4):595–601.
23. Ehde DM, et al. Chronic phantom sensations, phantom pain, residual limb pain, and other regional pain after lower limb amputation. Arch Phys Med Rehabil. 2000;81(8):1039–44.
24. Ehrsson HH, et al. Upper limb amputees can be induced to experience a rubber hand as their own. Brain. 2008;131(Pt 12):3443–52.
25. National Institue for Health and Care Excellence (NICE). Neuropathic pain in adults: pharmacological management in non-specialist settings. 2013. nice.org.uk/guidance/cg173.
26. Fang J, et al. Pharmacological interventions for phantom limb pain. Chin Med J. 2013;126(3):542–9.
27. Ferraro F, et al. Diagnosis and treatment of pain in plexopathy, radiculopathy, peripheral neuropathy and phantom limb pain. Evidence and recommendations from the Italian Consensus Conference on Pain on Neurorehabilitation. Eur J Phys Rehabil Med. 2016;52(6):855–66.
28. Finn SB, et al. A randomized, controlled trial of mirror therapy for upper extremity phantom limb pain in male amputees. Front Neurol. 2017;8:267.
29. Flor H. Phantom-limb pain: characteristics, causes, and treatment. Lancet Neurol. 2002;1(3):182–9.
30. Flor H, et al. Effect of sensory discrimination training on cortical reorganisation and phantom limb pain. Lancet. 2001;357(9270):1763–4.
31. Flor H, et al. Phantom-limb pain as a perceptual correlate of cortical reorganization following arm amputation. Nature. 1995;375(6531):482–4.
32. Flor H, Nikolajsen L, Staehelin J. Phantom limb pain: a case of maladaptive CNS plasticity? Nat Rev Neurosci. 2006;7(11):873–81.
33. Giuffrida O, Simpson L, Halligan PW. Contralateral stimulation, using TENS, of phantom limb pain: two confirmatory cases. Pain Med. 2010;11(1):133–41.
34. Goldet G, Howick J. Understanding GRADE: an introduction. J Evid Based Med. 2013;6:50–4.
35. Griessinger N, Sittl R, Likar R. Transdermal buprenorphine in clinical practice—a post-marketing surveillance study in 13,179 patients. Curr Med Res Opin. 2005;21(8):1147–56.
36. Hall GC, et al. An observational descriptive study of the epidemiology and treatment of neuropathic pain in a UK general population. BMC Fam Pract. 2013;14(28):1–10.
37. Hsu E, Cohen SP. Postamputation pain: epidemiology, mechanisms, and treatment. J Pain Res. 2013;6:121–36.
38. Hunter JP, Katz J, Davis KD. The effect of tactile and visual sensory inputs on phantom limb awareness. Brain. 2003;126(Pt 3):579–89.
39. Huse E, et al. The effect of opioids on phantom limb pain and cortical reorganization. Pain. 2001;90(1–2):47–55.
40. Huse E, Larbig W, Birbaumer N, Flor H. Kortikale Reorganisation und Schmerz: Empirische Befunde und therapeutische Implikationen am Beispiel des Phantomschmerzes. Der Schmerz. 2001;15:131–7.
41. Ichinose A, et al. Somatosensory feedback to the cheek during virtual visual feedback therapy enhances pain alleviation for phantom arms. Neurorehabil Neural Repair. 2017;31(8):717–25.
42. Johnson MI, Mulvey MR, Bagnall AM. Transcutaneous electrical nerve stimulation (TENS) for phantom pain and stump pain following amputation in adults. Cochrane Database Syst Rev. 2015;(8):Cd007264.
43. Kern KU, et al. [Capsaicin 8% cutaneous patches for phantom limb pain. Results from everyday practice (non-interventional study)] [Article in German]. Schmertz. 2014;28(4):374–83.
44. Kim J, et al. Motor cortex stimulation and neuropathic pain: how does motor cortex stimula-

tion affect pain-signaling pathways? J Neurosurg. 2016;124(3):866–76.
45. Kim SY, Kim YY. Mirror therapy for phantom limb pain. Korean J Pain. 2012;25(4):272–4.
46. Knotkova H, et al. Current and future options for the management of phantom-limb pain. J Pain Res. 2012;5:39–49.
47. Kooijman CM, et al. Phantom pain and phantom sensations in upper limb amputees: an epidemiological study. Pain. 2000;87(1):33–41.
48. Kuiken TA, et al. The use of targeted muscle reinnervation for improved myoelectric prosthesis control in a bilateral shoulder disarticulation amputee. Prosthetics Orthot Int. 2004;28(3):245–53.
49. Kuiken TA, et al. Targeted muscle reinnervation for the upper and lower extremity. Tech Orthop. 2017;32(2):109–16.
50. Lenggenhager B, Arnold CA, Giummarra MJ. Phantom limbs: pain, embodiment, and scientific advances in integrative therapies. Wiley Interdiscip Rev Cogn Sci. 2014;5(2):221–31.
51. Licina L, et al. Buprenorphine/naloxone therapy for opioid refractory neuropathic pain following traumatic amputation: a case series. Mil Med. 2013;178(7):e858–61.
52. Likar R. Transdermal buprenorfine in the management of persistent pain—safety aspects. Ther Clin Risk Manag. 2006;2(1):115–25.
53. Limakatso K, Corten L, Parker R. The effects of graded motor imagery and its components on phantom limb pain and disability in upper and lower limb amputees: a systematic review protocol. Syst Rev. 2016;5(1):145.
54. Lotze M, et al. Does use of a myoelectric prosthesis prevent cortical reorganization and phantom limb pain? Nat Neurosci. 1999;2(6):501–2.
55. Lotze M, et al. Phantom movements and pain. An fMRI study in upper limb amputees. Brain. 2001;124(Pt 11):2268–77.
56. Mason L, et al. Systematic review of topical capsaicin for the treatment of chronic pain. BMJ. 2004;1–5. https://doi.org/10.1136/bmj.38042.506748.EE. (Published 19 March 2004).
57. Mason L, et al. Systematic review of topical capsaicin for the treatment of chronic pain (Structured abstract). BMJ/Cochrane DARE, 2004/2014. Cochrane DARE. 2014;3:4.
58. McAuley J, van Groningen R, Green C. Spinal cord stimulation for intractable pain following limb amputation. Neuromodulation. 2013;16(6):530–6; discussion 536.
59. McCormick Z, et al. Phantom limb pain: a systematic neuroanatomical-based review of pharmacologic treatment. Pain Med. 2014;15:292–305.
60. Mehta S, et al. Gabapentinoids are effective in decreasing neuropathic pain and other secondary outcomes after spinal cord injury: a meta-analysis. Arch Phys Med Rehabil. 2014;95(2):2180–6.
61. Melzack R, Wall PD. Pain mechanisms: a new theory. Science. 1965;150(3699):971–9.
62. Mercier C, Sirigu A. Training with virtual visual feedback to alleviate phantom limb pain. Neurorehabil Neural Repair. 2009;23(6):587–94.
63. Moore RA, et al. Oral nonsteroidal anti-inflammatory drugs for neuropathic pain. Cochrane Database Syst Rev. 2015;10:CD010902.
64. Moore RA, et al. Pregabalin for acute and chronic pain in adults. Cochrane Database Syst Rev. 2009;2009(3). Art. No: CD007076. https://doi.org/10.1002/14651858.CD007076.pub2.
65. Morgenstern FS. The effects of sensory input and concentration on post-amputation phantom limb pain. J Neurol Neurosurg Psychiatry. 1964;27:58–65.
66. Moseley GL. Graded motor imagery is effective for long-standing complex regional pain syndrome: a randomised controlled trial. Pain. 2004;108(1–2):192–8.
67. Moseley GL. Graded motor imagery for pathologic pain: a randomized controlled trial. Neurology. 2006;67(12):2129–34.
68. Mou J, et al. Efficacy of Qutenza (capsaicin) 8% patch for neuropathic pain: a meta-analysis of the Qutenza Clinical Trials Database. Pain. 2013;154(9):1632–9.
69. Mulvey MR, Radford RE, Fawkner HJ, Hirst L, Neumann V, Johnson MI. Transcutaneous electrical nerve stimulation for phantom pain and stump pain in adult amputees. Pain Pract. 2013;13(4):289–96.
70. Murray CD, et al. The treatment of phantom limb pain using immersive virtual reality: three case studies. Disabil Rehabil. 2007;29(18):1465–9.
71. Ney JP, et al. Comparative efficacy of oral pharmaceuticals for the treatment of chronic peripheral neuropathic pain: meta-analysis and indirect treatment comparisons. Pain Med. 2013;14:706–19.
72. Nikolajsen L, et al. A randomized study of the effects of gabapentin on postamputation pain. Anesthesiology. 2006;105:1008–15.
73. Ortiz-Catalan M, et al. Phantom motor execution facilitated by machine learning and augmented reality as treatment for phantom limb pain: a single group, clinical trial in patients with chronic intractable phantom limb pain. Lancet. 2016;388(10062):2885–94.
74. Ortiz-Catalan M, et al. Treatment of phantom limb pain (PLP) based on augmented reality and gaming controlled by myoelectric pattern recognition: a case study of a chronic PLP patient. Front Neurosci. 2014;8:24.
75. Østlie K. Evidence-based guideline for rehabilitation after acquired upper-limb loss in Norway [Norwegian]. 2016. www.magicapp.org/public/guideline/Jn3zaL. Accessed 30 Mar 2016.
76. Østlie K, Granan LP. Retningslinje for rehabilitering etter ervervet overekstremitetsamputasjon [Norwegian]. Tidsskr Nor Laegeforen. 2017;137(1):15.
77. Osumi M, et al. Restoring movement representation and alleviating phantom limb pain through short-term neurorehabilitation with a virtual reality system. Eur J Pain. 2017;21(1):140–7.

78. Pereira EA, et al. Thalamic deep brain stimulation for neuropathic pain after amputation or brachial plexus avulsion. Neurosurg Focus. 2013;35(3):E7.
79. Pet MA, et al. Does targeted nerve implantation reduce neuroma pain in amputees? Clin Orthop Relat Res. 2014;472(10):2991–3001.
80. Prahm C, et al. PlayBionic: game-based interventions to encourage patient engagement and performance in prosthetic motor rehabilitation. PM&R. 2018;10(11):1252–60.
81. Prahm C, et al. Increasing motivation, effort and performance through game-based rehabilitation for upper limb myoelectric prosthesis control. In: 2017 International Conference on Virtual Rehabilitation (ICVR); 2017.
82. Prahm C, et al. Game-based rehabilitation for myoelektric prosthesis control. JMIR Ser Games. 2017;5(1):e3.
83. Preissler S, et al. Preliminary evidence for training-induced changes of morphology and phantom limb pain. Front Hum Neurosci. 2017;11:319.
84. Ramachandran VS, Hirstein W. The perception of phantom limbs. The D. O. Hebb lecture. Brain. 1998;121(Pt 9):1603–30.
85. Ramachandran VS, Rogers-Ramachandran D. Synaesthesia in phantom limbs induced with mirrors. Proc Biol Sci. 1996;263(1369):377–86.
86. Rosenquist EWK, Aronson MD, Park L. Neuropathic pain: pharmacologic approach. 2014. p. 1.
87. Rosenquist EWK, Aronson MD, Park L. Overview of the treatment of chronic pain. UpToDate. 2014. p. 32.
88. Rusy LM, Troshynski TJ, Weisman SJ. Gabapentin in phantom limb pain. Management in children and young adults: report of seven cases. J Pain Symptom Manag. 2001;21(1):78–82.
89. Saarto T, Wiffen PJ. Antidepressants for neuropathic pain. Cochrane Database Syst Rev. 2007; 2007(4). Art. No.: CD005454. https://doi.org/10.1002/14651858.CD005454.pub2.
90. Salminger S, et al. [Prosthetic reconstruction in high amputations of the upper extremity]. Orthopade. 2015;44(6):413–8.
91. Sano Y, et al. Tactile feedback for relief of deafferentation pain using virtual reality system: a pilot study. J Neuroeng Rehabil. 2016;13(1):61.
92. Sherman RA, Sherman CJ, Parker L. Chronic phantom and stump pain among American veterans: results of a survey. Pain. 1984;18(1):83–95.
93. Souza JM, et al. Targeted muscle reinnervation: a novel approach to postamputation neuroma pain. Clin Orthop Relat Res. 2014;472(10):2984–90.
94. Subedi B, Grossberg GT. Phantom limb pain: mechanisms and treatment approaches. Pain Res Treat. 2011;2011:864605.
95. Tilak M, et al. Mirror therapy and transcutaneous electrical nerve stimulation for management of phantom limb pain in amputees—a single blinded randomized controlled trial. Physiother Res Int. 2016;21(2):109–15.
96. Topper R, et al. Repetitive transcranial magnetic stimulation of the parietal cortex transiently ameliorates phantom limb pain-like syndrome. Clin Neurophysiol. 2003;114(8):1521–30.
97. Vandvik PO, Berg R, Vist G. En ny generasjon troverdige kliniske retningslinjer. Norsk Epidemiol. 2013;23(2):197–204.
98. Vasluian E, de Jong I, Janssen WGM, Poelma MJ, van Wijk I, Reinders-Messelink HA, van der Sluis CK. Opinions of youngsters with congenital below-elbow deficiency, and those of their parents and professionals concerning prosthetic use and rehabilitation treatment. PLoS One. 2013;8(6):e67101.
99. Wakolbinger R, et al. Home based Tactile Discrimination Training (TDT) reduces phantom limb pain. Pain Pract. 2018;18(6):709–15.
100. Wartan SW, et al. Phantom pain and sensation among British veteran amputees. Br J Anaesth. 1997;78:652–9.
101. Weiss T, et al. Decrease in phantom limb pain associated with prosthesis-induced increased use of an amputation stump in humans. Neurosci Lett. 1999;272(2):131–4.
102. Wiffen PJ, et al. Antiepileptic drugs for neuropathic pain and fibromyalgia—an overview of Cochrane reviews. Cochrane Database Syst Rev. 2013;2013(11). Art. No: CD010567. https://doi.org/10.1002/14651858.CD010567.pub2.
103. Wilder-Smith CH, Hill LT, Laurent S. Postamputation pain and sensory changes in treatment-naive patients: characteristics and responses to treatment with tramadol, amitryptiline, and placebo. Anesthesiology. 2005;103(3):619–28.
104. Wolff A, et al. 21. Phantom pain. Pain Pract. 2011;11(4):403–13.
105. Wu CL, et al. Morphine versus mexiletine for treatment of postamputation pain: a randomized, placebo-controlled, crossover trial. Anesthesiology. 2008;109(2):289–96.

Part V

Man–Machine Interfaces in Prosthetics

Control Strategies for Functional Upper Limb Prostheses

13

Janne Hahne, Cosima Prahm, Ivan Vujaklija, and Dario Farina

Electrically powered hand protheses are typically controlled with electromyographic (EMG) signals, acquired from muscles of the residual limb. In this chapter we will give an overview on classical EMG control as well as recent developments based on machine learning. Classical approaches utilize two EMG electrodes and allow to control only a single prosthetic function at a time. Machine learning-based approaches utilize more electrodes and can be divided into classification and regression. Classification-based approaches have become recently commercially available and allow a direct access to many prosthetic functions, while classification-based approaches allow for an independent simultaneous control of two degrees of freedom (DOF). Targeted muscle reinnervation is a surgical procedure to acquire additional control sites in the amputees and enables to directly control up to three DOF simultaneously.

Introduction

Electrically powered hand protheses are assistive devices that can help to compensate for the impact an amputation has to a person's life. First prototypes for electrically powered hand prostheses were developed after the 2nd World War in Germany [41] and the first device that became commercially available was released 1964 in the UDSSR [50]. From the very beginning, electromyographic (EMG) signals acquired form residual muscles have been the most important way to control the prostheses. Over many decades myoelectric-prostheses had only one actuated degree for freedom (DOF) for opening and closing the hand. Then a second DOF rotation of the wrist was introduced. In the last years, great advances were achieved in the development of highly functional electrically powered hand prostheses with a high number of actuated joints. Currently at least four manufacturers offer multifunctional hand prostheses with 6 to 11 actuated joints. However, the bottleneck for introducing advanced functionality is not the protheses hardware, but the techniques to read the intention from the user and control the prosthesis. In this chapter we will give an overview on the

J. Hahne (✉)
Applied Rehabilitation Technology Lab (ART—Lab), Department of Trauma Surgery, Orthopedic Surgery and Hand Surgery, University Medical Center Göttingen, Göttingen, Germany
e-mail: janne.hahne@bccn.uni-goettingen.de

C. Prahm
Department for Hand, Plastic, Reconstructive and Burn Surgery, BG Trauma Clinic Tübingen, Eberhard Karls University, Tübingen, Germany

I. Vujaklija
Department of Electrical Engineering and Automation, Aalto University, Espoo, Finland

D. Farina
Department of Bioengineering, Imperial College London, London, UK

O. C. Aszmann, D. Farina (eds.), *Bionic Limb Reconstruction*,
https://doi.org/10.1007/978-3-030-60746-3_13

commercially available control techniques as well as recent developments in research.

EMG Signal Acquisition

Physiological Background

Electromyographic signals are electric potentials in the range of 50 μV to 10 mV that are generated by skeletal muscles during their contraction [31]. Muscle fibers, which constitute the muscle, are innervated by terminal axonal branches of motor neurons. The motor neuron and its innervated muscle fibers constitute a motor unit (MU) [18]. Each action potential of the motor axon triggers a motor unit action potential (MUAP) that propagates toward both ends of the fibers and causes their contraction. The MUAPs of all MUs superimpose and form the electromyogram that can be measured on the surface of the skin. Since an increase in muscle force is mediated by increases in both the number of active MUs and their firing rates, the amplitude of the stochastic interference EMG signal proportionally increases with force [30].

Noninvasive EMG Acquisition

In clinical diagnostics and in electrophysiological research, disposable pre-gelled electrodes are often used in combination with monopolar EMG signal derivation. On the other hand, for the control of active prostheses, bipolar derivation with active electrodes is the commonly applied configuration. Here, the electric potential difference between two electrode contacts, typically located at a distance of 20 mm in the direction of the muscle fibers, is picked up and amplified by a biosignal amplifier [28]. Due to its high common mode rejection ratio and high input impedance, noise that is present on both bipolar electrode contacts are suppressed, as the EMG signal gets amplified [29]. In order to suppress motion artifacts and high-frequency contamination that tends to corrupt the EMG signals, commonly a band-pass filter with cut-off frequencies ranging from 5 to 30 Hz at the low end and 300–500 Hz at the top end is applied [31]. A notch filter is often used to eliminate the power line interference at 50 Hz or 60 Hz depending on the region.

For practical reasons dry metal electrodes are typically used in prosthetics. Due to their relatively large electrode-skin impedance, it is essential to keep the leads to the first amplification stage very short in order to prevent artifacts from saturating the signals. Therefore, active electrode modules are commonly applied, which integrate the electrodes with the amplifiers and the filters in a compact space. Many advanced control approaches use the raw EMG to extract multiple features per channel. However, electrode modules for conventional myoelectric control more often include a rectification and additional low-pass filtering of the signal and thus provide access to the EMG envelope.

Classical Control Approaches

Two-Channel Approaches

The most common myoelectric hand prosthesis control methods are based on two bipolar EMG signals picked up from a pair of antagonistic muscles or muscle groups available at the stump (Fig. 13.1). In case of an amputation, in most cases a phantom representation of the lost limb remains, and the residual muscles contract during phantom limb motions [47]. For individuals with transradial amputation, the wrist flexors and extensors are typically employed while the biceps and triceps brachii are commonly used in cases of transhumeral amputation. The pectoralis major or minor or infraspinatus or teres minor might be selected as source muscles for deriving control signals for amputations at the shoulder level. The electrode modules are integrated into the inner socket of the prosthesis and pressed against the skin with flexible polymer suspension that can partly compensate for stump volume variations [42].

Commonly, EMG activity of the flexor muscles is mapped to a closing function of the prosthetic hand, while activity on the extensors related to its opening. Most devices provide a

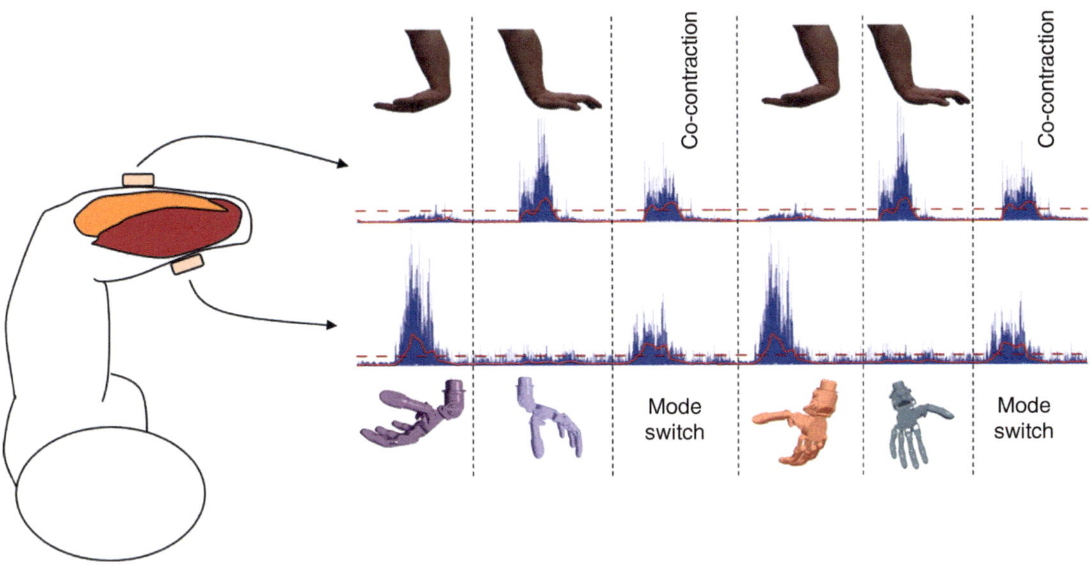

Fig. 13.1 Conventional two-channel proportional myoelectric control with antagonistic muscles. Only one DOF is controlled at a time and the active function is altered by performing a co-contraction. (© Ivan Vujaklija)

proportional control, i.e., the stronger the muscles contract, the faster the prosthesis moves, or a higher grip force is delivered. This control scheme, often referred to as direct control, was already applied in the first commercially available myoelectric prosthesis in the 1960s [50].

With the exception of targeted muscle reinnervation (TMR) patients, there are typically not enough separately addressable muscles to generate independent EMG signals, which could directly extend the prosthesis control to more degrees of freedom (DOFs) simply by increasing the number of channels. To control more than one DOF with only two EMG channels, several heuristics have been developed and are in commercial use. For a good overview on clinically available techniques, please refer to [34].

The most common approach for switching between the DOFs is the co-contraction-based triggering. Once the user contracts both muscle groups at the same time, the active DOF is switched, and the two EMG signals now control another function of the prosthesis, i.e., changing from grasping to wrist rotation. For above-elbow prostheses, even alternating between three functions (grasping, wrist rotation, and elbow flexion) is common. However, as only one function can be controlled at a time, cycling through DOFs is a rather cumbersome control method and thus limits the benefit of additional functions.

Another commercially popular approach distinguishes between grasping and rotation based on the slope with which the EMG signal increases. Slowly increasing EMG activity causes the opening or closing of the prosthetic hand, while a quickly increasing activity causes rotation of the wrist. Once a high or a low slope is detected, the prosthesis stays within the corresponding DOF until the user relaxes the muscles completely and therefore allows for a proportional control of the speed or the force by adapting the contraction force. Here, no explicit mode switching is required, but the DOFs still have to be activated sequentially.

Single-Channel Approaches

In cases when only one EMG channel can reliably be controlled, both directions of a single DOF have to be addressed using a single channel. The direction is selected either by the initial slope of the signal or by its absolute level (e.g., high level for opening, low level for closing). Once the controller detects the direction, it locks into that state until a rest phase is detected and thus allows for a proportional control via modulation of the EMG activity. Alternatively, in those cases where even the proportional modulation of the EMG amplitude is not an option (commonly in kids), a hand can be open using a single channel, and the closing is automatically done by the prosthesis when the user relaxes. This approach is commonly referred to as "cookie crusher" [27].

Non-EMG Approaches

In a clinical setting, control inputs other than EMG are also used as an alternatively or in addition to it. In prostheses with individually actuated fingers, which are usually controlled with the two-channel approach, grip patterns can be preconfigured and selected by pressing a button on the prosthesis with the other hand, or by moving the prosthesis into a certain direction after an EMG-trigger-signal. Also RFID tags placed on certain objects can be used to automatically select grip patterns, when the hand approaches the tag [52].

Control Approaches Following TMR

In TMR, additional EMG sites are obtained by surgically reconnecting the still intact nerves of the lost limb to other muscles in the vicinity [26]. In this way, intuitively controlled muscles that react to phantom limb motions of the lost limb are obtained and used for prosthesis control. Since the muscles are spatially well separated and can be actuated independently by the user, the two-channel direct control approach can be extended to multiple DOFs (Fig. 13.2). In a typical TMR prosthesis following a shoulder disarticulation, up to six EMG signals can be detected and grouped in three pairs of commands. Thus, independent, proportional, and simultaneous control of three DOFs (elbow flexion/extension, wrist rotation, and hand opening/closing) is possible [32].

In those TMR cases where fewer control signals are available, a combination of other conventional techniques can be adopted. For instance, when only five independent EMG channels are present, a pair of electrodes could be mapped to hand open/close and another one to elbow flexion and extension. The signal of the remaining electrode could then be split so that the initial slope maps either pronation or supination and vice versa. With four available channels, two electrodes can be used for elbow flexion/extension exclusively and the other two channels to either control grasping or wrist rotation, which can be switched through co-contraction.

Machine Learning-Based Approaches

To overcome the limitations of the classical control approaches, significant research has been conducted in the past decades with the goal to employ machine learning techniques for extracting more control information from a larger number of EMG signals. Most of these advanced approaches follow an established control chain containing the processing blocks shown in

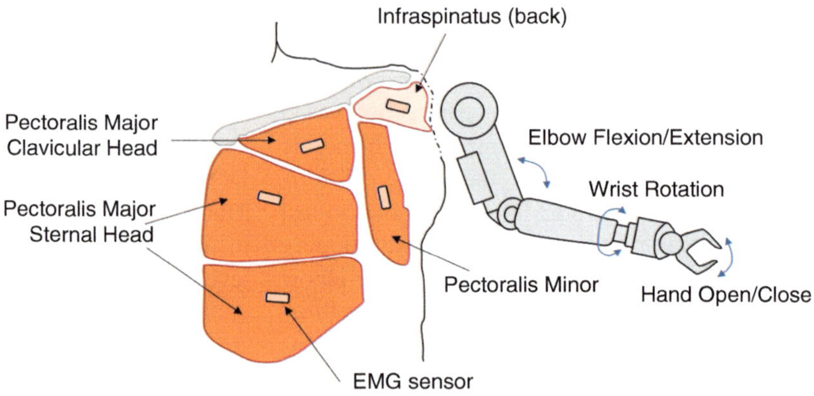

Fig. 13.2 A schematic of a shoulder disarticulation TMR fitting (ventral side). (© Ivan Vujaklija)

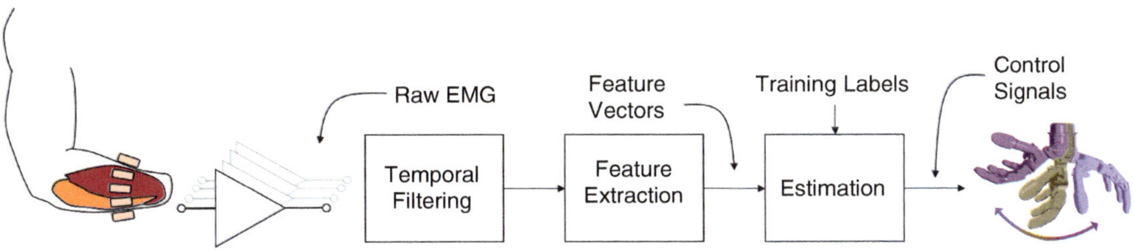

Fig. 13.3 Block diagram describing the signal processing chain in most machine learning approaches for prosthesis control. (© Ivan Vujaklija)

Fig. 13.3. Instead of just two, typically six to ten EMG channels are used, either specifically placed on certain muscles or equally distributed over the area of interest. In order to extract as much information from the signals, the raw EMG is typically used. The signals are digitized and processed in blocks (windows) of 100–200 ms in duration as to satisfy the optimal controller delay [10]. After applying similar noise removal filters as in the conventional control, certain features are extracted from the filtered EMG to condense and describe the information that can be used by the control algorithm. A large number of different features and their combinations (feature sets) can be extracted in time and/or frequency domain. The most common ones include root mean square (RMS), mean absolute value (MAV), slope sign changes (SSC), zero crossings (ZC), wavelength (WL), auto recursive coefficients, short-time Fourier transform features, wavelets, and more [19]. After feature selection, a machine learning algorithm, commonly a classifier or a regressor, interprets the signal features and transforms them into control signals to be used by the prosthesis. Most approaches employ supervised algorithms, i.e., they have to be trained with samples of labeled training data prior to their application. To obtain these training samples, data is recorded for which the type of contraction is known, i.e., by relying on visual cues [1] or by performing bilateral symmetric motions and measuring kinematic or forces of the contralateral, unaffected side [35]. In both cases (visual cues and bilateral motions), a certain error in the labels may be introduced due to variability in task execution.

Classification

In classification-based approaches, an algorithm is trained with a pre-recorded feature samples of all motions the controller should be able to detect. It then compares the current feature vector with the model generated from the training data and decides for a certain motion (class). To achieve a high classification accuracy, several types of classifiers for myoelectric prosthetic control have been examined, such as linear discriminant analysis (LDA) [8], artificial neural networks (ANN) [24], support vector machines (SVMs) [49], k-nearest neighbors [24], and many more.

A classifier only estimates which motion is active, but it does not provide any information on the strength of the activation. It can therefore be combined with a parallel signal path to estimate the activation level (e.g., from the mean amplitude of all channels) and enable the clinically required proportional control of the velocity [4]. Compared to the co-contraction control, the mode switching is omitted, but different DOFs still have to be executed one by one, which requires complex motions to be separated into sequentially executed sub-motions. In the last years, however, EMG pattern recognition has been extended to concurrent classification of motion intent by introducing additional classes for all motion combinations that should be activated simultaneously [36]. In this way, the pattern recognition approach enables simultaneous control and thus promotes a more natural interaction with the environment. However, with increasing number of classes, the classification accuracy decreases, which increases the risk of false motions.

Although classification-based control approaches for myoelectric control have been proposed already several decades ago [13], their clinical impact has been so far limited. Most commercial prostheses still use the established two-channel control approaches. As of recent Coapt LLC [5] offers an FDA approved controller with a classification-based control as an extension to most common hand and wrist devices. In 2018, Otto Bock has introduced their own classification-based controller to the market [46]. The reasons for a limited transfer into clinical applications are related to reliability problems under real-world conditions. Factors such as a change in arm position [11], electrode shifts [54], and time between training and application [51] or changing skin conditions, e.g., due to sweat [21], alter the signal patterns and cause significantly decreasing classification accuracies.

Regression

To overcome some of the limitations that classification-based approaches present, regression-based techniques have been investigated to achieve an independent simultaneous and proportional control of multiple DOFs [22]. Similar to classification, regression is also a machine learning technique and needs to be trained with some calibration data. The essential difference to classification is that a regressor does not estimate a specific class (movement) but instead a continuous physical value (force, speed, position, etc.) for each DOF individually. In this way the activation ratio can be controlled independently in all DOF, which allows, e.g., to open the hand quickly while performing a slow rotation at the same time. Various nonlinear and linear techniques have been investigated for regression-based myoelectric control. Nielsen et al. demonstrated successful control of two DOF with artificial neuronal networks [35] trained on bilateral mirrored motions, and Muceli et al. extended this approach to three DOF [33]. Ameri et al. employed nonlinear kernel-based support vector regression [2] and investigated different strategies for obtaining the labels for supervised training, including visual cues [1]. Gijsberts et al. demonstrated an effective approximation of the kernel approach by Random Fourier Features [12]. Matrix factorization techniques have been able to successfully estimate anywhere from two DoF [23] up to seven DoF [20] control. Finally, regression control established using autoencoders has shown to outperform classic control when driving multiple wrist DOFs [53].

An extensive comparison of linear and nonlinear regression techniques has been made [14] showing that simplest of the approaches still seem to be sufficiently effective in delivering satisfactory user performance. This well may be the consequence of continuous feedback that regression-based approaches offer. Unlike classification where due to discrete nature of the estimation the users are not fully aware in which way the misestimation occurred, regression approaches, with their unbounded solution space, allow implicit adaptation to the user. This way, they can actively compensate for false estimations of the algorithm despite the commonly occurring disturbances [15, 48]. As recently shown in five prosthesis users, this leads to relative high robustness against potential sources of non-stationarities such as changing arm position or donning and doffing the prosthesis [16]. However, the additional capabilities of simultaneous and proportional control are followed by the risk of unintended co-activations in other DOFs. Thresholds to suppress small activations can be introduced, but this problem increases with increasing number of DOFs.

Modeling

It has been shown that simultaneous and proportional control can also be achieved through musculoskeletal modeling that estimates joint moments and joint torques from the muscle activations [7, 43, 44]. This approach has been successfully demonstrated in both upper and lower limb prostheses [6, 45]. Instead of focusing on

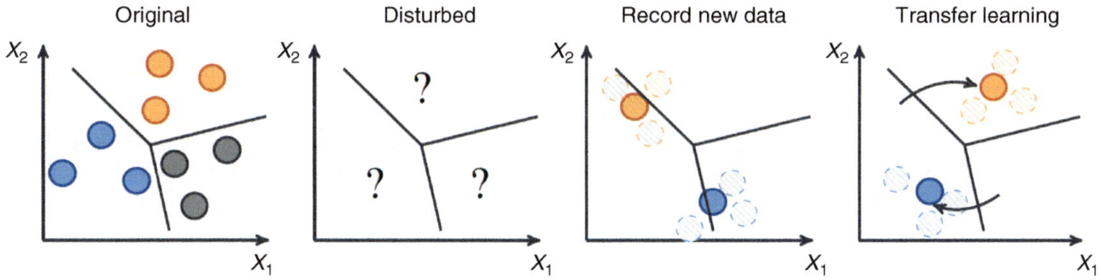

Fig. 13.4 Overview of the transfer learning process [40]: a pattern recognition model (original) is trained on recorded data, through a shift in electrodes, the incoming data is thus disturbed and cannot be accurately recognized by the original model. This occurrence is counteracted by re-recording only a few new data points with this shifted position to estimate new data, so that the original model can be applied again. Thereby virtually shifting the electrodes back into place, when in reality they are not. This transformation of data is called transfer learning. (Used with permission from IEEE)

the explicit properties of the data or the correlations, forward musculoskeletal models recreate the biological process of motion generation. They incorporate the physiological and biomechanical constraints in order to estimate natural limb motions.

In particular, since TMR allows the detection of the neural activity of all nerves involved into the task, including missing muscles, musculoskeletal models allow rather detailed reconstruction of the internal biomechanical representation of missing limbs [9].

Transfer Learning

Various sources of disturbance, such as posture change and resulting electrode shifts can impede the user in everyday prosthesis control [17, 54]. Several approaches have been proposed to improve robustness, such as implanted EMG electrodes instead of surface electrodes [14, 38], high-density EMG surface electrode grids [33, 51] more sophisticated feature extraction [25], and post hoc error detection within the algorithm [3]. However, for the most commonly used surface EMG control, a quick and easy approach to counteract electrode shift is the concept of transfer learning, an approach which adapts the machine learning model to the disturbed data such that the original model is applicable again [37, 39, 40]. Thereby, a pattern recognition model is trained on recorded data (original) (see Fig. 13.4). The colored circles indicate a different class each. Then, incoming data is disturbed through electrode shift, and the learnt model is not applicably anymore (disturbed). The disturbance is estimated by recording only few new instances from only few selected classes in the disturbed condition (record new data). Grey circles indicate possible future positions of transformed data. Finally, the algorithms learns an updated model based on the newly acquired data to virtually transform the data to its original domain so that the model can be employed again (transfer learning) [40].

References

1. Ameri A, Kamavuako EN, Scheme EJ, Englehart KB, Parker PA. Real-time, simultaneous myoelectric control using visual target-based training paradigm. Biomed Signal Process Control. 2014;13:8–14. https://doi.org/10.1016/j.bspc.2014.03.006.
2. Ameri A, Kamavuako EN, Scheme EJ, Englehart KB, Parker PA. Support vector regression for improved real-time, simultaneous myoelectric control. IEEE Trans Neural Syst Rehabil Eng. 2014;22:1198–209. https://doi.org/10.1109/TNSRE.2014.2323576.
3. Amsuss S, Goebel PM, Jiang N, Graimann B, Paredes L, Farina D. Self-correcting pattern recognition system of surface EMG signals for upper limb prosthesis control. IEEE Trans Biomed Eng. 2014;61:1167–76. https://doi.org/10.1109/TBME.2013.2296274.
4. Castellini C, van der Smagt P. Surface EMG in advanced hand prosthetics. Biol Cybern. 2009;100:35–47. https://doi.org/10.1007/s00422-008-0278-1.

5. COAPT LLC, Coapt, 2020. [Online]. Available: https://www.coaptengineering.com/.
6. Crouch DL, Huang H. Lumped-parameter electromyogram-driven musculoskeletal hand model: a potential platform for real-time prosthesis control. J Biomech. 2016;49:3901–7. https://doi.org/10.1016/j.jbiomech.2016.10.035.
7. Crouch DL, Huang HH. Musculoskeletal model-based control interface mimics physiologic hand dynamics during path tracing task. J Neural Eng. 2017;14(3):036008. https://doi.org/10.1088/1741-2552/aa61bc.
8. Englehart K, Hudgins B. A robust, real-time control scheme for multifunction myoelectric control. Biomed Eng IEEE Trans. 2003;50:848–54.
9. Farina D, Vujaklija I, Sartori M, Kapelner T, Negro F, Jiang N, Bergmeister K, Andalib A, Principe J, Aszmann OC. Man/machine interface based on the discharge timings of spinal motor neurons after targeted muscle reinnervation. Nat Biomed Eng. 2017;1:0025. https://doi.org/10.1038/s41551-016-0025.
10. Farrell TR, Weir RF. The optimal controller delay for myoelectric prostheses. Neural Syst Rehabil Eng IEEE Trans [see also IEEE Trans Rehabil Eng]. 2007;15:111–8.
11. Fougner A, Scheme E, Chan A, Englehart K, Stavdahl O. Resolving the limb position effect in myoelectric pattern recognition. Neural Syst Rehabil Eng IEEE Trans. 2011;9:644–51.
12. Gijsberts A, Bohra R, González DS, Werner A, Nowak M, Caputo B, Roa MA, Castellini CPD, Sierra González D, Werner A, Nowak M, Caputo B, Roa MA, Castellini CPD. Stable myoelectric control of a hand prosthesis using non-linear incremental learning. Front Neurorobot. 2014;8:1–8. https://doi.org/10.3389/fnbot.2014.00008.
13. Graupe D, Cline WK. Functional separation of EMG signals via ARMA identification methods for prosthesis control purposes. Syst Man Cybern IEEE Trans. 1975;SMC-5(2):252–9.
14. Hahne JM, Bießmann F, Jiang N, Rehbaum H, Farina D, Meinecke FC, Muller K-R, Parra LC. Linear and nonlinear regression techniques for simultaneous and proportional myoelectric control. IEEE Trans Neural Syst Rehabil Eng. 2014;22:269–79. https://doi.org/10.1109/TNSRE.2014.2305520.
15. Hahne JM, Markovic M, Farina D. User adaptation in myoelectric man-machine interfaces. Sci Rep. 2017;7:4437. https://doi.org/10.1038/s41598-017-04255-x.
16. Hahne JM, Schweisfurth MA, Koppe M, Farina D. Simultaneous control of multiple functions of bionic hand prostheses: performance and robustness in end users. Sci Robot. 2018;3:eaat3630. https://doi.org/10.1126/scirobotics.aat3630.
17. Hargrove L, Englehart K, Hudgins B. A training strategy to reduce classification degradation due to electrode displacements in pattern recognition based myoelectric control. Biomed Signal Process Control. 2008;3(2):175–80. https://doi.org/10.1016/j.bspc.2007.11.005.
18. Heckman CJ, Enoka RM. Motor unit. In: Comprehensive physiology. Hoboken, NJ: Wiley; 2012. p. 2629–82.
19. Hudgins B, Parker P, Scott RN. A new strategy for multifunction myoelectric control. IEEE Trans Biomed Eng. 1993;40:82–94. https://doi.org/10.1109/10.204774.
20. Ison M, Vujaklija I, Whitsell B, Farina D, Artemiadis P. High-density electromyography and motor skill learning for robust long-term control of a 7-DoF robot arm. IEEE Trans Neural Syst Rehabil Eng. 2015:1–10.
21. Jiang N, Dosen S, Muller KR, Farina D. Myoelectric control of artificial limbs: is there a need to change focus? [in the spotlight]. IEEE Signal Process Mag. 2012;29:150–2. https://doi.org/10.1109/msp.2012.2203480.
22. Jiang N, Englehart KB, Parker PA. Extracting simultaneous and proportional neural control information for multiple-dof prostheses from the surface electromyographic signal. IEEE Trans Biomed Eng. 2009;56:1070–80. https://doi.org/10.1109/TBME.2008.2007967.
23. Jiang N, Rehbaum H, Vujaklija I, Graimann B, Farina D. Intuitive, online, simultaneous, and proportional myoelectric control over two degrees-of-freedom in upper limb amputees. IEEE Trans Neural Syst Rehabil Eng. 2014;22:501–10. https://doi.org/10.1109/TNSRE.2013.2278411.
24. Kelly MF, Parker PA, Scott RN. The application of neural networks to myoelectric signal analysis: a preliminary study. IEEE Trans Biomed Eng. 1990;37:221–30. https://doi.org/10.1109/10.52324.
25. Khushaba RN, Takruri M, Miro JV, Kodagoda S. Towards limb position invariant myoelectric pattern recognition using time-dependent spectral features. Neural Netw. 2014;55:42–58. https://doi.org/10.1016/j.neunet.2014.03.010.
26. Kuiken TA, Dumanian GA, Lipschutz RD, Miller LA, Stubblefield KA. The use of targeted muscle reinnervation for improved myoelectric prosthesis control in a bilateral shoulder disarticulation amputee. Prosthetics Orthot Int. 2004;28:245–53. https://doi.org/10.3109/03093640409167756.
27. Meredith JM, Uellendahl JE, Keagy RD. Successful voluntary grasp and release using the cookie crusher myoelectric hand in 2-year-olds. Am J Occup Ther. 1993;47:825–9. https://doi.org/10.5014/ajot.47.9.825.
28. Merletti R. Surface electromyography: the SENIAM project. Eur Med Phys. 2000;36:167–9.
29. Merletti R, Botter A, Troiano A, Merlo E, Minetto MA. Technology and instrumentation for detection and conditioning of the surface electromyographic signal: state of the art. Clin Biomech. 2009;24:122–34. https://doi.org/10.1016/j.clinbiomech.2008.08.006.

30. Merletti R, Farina D. Surface electromyography: physiology, engineering and applications. Wiley-IEEE Press; 2015.
31. Merletti R, Parker P. Electromyography: physiology engineering, and noninvasive applications. Hoboken: Wiley; 2004.
32. Miller LA, Lipschutz RD, Stubblefield KA, Lock BA, Huang H, Williams TW, Weir RF, Kuiken TA. Control of a six degree of freedom prosthetic arm after targeted muscle reinnervation surgery. Arch Phys Med Rehabil. 2008;89:2057–65. https://doi.org/10.1016/j.apmr.2008.05.016.
33. Muceli S, Jiang N, Farina D. Extracting signals robust to electrode number and shift for online simultaneous and proportional myoelectric control by factorization algorithms. IEEE Trans Neural Syst Rehabil Eng. 2014;22:623–33. https://doi.org/10.1109/TNSRE.2013.2282898.
34. Muzumdar A. Powered upper limb prostheses: control, implementation and clinical application. Berlin: Springer; 2004.
35. Nielsen JLG, Holmgaard S, Jiang N, Englehart KB, Farina D, Parker PA. Simultaneous and proportional force estimation for multifunction myoelectric prostheses using mirrored bilateral training. IEEE Trans Biomed Eng. 2011;58:681–8. https://doi.org/10.1109/TBME.2010.2068298.
36. Ortiz-Catalan M, Hkansson B, Brnemark R. Real-time and simultaneous control of artificial limbs based on pattern recognition algorithms. IEEE Trans Neural Syst Rehabil Eng. 2014;22:756–64. https://doi.org/10.1109/TNSRE.2014.2305097.
37. Paaßen B, Schulz A, Hahne J, Hammer B. Expectation maximization transfer learning and its application for bionic hand prostheses. Neurocomputing. 2018;298:122–33. https://doi.org/10.1016/j.neucom.2017.11.072.
38. Pasquina PF, Evangelista M, Carvalho AJ, Lockhart J, Griffin S, Nanos G, McKay P, Hansen M, Ipsen D, Vandersea J, Butkus J, Miller M, Murphy I, Hankin D, et al. First-in-man demonstration of a fully implanted myoelectric sensors system to control an advanced electromechanical prosthetic hand. J Neurosci Methods. 2015;244:85–93. https://doi.org/10.1016/j.jneumeth.2014.07.016.
39. Prahm C, Paassen B, Schulz A, Hammer B, Aszmann O. Transfer learning for rapid re-calibration of a myoelectric prosthesis after electrode shift. In: Converging clinical and engineering research on neurorehabilitation {II}; 2017. p. 153–7.
40. Prahm C, Schulz A, Paaben B, Schoisswohl J, Kaniusas E, Dorffner G, Hammer B, Aszmann O. Counteracting electrode shifts in upper-limb prosthesis control via transfer learning. IEEE Trans Neural Syst Rehabil Eng. 2019; https://doi.org/10.1109/TNSRE.2019.2907200.
41. Reiter R. Eine neue elektrokunsthand. Grenzgeb Med. 1948;1(4):133–5.
42. RSL Steeper. Upper limb prosthetic components. 2020 [Online] Available: https://www.steepergroup.com.
43. Sartori M, Durandau G, Došen S, Farina D. Robust simultaneous myoelectric control of multiple degrees of freedom in wrist-hand prostheses by real-time neuromusculoskeletal modeling. J Neural Eng. 2018;15:066026. https://doi.org/10.1088/1741-2552/aae26b.
44. Sartori M, Farina D. Neural data-driven musculoskeletal modeling for personalized neurorehabilitation technologies. IEEE Trans Biomed Eng. 2016;63:879–93. https://doi.org/10.1109/TBME.2016.2538296.
45. Sartori M, Reggiani M, Farina D, Lloyd DG. EMG-driven forward-dynamic estimation of muscle force and joint moment about multiple degrees of freedom in the human lower extremity. PLoS One. 2012;7:e52618. https://doi.org/10.1371/journal.pone.0052618.
46. Schäfer M, Muders F, Kunz S, Laassidi K. Experience with the use of a novel system for pattern recognition in forearm Prosthesesk. Orthopädietechnik. 2019:18–22.
47. Scott RN, Parker PA. Myoelectric prostheses: state of the art. J Med Eng Technol. 1988;12:143–51.
48. Shehata AW, Scheme EJ, Sensinger JW. Evaluating internal model strength and performance of myoelectric prosthesis control strategies. IEEE Trans Neural Syst Rehabil Eng. 2018;26:1046–55. https://doi.org/10.1109/TNSRE.2018.2826981.
49. Shenoy P, Miller KJ, Crawford B, Rao RPN. Online electromyographic control of a robotic prosthesis. Biomed Eng IEEE Trans. 2008;55:1128–35. https://doi.org/10.1109/TBME.2007.909536.
50. Sherman ED. A Russian bioelectric-controlled prosthesis. Can Med Assoc J. 1964;91:1268–70.
51. Vidovic MM-C, Hwang H-J, Amsuss S, Hahne JM, Farina D, Muller K-R. Improving the robustness of myoelectric pattern recognition for upper limb prostheses by covariate shift adaptation. IEEE Trans Neural Syst Rehabil Eng. 2016;24(9):961–70. https://doi.org/10.1109/TNSRE.2015.2492619.
52. Vujaklija I, Farina D, Aszmann O. New developments in prosthetic arm systems. Orthop Res Rev. 2016;8:31–9. https://doi.org/10.2147/ORR.S71468.
53. Vujaklija I, Shalchyan V, Kamavuako EN, Jiang N, Marateb HR, Farina D. Online mapping of EMG signals into kinematics by autoencoding. J Neuroeng Rehabil. 2018;15:21. https://doi.org/10.1186/s12984-018-0363-1.
54. Young AJ, Hargrove LJ, Kuiken TA. The effects of electrode size and orientation on the sensitivity of myoelectric pattern recognition systems to electrode shift. IEEE Trans Biomed Eng. 2011;58:2537–44. https://doi.org/10.1109/TBME.2011.2159216.

Implantable Myoelectric Sensors for Prosthetic Control

14

Clemens Gstoettner, Stefan Salminger, Konstantin Bergmeister, Anna Willensdorfer, Martin Aman, and Oskar C. Aszmann

While prosthetic technology has been continuously advancing over the last decades, the bottleneck of translation into intuitive and natural prosthetic control has been the functional interface between user and prosthesis. Currently used surface electrodes entail various shortcomings, ranging from low selectivity to frequent signal instability. Most of these limitations can be overcome by implantation of myoelectric sensors, thereby moving them closer to the biological signal source. Different implantable solutions for prosthetic interfacing have been developed and tested in animal as well as human studies. This chapter will give a short overview of the current limitations and go on to present promising implantable solutions as well as a future outlook.

C. Gstoettner (✉) · A. Willensdorfer
Clinical Laboratory for Bionic Extremity Reconstruction, Department of Surgery, Medical University of Vienna, Vienna, Austria
e-mail: clemens.gstoettner@meduniwien.ac.at

S. Salminger · O. C. Aszmann
Clinical Laboratory for Bionic Extremity Reconstruction, Department of Surgery, Medical University of Vienna, Vienna, Austria

Division of Plastic and Reconstructive Surgery, Department of Surgery, Medical University of Vienna, Vienna, Austria

K. Bergmeister
Clinical Laboratory for Bionic Extremity Reconstruction, Department of Surgery, Medical University of Vienna, Vienna, Austria

Karl Landsteiner University of Health Sciences, Department of Plastic, Aesthetic and Reconstructive Surgery, University Hospital St. Pölten, Krems, Austria

Department of Plastic, Aesthetic and Reconstructive Surgery, University Hospital St. Poelten, St. Poelten, Austria

M. Aman
Clinical Laboratory for Bionic Extremity Reconstruction, Department of Surgery, Medical University of Vienna, Vienna, Austria

Department of Hand, Plastic and Reconstructive Surgery, Burn Center-Hand and Plastic Surgery, University of Heidelberg, Heidelberg, Germany

BG Trauma Center Ludwigshafen, Ludwigshafen am Rhein, Germany

O. C. Aszmann, D. Farina (eds.), *Bionic Limb Reconstruction*,
https://doi.org/10.1007/978-3-030-60746-3_14

Current Solutions and Their Shortcomings

While there have been major advances in prosthetic technology and amputation surgery in the last decades, the basic mechanisms of myoelectric signal pickup have remained largely unchanged since its first use roughly 60 years ago [20]. Currently, the control of myoelectric prosthetic devices is achieved via transcutaneous detection of EMG signals using surface electrodes. These are placed in the socket of the device and positioned on the skin over the corresponding muscle. Surface EMG (sEMG) activity can thus be registered and used as control input for prosthetic movements, employing either direct approaches or signal processing algorithms such as pattern recognition or regression. Transcutaneous signal transmission bears several well-known limitations, which negatively influence reliability and overall performance of prosthetic devices, increasing frustration and device abandonment rates among users [4]. An important factor to consider is the amount of tissue between electrode and muscle, comprising mostly skin and fat, which decreases EMG signal amplitude and promotes signal crosstalk between recording sites [8]. Particularly in overweight patients, this leads to a significant decrease in EMG quality, limiting sensitivity and selectivity of each electrode and constraining the number and quality of available sites for myoelectric control. Furthermore, changes of skin electrode position are unavoidable during movement of the stump, especially when lifting heavy objects, or after donning and doffing the device. Other factors which contribute to signal instability are variability of stump size, relative movement of the muscle with respect to the electrode and changes in electrode impedance due to sweating. Given the low signal intensity, higher threshold values need to be used to discriminate between volitional EMG signals and background noise or artefacts. This results in the need for stronger muscular contractions, limiting the accuracy of proportional myoelectric control and promoting fatigue.

The shortcomings of surface electrodes become particularly apparent in patients who have received targeted muscle reinnervation surgery to increase the number of myosignals [18]. Since up to six independent muscle signals have to be registered within the limited surface area of a residual limb, electrode placement becomes increasingly difficult and time-consuming. Crosstalk between signals and incorrect placement of the electrodes often limits control performance and adds to the frustration of patient and prosthetist during the process of prosthetic rehabilitation. Finally, research on the use of cognitive nerve transfers for muscle reinnervation in amputees has shown that the resulting biosignals offer a complex array of compacted information within a relatively small space [5]. In order to fully harvest this potential, high-fidelity signal transmission is necessary (see section "Future Outlook") [3].

While improving efferent control is a major focus in prosthetic research, the inclusion of sensory feedback to close the loop is another important goal which has so far remained elusive in clinical practice. Due to the lack of touch and proprioception, the user needs to rely on visual feedback to guide prosthetic control, which is slow, unintuitive and impedes natural use as well as device embodiment. The strong predominance of afferent fibres within the human brachial plexus, of a factor of approximately 10:1 in relation to motor fibres, underlines the great importance of sensory perceptions for effective interaction with our surroundings [6]. The next chapter will go into further detail regarding the complex topic of sensory feedback.

Developing Implantable Interfaces

Research and clinical experience over the last years have suggested that the necessary way forward to attain more reliable and selective myoelectric control is implantation of EMG sensors. Implantable electrodes can either be directly attached to the surface of a muscle (epimysial) or placed within the muscle belly (intramuscular). This approach largely eliminates the limitations associated with sEMG pickup, while offering the possibility to access a higher number of independent signals [13]. Close proximity to the signal source greatly increases the amplitude of the biosignal, removes crosstalk and effectively prevents changes in impedance and signal quality. Furthermore, implanted electrodes can also be used to stimulate nerves which are cognitively related to the lost hand. In combination with pressure sensors incorporated into prosthetic fingers, this approach can be used to elicit sensory feedback when grasping objects.

However, while these are promising perspectives, development of an implantable medical device comes with its own challenges. First of all, any long-term implant should be designed without the need for percutaneous leads, which are susceptible to infection and may cause additional harm through dislodging, while also being psychologically unacceptable for most patients [19]. Data transmission and powering of the implant must therefore be achieved wirelessly, e.g. using radiofrequency or light-related transmission. Further considerations that need to be addressed when developing an implant for clinical use are biocompatibility of all materials used, chemical and mechanical stability, injury during implantation and chronic functionality. Before moving into human application, these factors need to be evaluated thoroughly in animal models, generally moving from biocompatibility studies in small animals to long-term functionality studies in large animals [1]. Materials used for the implant surface should be well-known and inert, such as silicone, titanium or ceramic for the casing and platin-iridium or steel for the electrode contacts. Any areas with risk of implant failure or wire breakage, such as connections between the central implant and a cable, need to receive special attention during the design process. Once material tests and animal experiments are conducted and confirm device stability and chronic functionality, human studies can be planned, which involves an extensive regulatory process and rigorous documentation. A limited number of implantable systems to improve prosthetic interfacing have been developed and tested so far, none of which have yet become commercially available.

Overview of Implantable Systems

IMES

The IMES ("Implantable MyoElectric Sensors") system is one of the few which has been chronically tested in humans. Originally it was developed by the Alfred Mann Foundation in the US. The IMES are small implants (16 mm long and 2.5 mm in diameter) with a ceramic housing of cylindrical shape and metal end caps acting as electrodes for recording intramuscular EMG [15]. Up to six individual sensors can be placed in different muscles, in order to wirelessly transmit EMG data to the prosthesis. Data transmission as well as power supply of the sensors is achieved through a circumferential external coil, which has to be integrated into the prosthetic socket (see Fig. 14.1). Recently, the first long-term implantation of IMES in conjunction with TMR in above-elbow amputees has been reported, demonstrating that the intramuscular sensors can chronically register and transmit EMG after selective nerve transfers for establishing natural prosthetic control [17]. The results over a period of more than 2.5 years showed substantial functional improvements compared to standard surface EMG control without any events of disconnection or malfunction of the system. Due to the intramuscular placement, signals are independent from the position of the prosthesis and are therefore not subject to disturbances during postural changes or after donning and doffing. For the same reason, these signals can be detected early after nerve transfer surgery, which leads to a significant decrease in rehabilitation time. The fact that these sensors individually transmit signals without a central transmission unit close to the surface allows a certain freedom of wireless placement. However, the distance that needs to be covered requires energy supply which cannot be integrated into the prosthesis. Patients therefore need to wear an additional external belt-worn device, which is generally perceived as bothersome by its users. Another drawback is that the current IMES system is not compatible with metal implants at the stump region. Thus, surgical procedures such as angulation osteotomy or osseointegration cannot

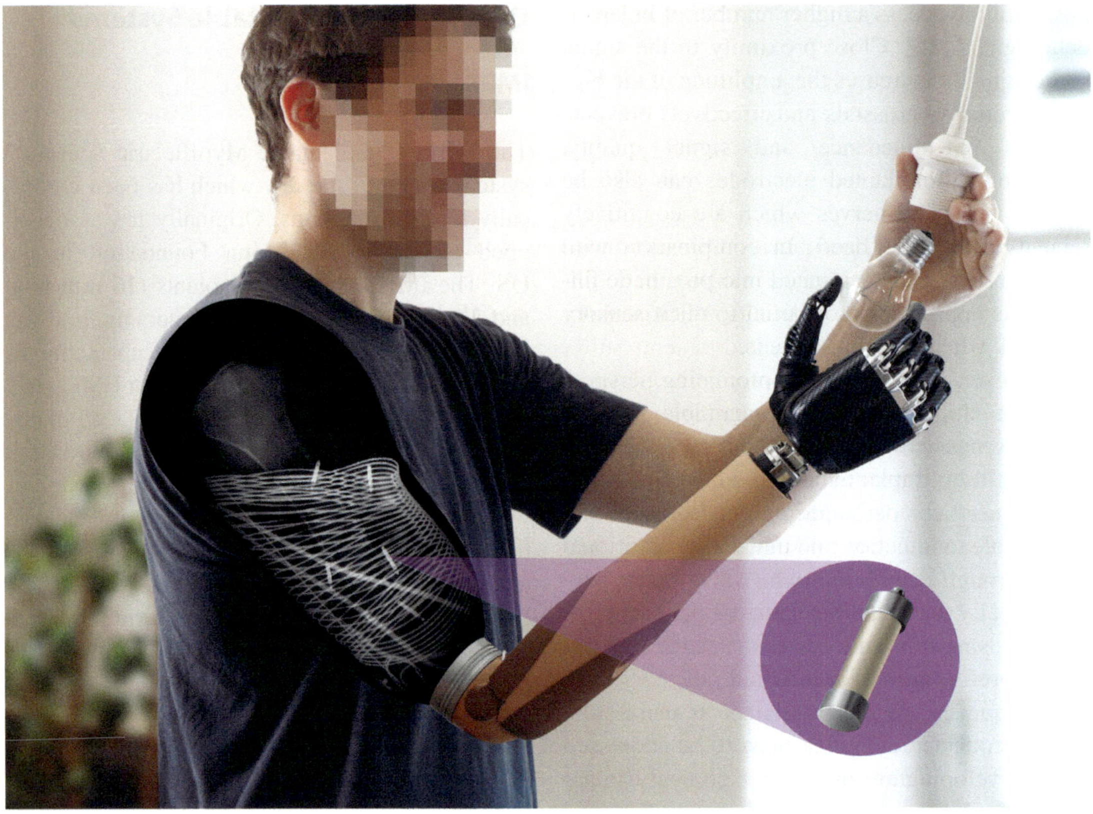

Fig. 14.1 A patient using the IMES system to handle a light bulb. An X-ray of the stump is shown, which includes the individual sensors of the system as well as the circumferential coil which is embedded into the prosthetic socket. The coil is used for wireless communication with the implanted sensors and also provides power supply. The purple circle highlights an illustration of an intramuscular sensor. (© Aron Cserveny for Oskar Aszmann)

be currently combined with the implantable system. Additionally, as the coil has to be placed circumferentially around the stump, short above-elbow amputations or shoulder disarticulations cannot be treated with this system. Finally, the IMES sensors are unidirectional and cannot serve as an afferent signal interface.

eOPRA/OHMG

Osseointegration has emerged as an increasingly accepted approach to improve the mechanical connection between stump and prosthesis after limb amputation. Through an intramedullary titanium implant which is connected to a transcutaneous abutment, a direct and durable link between skeleton and prosthetic device is achieved. Conventional sockets are thus not necessary, and full range of motion can be maintained in the proximal joints. The OPRA (Integrum, Sweden) system was the first of this kind, having been developed by Prof. Rickard Brånemark. More recently, the eOPRA (also referred to as osseointegrated human-machine gateway, OHMG) was presented, which harnesses the transcutaneous port of the osseointegrated implant to establish a connection to permanently implanted neuromuscular electrodes. Wires are tunnelled through the implant, so that the eOPRA neither requires telemetry nor implanted active electronic components (see Fig. 14.2). However, it relies on connectors between the wires and the electrodes, which may increase the risk of implant failure due to fluid intrusion or wire breakage. Epimysial electrodes

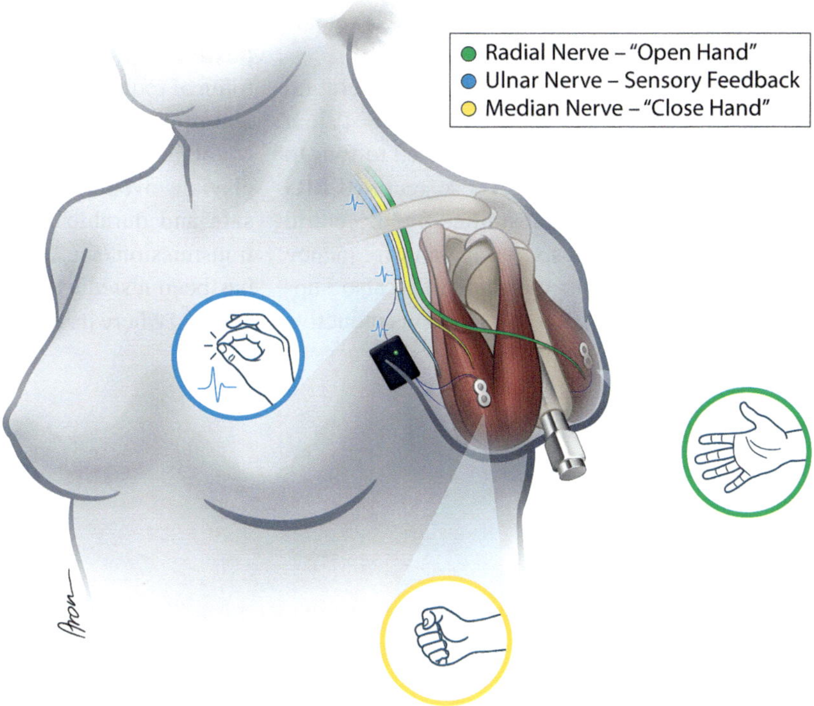

Fig. 14.2 The eOPRA combines the use of an osseointegrated titanium implant for prosthesis attachment with implanted neuromuscular electrodes for bidirectional functional interfacing. The electrode wires are tunnelled through the transcutaneous titanium implant, thus no telemetry is needed. The figure shows two epimysial electrodes placed on reinnervated muscles in a transhumeral amputee, as well as the respective cognitive signals for prosthetic control. A cuff electrode is placed around the ulnar nerve for sensory feedback. (© Aron Cserveny for Oskar Aszmann)

are used for recording EMG signals, and nerve cuff electrodes are employed for sensory feedback, effectively closing the loop of upper limb prosthetic control. The prosthetic device can thus be connected to the patient's bone, nerves and muscles. The eOPRA has been used in human studies on four transhumeral amputees for up to 7 years, demonstrating chronic functionality and functional benefits for patients in daily life [14]. In two of the patients which also underwent targeted muscle reinnervation, the first EMG signals could be registered as early as 1 month after surgery through the epimysial electrodes. Similar to the results of the IMES study, this facilitates an accelerated rehabilitation protocol after TMR. For further details, please see Chap. 6. Advantages compared to the IMES system include implementation without the need for a socket and inclusion of a feedback mechanism. However, similarly to the IMES, adequate stump length is necessary for the osseointegration, and glenohumeral amputees are therefore not suitable. Also, while chronic sensory feedback through nerve stimulation presents an important step forward, this approach is currently not able to restore natural touch perception. The sensations elicited through peripheral nerve stimulation are frequently described by patients as electrical, vibratory or twitching [16].

Flat Coil Technology with Central Implant

In order to offer implantable solutions which are applicable at all levels of amputation, including glenohumeral amputees, different design approaches need to be chosen. Instead of a circumferential coil, as is used in the IMES system, a flat coil technology can be implemented in order to establish a wireless communication and power supply link. This approach has been frequently used in other biomedical devices, such as cochlear or retinal implants, and allows for flexible positioning within the body. Two such systems have been developed to detect EMG signals for prosthetic control.

The MyoPlant was developed in a cooperation between academic and industry partners [9]. It consists of a central implant connected to four to eight epimysial electrodes. These record EMG signals, which are then transmitted by the central unit to the prosthesis via radiofrequency transmission. Power to the implant is also supplied wirelessly, using a transcutaneous inductive link from an external battery. A major advantage of this fully implantable flat coil design is that placement of the central unit during surgery is flexible and can thus accommodate various anatomical requirements. Regarding a potential clinical application in glenohumeral amputees, the MyoPlant system can be placed just below the clavicle over the pectoral muscle, to provide a safe and durable link for transcutaneous signal transmission (see Fig. 14.3). So far, the MyoPlant has been tested extensively in several preclinical studies, where its long-term stability was assessed

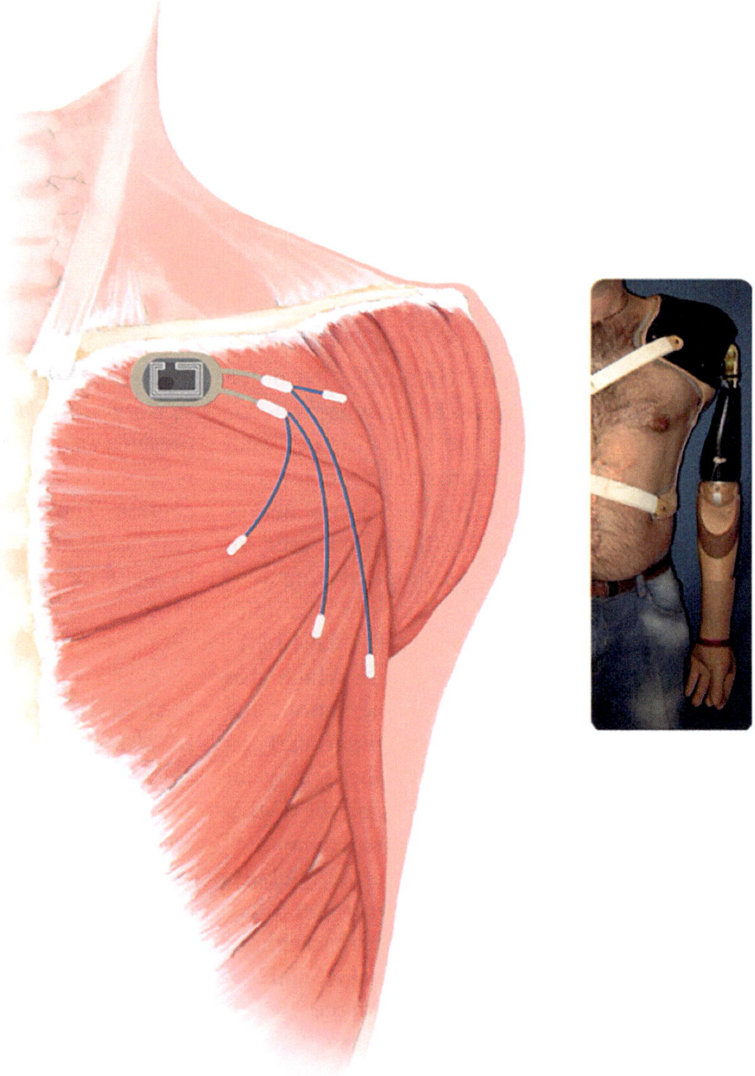

Fig. 14.3 Possible human application of the MyoPlant system. The central unit is placed identically to cardiac pacemakers and the electrodes are implanted epimysial to the targeted muscles. In this scenario, the electrodes receive signals from three separate pectoralis parts and the latissimus dorsi muscle after TMR surgery before being transmitted by the central unit wirelessly to the prosthesis. The system can be equipped with up to eight electrodes to increase the number of muscle signals. (Figure reprinted with permission from Wolters Kluwer [2])

in rats, sheep and rhesus monkeys [2, 10]. Chronic implantation in large animals demonstrated its reliable ability to record EMG data with low crosstalk between agonist and antagonist muscles. A drawback of the MyoPlant system is that it relies on connectors between central unit and electrodes, which represents a predilection site for wire breakage. Also, sensory feedback is not featured in the current design.

The MIRA ("Myoelectric Implantable Recording Array") is a similar solution, which was developed by Ripple Neuro Inc. (Utah, USA). Its central unit is connected to eight silicone leads, each of which carries four electrode contacts (see Fig. 14.4). The leads are placed intramuscularly, in order to gather EMG information from different locations within each muscle. A polypropylene anchor is placed at the end of each intramuscular lead, which ensures fixation within the muscle. In total, the MIRA can record myosignals from 32 individual channels. The central unit is placed directly under the skin and transmits the gathered data to an externally aligned transceiver disc via infrared communication. Power supply is provided through inductive coupling. Similar in its basic design to the MyoPlant, it can also be placed in various anatomic locations and does not require a defined stump length. Also, it offers the advantage of being one complete unit, while the MyoPlant requires assembly of different components via damage-prone connectors. The MIRA has been tested extensively regarding material safety and biocompatibility. Large-animal studies have been performed in dogs and sheep for up to 2 years, demonstrating its ability to chronically transmit stable and highly selective EMG signals from individual muscles (unpublished data). First in human testing is currently being planned. A drawback of the current system is that it does not include sensory feedback. However, the next generation of the MIRA is being designed to include the option for bidirectional transmission. Also, given its high number of channels, the MIRA may be used together with advanced control algorithms, which are able to determine the firing pattern of individual spinal motor units from the activity of reinnervated muscles (see below).

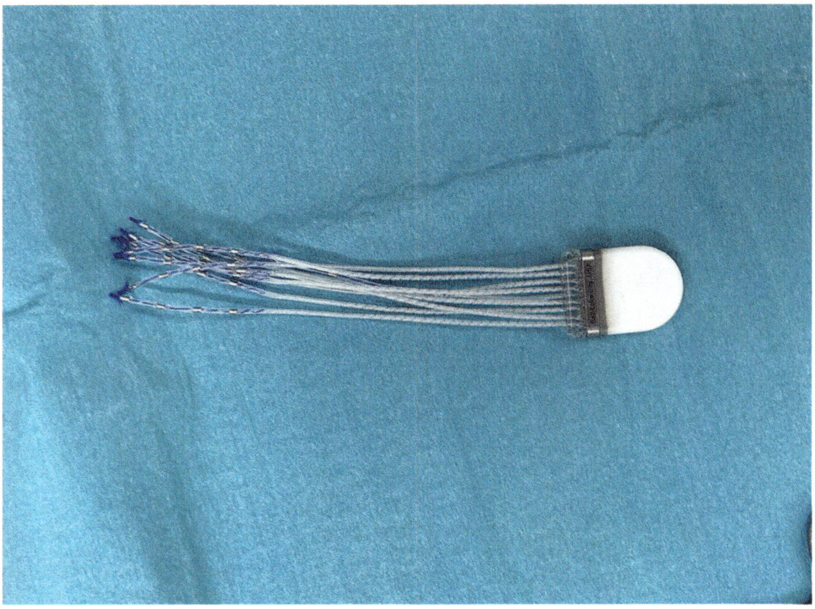

Fig. 14.4 The MIRA implant (Ripple Neuro Inc., Utah, USA) consists of a ceramic-encased central unit which is connected to eight intramuscular silicone leads. Each of the leads carries four steel electrode contacts as well as an anchor for intramuscular fixation. The central unit gathers the EMG signals and sends them to an external transceiver via infrared transmission

Future Outlook: Implantable High-Density EMG Electrodes

With the technological advances described above, an important step forward has been made towards implantation of EMG-triggered systems in clinical practice. This improves signal acquisition and reduces disadvantages of surface EMG as mentioned before. However, most of the systems described only offer between four and eight individual signals, which translates into a maximum of four intuitive degrees of freedom. While these systems lead to a more natural control of prosthetic devices which requires less effort by the user, patients will still be limited to a low number of individual movements, far from replacing the numerous intricate motions of the human hand.

As described in previous chapters, TMR provides additional axonal input to each reinnervated muscle, since highly capable nerves such as the median or the ulnar are used to drive cognitively simple muscles such as the biceps or the triceps. Research has shown, that high-density electrode grids placed over such muscles can be combined with advanced decoding algorithms in order to estimate the actual neural drive of the motor nerve used for reinnervation [5]. So far, this represents the only existing option to access the activity of individual motor neurons and is therefore a major milestone towards highly advanced and natural prosthetic control. Such high-density interfaces, however, still face the well-known problems of superficial electrodes. Therefore, implantation of these multichannel systems will allow to fully harvest the potential of motor unit decoding [7]. The concept of motor unit identification using different implantable muscle electrodes has so far been validated in the acute setting, in animal as well as human subjects [11, 12]. Current studies are ongoing to further develop this approach, evaluating its full potential and biosafety during long-term implantation of high-density electrode grids, after which translation into human subjects will be conducted. In its final application, this implantable system also aims to include natural sensory feedback, through mechanical stimulation of reinnervated dermal matrices. Through a combination of muscle reinnervation and dermal matrix implantation, a cognitive representation of the missing hand will be created within the stump. Together with advanced implanted electronic systems as well as specific mathematical decoding algorithms, this approach aims to create a truly natural substitution of the lost limb (see Fig. 14.5).

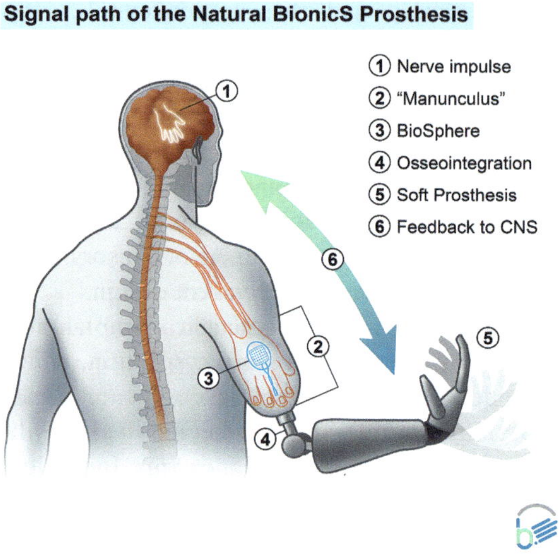

Fig. 14.5 A schematic illustration of truly natural bionic limb replacement, representing the goal of our current research efforts. The neural impulses (1) directed to the missing hand travel from the brain to the "manunculus" within the stump (2). The manunculus describes a surgical concept, which aims to recreate the cognitive representation of the missing hand, employing a combination of targeted nerve transfers and dermal matrix transplantation. In a biosphere (3), which consists of different afferent as well as efferent electrodes, the biological signals are acquired and sent to the prosthetic device, which is attached to the stump via osseointegration (4). Soft robotic hands (5) will be used, which have recently been developed in order to more closely mimic the natural characteristics of the human hand. Sensors within the device send feedback impulses through the biosphere back to the brain (6), in order to close the loop of prosthetic control and promote natural use as well as device embodiment. (© Aron Cserveny for Oskar Aszmann)

Conclusion

The use of implantable EMG sensors enables the extraction and transmission of selective, high-quality myosignals, without crosstalk or other signal disturbances. While the risk of infection and additional surgery needs to be considered, implantable devices nonetheless represent a major improvement compared to current surface electrodes. Both the IMES and the eOPRA have been used in combination with selective nerve transfers in above-elbow amputees, leading to long-term intuitive and dexterous control of robotic arms. Due to the placement of sensors close to the muscle, EMG signals can be detected early after nerve transfer surgery with significant decrease in rehabilitation time. The MyoPlant and the MIRA have so far only been tested preclinically but offer the advantage of flexible placement during surgery and can thus be used in glenohumeral amputees as well. The use of implantable sensors will have an impact on the surgical procedure of TMR, since these sensors can record from deep and small muscles which may thus become new targets for nerve transfers. Future systems should aim to combine the benefits of implantable high-density electrode grids with solutions for natural sensory feedback, in order to close the loop and promote intuitive prosthetic use as well as device embodiment.

References

1. Aman M, Bergmeister KD, Festin C, Sporer ME, Russold MF, Gstoettner C, Podesser BK, Gail A, Farina D, Cederna P, Aszmann OC. Experimental testing of bionic peripheral nerve and muscle interfaces: animal model considerations. Front Neurosci. 2019;13:1442. https://doi.org/10.3389/fnins.2019.01442.
2. Bergmeister KD, Hader M, Lewis S, Russold M-F, Schiestl M, Manzano-Szalai K, Roche AD, Salminger S, Dietl H, Aszmann OC. Prosthesis control with an implantable multichannel wireless electromyography system for high-level amputees: a large-animal study. Plast Reconstr Surg. 2016;137:153–62. https://doi.org/10.1097/PRS.0000000000001926.
3. Bergmeister KD, Vujaklija I, Muceli S, Sturma A, Hruby LA, Prahm C, Riedl O, Salminger S, Manzano-Szalai K, Aman M, Russold M-F, Hofer C, Principe J, Farina D, Aszmann OC. Broadband prosthetic interfaces: combining nerve transfers and implantable multichannel EMG technology to decode spinal motor neuron activity. Front Neurosci. 2017;11:421. https://doi.org/10.3389/fnins.2017.00421.
4. Farina D, Jiang N, Rehbaum H, Holobar A, Graimann B, Dietl H, Aszmann OC. The extraction of neural information from the surface EMG for the control of upper-limb prostheses: emerging avenues and challenges. IEEE Trans Neural Syst Rehabil Eng. 2014;22:797–809. https://doi.org/10.1109/TNSRE.2014.2305111.
5. Farina D, Vujaklija I, Sartori M, Kapelner T, Negro F, Jiang N, Bergmeister K, Andalib A, Principe J, Aszmann OC. Man/machine interface based on the discharge timings of spinal motor neurons after targeted muscle reinnervation. Nat Biomed Eng. 2017;1:0025. https://doi.org/10.1038/s41551-016-0025.
6. Gesslbauer B, Hruby LA, Roche AD, Farina D, Blumer R, Aszmann OC. Axonal components of nerves innervating the human arm: arm nerve axonal components. Ann Neurol. 2017;82:396–408. https://doi.org/10.1002/ana.25018.
7. Kapelner T, Jiang N, Holobar A, Vujaklija I, Roche AD, Farina D, Aszmann OC. Motor unit characteristics after targeted muscle reinnervation. PLoS One. 2016;11:e0149772. https://doi.org/10.1371/journal.pone.0149772.
8. Kuiken TA, Lowery MM, Stoykov NS. The effect of subcutaneous fat on myoelectric signal amplitude and cross-talk. Prosthetics Orthot Int. 2003;27:48–54. https://doi.org/10.3109/03093640309167976.
9. Lewis S, Russold M, Dietl H, Ruff R, Audi JMC, Hoffmann K-P, Abu-Saleh L, Schroeder D, Krautschneider WH, Westendorff S, Gail A, Meiners T, Kaniusas E. Fully implantable multi-channel measurement system for acquisition of muscle activity. IEEE Trans Instrum Meas. 2013;62:1972–81. https://doi.org/10.1109/TIM.2013.2253992.
10. Morel P, Ferrea E, Taghizadeh-Sarshouri B, Audí JMC, Ruff R, Hoffmann K-P, Lewis S, Russold M, Dietl H, Abu-Saleh L, Schroeder D, Krautschneider W, Meiners T, Gail A. Long-term decoding of movement force and direction with a wireless myoelectric implant. J Neural Eng. 2016;13:016002. https://doi.org/10.1088/1741-2560/13/1/016002.
11. Muceli S, Bergmeister KD, Hoffmann K-P, Aman M, Vukajlija I, Aszmann OC, Farina D. Decoding motor neuron activity from epimysial thin-film electrode recordings following targeted muscle reinnervation. J Neural Eng. 2019;16:016010. https://doi.org/10.1088/1741-2552/aaed85.
12. Muceli S, Poppendieck W, Negro F, Yoshida K, Hoffmann KP, Butler JE, Gandevia SC, Farina D. Accurate and representative decoding of the neural drive to muscles in humans with multi-channel intramuscular thin-film electrodes: multi-channel intramuscular EMG electrode. J Physiol. 2015;593:3789–804. https://doi.org/10.1113/JP270902.
13. Ortiz-Catalan M, Brånemark R, Håkansson B, Delbeke J. On the viability of implantable electrodes for the natural control of artificial limbs: review and discussion. Biomed Eng Online. 2012;11:33. https://doi.org/10.1186/1475-925X-11-33.
14. Ortiz-Catalan M, Mastinu E, Sassu P, Aszmann O, Brånemark R. Self-contained neuromusculoskeletal

arm prostheses. N Engl J Med. 2020;382:1732–8. https://doi.org/10.1056/NEJMoa1917537.
15. Pasquina PF, Evangelista M, Carvalho AJ, Lockhart J, Griffin S, Nanos G, McKay P, Hansen M, Ipsen D, Vandersea J, Butkus J, Miller M, Murphy I, Hankin D. First-in-man demonstration of a fully implanted myoelectric sensors system to control an advanced electromechanical prosthetic hand. J Neurosci Methods. 2014;244:85–93.
16. Petrini FM, Valle G, Strauss I, Granata G, Di Iorio R, D'Anna E, Čvančara P, Mueller M, Carpaneto J, Clemente F, Controzzi M, Bisoni L, Carboni C, Barbaro M, Iodice F, Andreu D, Hiairrassary A, Divoux J-L, Cipriani C, Guiraud D, Raffo L, Fernandez E, Stieglitz T, Raspopovic S, Rossini PM, Micera S. Six-month assessment of a hand prosthesis with intraneural tactile feedback. Ann Neurol. 2019;85:137–54. https://doi.org/10.1002/ana.25384.
17. Salminger S, Sturma A, Hofer C, Evangelista M, Perrin M, Bergmeister KD, Roche AD, Hasenoehrl T, Dietl H, Farina D, Aszmann OC. Long-term implant of intramuscular sensors and nerve transfers for wireless control of robotic arms in above-elbow amputees. Sci Robot. 2019;4:eaaw6306. https://doi.org/10.1126/scirobotics.aaw6306.
18. Salminger S, Sturma A, Roche AD, Mayer JA, Gstoettner C, Aszmann OC. Outcomes, challenges and pitfalls after targeted muscle reinnervation in high level amputees. Is it worth the effort? Plast Reconstr Surg. 2019;144(6):1037e–43e. https://doi.org/10.1097/PRS.0000000000006277.
19. Weir RF f, Troyk PR, DeMichele GA, Kerns DA, Schorsch JF, Maas H. Implantable myoelectric sensors (IMESs) for intramuscular electromyogram recording. IEEE Trans Biomed Eng. 2009;56:159–71. https://doi.org/10.1109/TBME.2008.2005942.
20. Zuo KJ, Olson JL. The evolution of functional hand replacement: from iron prostheses to hand transplantation. Plast Surg Oakv Ont. 2014;22:44–51.

Prosthetic Feedback Systems

15

Strahinja Dosen, Cosima Prahm, Sebastian Amsüss, Ivan Vujaklija, and Dario Farina

To fully replace the missing limb, a myoelectric prosthesis needs to provide a bidirectional communication between user's brain and its bionic limb. And indeed, modern prosthetic hands are advanced mechatronic systems that approach the design and capabilities of biological hands both morphologically (size, shape, and weight) and functionally (degrees of freedom). In addition, these hands are controlled intuitively by mapping muscles' activations to prosthesis functions using direct control or pattern classification. However, commercial systems do not yet provide somatosensory feedback to their users. In this chapter, we provide an overview of the methods and techniques that can be used to stimulate the sensory motor structures of an amputee subject in order to restore the missing sensations. We then discuss the prosthesis variables that are most often transmitted through the stimulation as well as the encoding schemes that can be used to map those variables to the stimulation parameters. The contradictory evidence about the impact of feedback on the prosthesis performance is presented next, illustrating that designing, implementing, and assessing effective feedback interfaces is indeed a challenging task. Finally, the chapter ends with discussion and recommendation for further research that will hopefully lead to a successful solution for closed-loop prosthesis control.

S. Dosen (✉)
Department of Health Science and Technology, The Faculty of Medicine, Aalborg University, Aalborg, Denmark
e-mail: sdosen@hst.aau.dk

C. Prahm
Department of Hand, Plastic, Reconstructive and Burn Surgery, BG Trauma Clinic Tübingen, Eberhard Karls University, Tübingen, Germany

S. Amsüss
Otto Bock Healthcare GmbH, Vienna, Austria

I. Vujaklija
Department of Electrical Engineering and Automation, Aalto University, Espoo, Finland

D. Farina
Department of Bioengineering, Imperial College London, London, UK

Introduction

Human hands serve a multitude of functions, from performing dexterous grasping and manipulation, and haptically exploring the environment, to communicating and bonding with others, through gestures and social and affective touch [1]. These functions are implemented through a skillful control by the nervous system that orchestrates an intricate set of muscle actuators and receives sensory feedback from a dense network of sensors. The skin mechanoreceptors provide rich information on the hand motion and interaction with the environment, including contact events and location, pressure distribution, shape, texture, vibrations, and slip [2], and this informa-

O. C. Aszmann, D. Farina (eds.), *Bionic Limb Reconstruction*,
https://doi.org/10.1007/978-3-030-60746-3_15

tion is critically important for all hand functions. When a person loses a hand due to an amputation, this has a profound impact on his/her quality of life. The lost motor functions can be restored to a certain degree using myoelectric prostheses. These systems are controlled by translating the electrical activity of the user muscles into prosthesis commands [3]. The control is intuitive, as the same muscles that were used to move the hand before amputation (hand flexors and extensors) are now mapped to analogous functions in the prosthesis (hand close and open). However, commonly used commercial prostheses do not restore the lost sensory functions. One three recently presented systems claim to provide feedback to the user, yet its clinical effectivness is still to be demonstrated [4–6]. The surveys among the users of myoelectric prostheses have indicated that restoring sensory feedback is indeed an important requirement [7–10]. The users expressed interest in receiving feedback on the prosthesis grasping force, hand aperture, and object contact and release using vibration, pressure, and electrical stimulation [9].

To restore the missing sensations, the prosthesis is equipped with artificial exteroceptive and proprioceptive sensors, the sensor information (joint angles and grasping force) is translated into stimulation parameters, and the stimulation is delivered to the sensory-motor structures that are still available after amputation [11]. This can be implemented using noninvasive methods, as when electro- and vibrotactile stimulation is applied to the surface of the skin [12, 13]. Noninvasive feedback can also be provided acoustically [14, 15] or visually [16, 17]. Alternatively, the stimulation can be delivered using an implanted interface, directly activating peripheral nerves [18, 19] and/or sensory areas in the central nervous system [20]. The advantage of the invasive approach is that it activates the same sensory structures that were used before the amputation to transmit the feedback, and thereby the stimulation could potentially elicit natural and somatotopic sensations, for example, a touch to a prosthetic finger can be perceived as a touch to a user phantom finger. However, this method of delivering feedback requires a surgical procedure that entails potential risks related to post-surgery complications. In addition, some amputee patients, especially those with amputations due to trauma, might be hesitant to undergo additional surgery, and hence they prefer noninvasive solutions [21].

The idea of providing the users of upper limb prostheses with a feeling of touch by implementing sensory feedback mechanisms into artificial limbs is not novel. In fact, this was an active area of research in the 1970s and 1980s [22]. The following years were marked by a rather slow research progress, but recent years have shown an enormous gain in interest and research activities in the area of tactile displays and the integration of haptic signals into prosthetic devices [23, 24]. User demands as well as neurophysiological characteristics of the human body are increasingly taken into consideration during the process of tactile display design [25]. An important point for the design of artificial feedback is to recognize that the feedback is one component within a complex framework of human motor control (Fig. 15.1). The user controls a prosthesis by generating myoelectric signals, which are in most cases recorded noninvasively from the surface of the skin. The control signals are therefore characterized with noise, which increases with the intensity of contraction [26], and this produces an uncertainty in controlling the prosthesis [3]. The user receives feedback through electrical or mechanical stimulation, which he/she needs to interpret in order to understand the feedback information. This process is also characterized with uncertainty [27] that depends on the characteristics of the feedback interface (e.g., number and discriminability of stimulation patterns). In addition, the user can exploit other sources of feedback information that are available intrinsically in the prosthesis. For example, he/she can hear the sound of the prosthesis motor and from that infer the prosthesis state (e.g., closing and opening speed or change in force) [28]. Finally, the user is a smart controller that can learn the behavior of the controlled system and use feedforward control (anticipation and prediction) to operate the prosthesis, without the need to rely on the feedback [29–31]. Therefore, an effective interface for supplemental somatosensory feedback needs to be designed considering the system

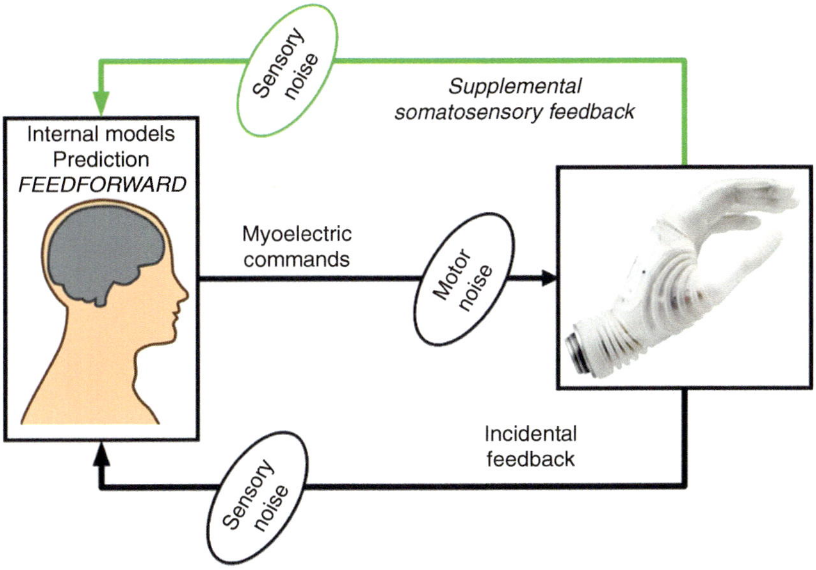

Fig. 15.1 A framework for artificial feedback in prosthetics. The supplemental feedback is one component of a motor control loop that includes a user, noisy myoelectric commands, and intrinsic feedback sources (e.g., motor sound, visual observation). The user is an advanced controller capable of learning and adapting to the environment and operating predictively. To improve prosthesis utility, an effective feedback interface needs to make an impact to the system as a whole

as a whole. For example, the artificial feedback needs to provide more information than what is already available intrinsically in the prosthesis; otherwise, the impact of closing the loop might be very limited or even completely absent.

In this chapter, we begin by providing an overview of the technology to deliver artificial feedback. Then, we summarize the choice of feedback variables and explain how they can be translated into stimulation profiles to be delivered to the user. Next, we discuss the impact of feedback at the functional and psychological level, and finally, we give a discussion of the present state of the art and an outlook into the future developments.

Technology for Artificial Feedback

Electrotactile Stimulation

In electrotactile (electrocutaneous) stimulation [13], low-amplitude electrical pulses are delivered via surface electrodes placed on the skin of the residual limb to depolarize cutaneous afferents and produce tactile sensations (Fig. 15.2a, c). The sensation quality and intensity can be regulated by modulating stimulation parameters. A single pulse is felt as a discrete tap on the skin. With increasing frequency, the individual sensations gradually fuse creating a perception of vibration, tingling, or constant pressure at the surface of the skin [34, 35]. The quantity of charge injected with each pulse (duration × amplitude) determines the number of recruited cutaneous nerve fibers and therefore the intensity of sensation. When the amount of charge exceeds a certain limit, known as discomfort or pain threshold, the stimulation can activate small nociceptive fibers, producing thereby a painful sensation. Typically, the stimulation is delivered using concentric electrodes to establish focused and superficial current flow leading to a localized sensation. The electrode size determines the area of the skin activated by the stimulation, and in general, increasing the area improves the perception [36]. The current flowing deeper into the tissue can activate sensory and motor nerves and elicit muscle contractions and referred sensations (e.g., hand/forearm paresthesia). By properly placing electrodes on the surface of the skin above targeted nerves, this approach can be employed to activate sensory nerves noninvasively [37, 38] and thereby produce somatotopic sensations in the phantom limb akin to those that are generated using invasive interfaces.

Several notable advantages make the electrotactile stimulation an attractive technology for

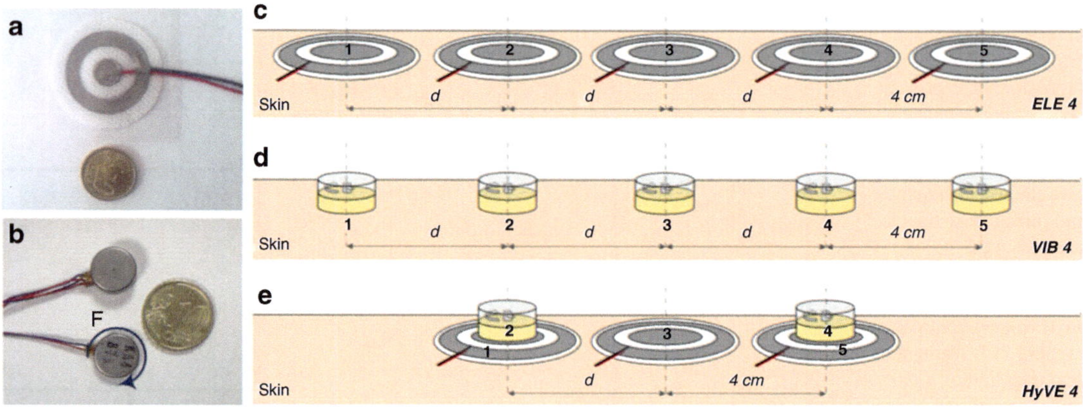

Fig. 15.2 Electro- and vibrotactile stimulation. (**a**) A concentric stimulation electrode producing localized current flow and focused tactile sensation; (**b**) a coin motor generating vibrations tangential to the skin; (**c**) and (**d**) arrays of electrodes and vibrators for multichannel stimulation; (**e**) a hybrid stimulation interface (HyVE) delivering vibrations and electrical stimulation through the same area of the skin. (Adapted from [32, 33])

implementation of feedback in prosthetics. Contrary to vibration motors, there are no moving mechanical elements, which implies low power consumption and fast responses to control inputs. The stimulation interface comprises thin, flexible electrodes and low-power electronics, which allows producing compact solutions integrating many electrodes within a limited area [39]. Therefore, advanced multichannel feedback interfaces can be embedded within the confined space of a prosthetic socket. The stimulation parameters can be modulated independently and simultaneously, which allows flexibility in shaping stimulation profiles to encode feedback information and regulate tactile sensations [36]. However, the stimulation can be uncomfortable if the parameters are set too high. Furthermore, some users can be initially concerned about the approach as they associate it with electrical shocks. Finally, the electrical stimulation is inherently unselective, and therefore, it is not possible to target specific afferents and tactile perception channels [40].

Vibrotactile Stimulation

The skin of the residual limb can be stimulated by delivering vibrations to directly activate skin mechanoreceptors [41, 42] (Fig. 15.2b, d). Depending on the mechanical construction of the vibration device, the oscillations can be perpendicular or tangential to the skin. The simplest solution is a pager vibrator (coin motor) commonly embedded in mobile phones, in which vibrations are produced by rotating a mass positioned eccentrically to the motor shaft (motion tangential to the skin). This design intrinsically couples the stimulation parameters, and the motor has only one control input (motor speed) to simultaneously adjust both the intensity and frequency of vibrations [43]. Several coin motors can be integrated into a common casing to generate more complex stimulation patterns through mechanical superposition of elementary oscillations produced by single motors [44]. Voice-coil technology, originally developed for loudspeakers, allows more flexibility in producing vibrations. The motor integrates a solenoid attracting/retracting a mass connected to the motor base by a spring. In principle, in this design the intensity and frequency of vibrations can be modulated independently by adjusting the amplitude and frequency of the sinusoidal input voltage driving the motor. However, there is still a significant interaction between the parameters due to the resonance properties of a mass-spring system. Typically, such motors have a strongly peaked frequency profile, and therefore, modu-

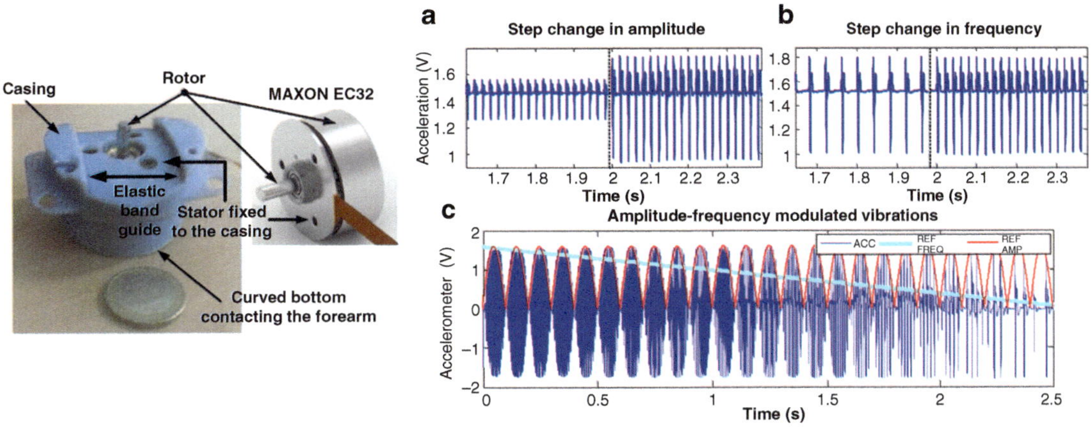

Fig. 15.3 A method to produce vibrations with independently adjustable parameters. The vibrations are generated by driving a DC motor using a train of pulses to produce acceleration ticks. The vibration amplitude and frequency can be adjusted by modulating (**a**) pulse amplitude and (**b**) frequency, respectively. The parameters are thereby decoupled and can be modulated simultaneously, producing complex vibration patterns, as demonstrated in (**c**). The light blue and red lines in (**c**) are command inputs into the motor, and the dark blue signal represents generated oscillations. The vibrations were measured using an accelerometer placed on the vibration motor (left). (Adapted from [43])

lating the frequency can have a significant impact on vibration intensity and vice versa [45]. An additional drawback is that voice-coil motors are noisy, which can be experienced as intrusive by amputees. Recently, a solution for generating rich patterns of amplitude- and frequency-modulated vibrations, where the parameters are fully decoupled, has been presented [43] (Fig. 15.3). The method uses a standard brushless DC motor and a unique control approach, where the oscillations are produced by generating a sequence of ticks that are analogous to pulses in electrical stimulation (Fig. 15.3a, b). The vibration intensity and frequency can therefore be controlled independently by adjusting the rate and amplitude of the generated ticks (Fig. 15.3c).

The advantage of vibrotactile stimulation is in the ease of application since the stimulation is comfortable and cannot produce pain as in electrotactile stimulation. Vibrotactile displays are an affordable solution because low-energy-consuming motors are available off the shelf. However, due to rotating masses, there is a delay in responding to vibration commands, and compared to electrodes, the stimulation units are bulkier and power consumption is higher.

Force and Torque Applicators

In this approach, the feedback is delivered by applying low-amplitude forces and/or torques to the residual limb. Linear motors can be used to push into the skin [46], while rotational actuators can apply torque around the joint [47] or stretch the skin rotationally and/or longitudinally [48]. Schoepp et al. [49] presented a design where a force applicator is actuated through a Bowden cable to allow flexibility in placement and minimize the vertical profile. A recent example for the skin stretch feedback is a servo motor controlling a rocker mechanism covered with rubber to make a tight, high-friction contact with the skin [50, 51]. Another design is based on an ultrasonic motor rotating a miniature end-effector at the surface of the skin, whereby both the position and rotational velocity can be controlled [48, 52]. When the end-effector moves, the skin is gently stretched. A passive mechanism delivering linear skin stretch was presented recently [53]. The prostheses fingers were passively connected to the contact pads attached to the forearm, and therefore, when the fingers move they pull on the pads. Sophisticated mechanisms capable of delivering three-dimensional force feedback have been developed [54], but they have not been translated to prosthetic applications. Finally, electrically actuated braces and pneumatic cuffs [55] can be used to squeeze the forearm delivering the feedback in the form of pressure around the residual limb. An interesting design has been presented in [56] demonstrating a brace with two DC motors to produce normal as well as tangential forces creating pressure and skin stretch around the upper arm, respectively.

The biggest advantage of mechanotactile devices is that they can deliver modality-matched feedback, i.e., the information on the prosthesis grasping force is transmitted to the user by applying force to the residual limb. This is a more intuitive method of providing feedback compared to delivering vibrations or creating tingling sensations using electrotactile interfaces. However, mechanical solutions for the modality-matched feedback are also far more complex compared to miniature vibration motors and compact electrotactile stimulators.

Implantable Interfaces

Sensory feedback can be also restored using invasive techniques (Fig. 15.4). In this approach, implanted electrodes are used to deliver electrical pulses directly to sensory neural structures within the body. The stimulation can be delivered at different levels along the human neuraxis, namely, to peripheral nerves [19, 58], dorsal root ganglions [59], dorsal column nuclei in the brain stem [60], or sensory cortex [20, 61]. The advantage of delivering the stimulation distally (peripheral nerves) is that neural information coding is well understood at this level of the CNS (rate and population coding) and can be therefore reproduced by the stimulation interface. The higher the stimulation target, the less is known regarding the information representation used by the nervous system, but there is also a possibility of producing more complex and higher-level percepts. A variety of electrode interfaces has been used to stimulate the sensory structures. In extraneural systems, the electrode does not penetrate into neural tissue. Examples are traditional cuff electrodes [19] encircling the nerve and a recent flat interface nerve electrodes (FINE) [62–64]. The FINE electrode flattens the nerve to spatially separate the fibers and improve the stimulation selectivity. Intraneural electrodes pierce through the epineurium. Longitudinal (LIFE) and transversal (TIME) intrafascicular electrodes belong to this category [65–68]. As it travels through the nerve, the TIME can activate different areas (bundles) along the nerve cross section. Utah

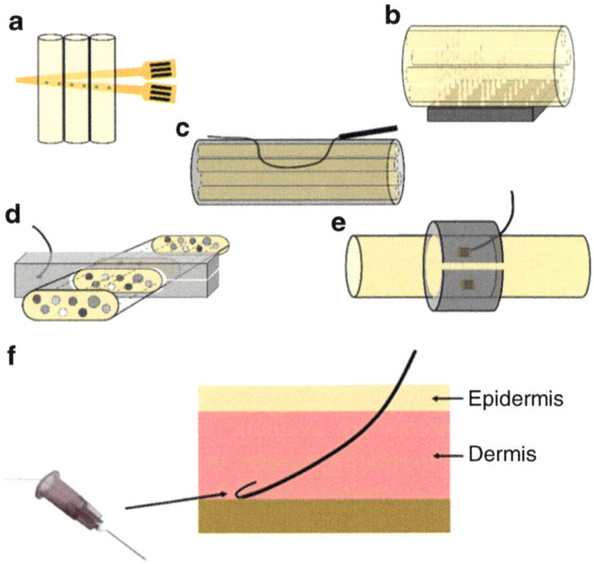

Fig. 15.4 Implantable interfaces that can be used to deliver electrical stimulation to evoke tactile sensations: (**a**) Transversal intrafascicular multichannel electrode (TIME), (**b**) Utah electrode array, (**c**) longitudinal intrafascicular electrode (LIFE), (**d**) flat interface nerve electrode (FINE), (**e**) cuff electrode, and (**f**) subdermal electrode. (Adapted from [57])

slanted electrode array is a matrix of fine needle electrodes that penetrate the nerve [69, 70]. The needles are of different length and therefore reach into different cross sections of the nerve.

The advantage of the implanted interfaces is permanent placement and the possibility, at least in theory, of producing natural sensations, both in quality and location to provide thereby a homologous and somatotopic feedback. The implanted electrodes activate the same sensory structures that were responsible for eliciting natural sensations before amputation. The sensation quality depends on the type of sensory fibers that are recruited by the stimulation. As demonstrated in [71] and more recently in [72], a multi-contact electrode can activate different fiber types and elicit qualitatively different tactile sensations. In addition, as the sensory fibers are organized somatotopically within the nerve, changing active contact can elicit tactile sensations experienced as coming from different areas of the missing hand. Finally, the intensity and frequency of stimulation determine the number of recruited fibers and their frequency of firing, modulating the sensation quality and intensity, analogously to the stimulation through the surface electrodes (see section Electrotactile Stimulation). Therefore, the implanted feedback interfaces can modulate location, intensity, and quality of the elicited tactile sensations. Recently, first studies that report on successful long-term use of such interfaces have been published [73–76]. Subdermal electrodes have been proposed for the restoration of sensory feedback in [77, 78]. They provide some of the benefits of the implantable interfacing (low power, small size) with minimal invasiveness, as they can be applied using a hypodermic needle to insert the electrode just beneath the skin.

Hybrid Methods

Multimodal interfaces combine several stimulation techniques to elicit rich and versatile tactile sensations. In hybrid vibro-electrotactile stimulation (HyVE), a vibration motor is placed on the top of a thin electrode to deliver electrical pulses and mechanical vibrations to the same area of the skin [32, 33] (Fig. 15.2e). It was demonstrated that human subjects could differentiate the properties of the two stimuli when they were delivered simultaneously [32]. Therefore, a single HyVE stimulator implements two information streams transmitted in parallel through the same area of the skin. Furthermore, in a psychometric assessment [33], a multichannel HyVE interface outperformed a vibration interface similar in size and performed similarly to an electrotactile interface larger in size. Another recent example of a hybrid system is a compact device integrating two vibrotactile motors and one mechanotactile stimulator to deliver vibration and force stimulation, respectively [79]. Vibro- and mechanotactile stimulation have been combined in [80], and psychometric tests in amputee subjects indicated an advantage of multimodal versus single-modal stimulation. A miniature haptic display described in [81], comprising a six-bar mechanism and a brushless DC motor, is probably the most advanced solution presented so far. The device can deliver contact (tapping), pressure, vibration, shear force, and temperature stimulation. Nevertheless, hybrid methods are still rarely used for prosthesis control likely due to technical challenges and potential information overload of the prosthesis user.

Feedback Variables and Information Coding

Ideally, the artificial feedback should mimic the natural connection between the hand and the brain and therefore provide complete exteroceptive and proprioceptive information. However, the technologies for tactile stimulation can transmit only a fraction of this information due to a low communication bandwidth. Normally, the subjects need training to perceive and interpret tactile sensations elicited by the feedback interface and to associate those with prosthesis state. Most often, only the prosthesis grasping force is selected as the information to be fed back to the prosthesis user [23, 82], because it is critical for grasp stability and difficult to estimate visually. If the force is too low, the object can slip from the grasp, while an excessive force can damage the object or the prosthesis. Modern prostheses are strong and can generate forces of up to 100 N [83], and therefore, grasping a delicate and brittle object (an egg) can be indeed a challenging task. In this case, there are no visual cues on the exerted force, as the object is rigid, and it snaps suddenly when the force crosses the breaking threshold.

The simplest approach to communicate grasping force is to use a single channel of vibro- and electrotactile stimulation and modulate the stimulation parameters proportionally to the force [84–87]. For example, a higher intensity or frequency of stimulation can indicate higher force. As demonstrated in a recent study [69], the mapping function from the measured force to the stimulation parameters can be important for the effectiveness of feedback. The stimulation intensity/frequency can increase linearly or exponentially with force [88, 89]. The latter follows the Weber law of human perception and translates the same change in force to a larger change in stimulation parameters as the baseline force increases. George et al. [69] presented a biomimetic encoding where frequency and/or amplitude was modulated with the stimulus position, velocity, and acceleration, and they have demonstrated that this approach led to a better performance compared to the "classic" linear mapping during an active exploration task using a myoelectric prosthesis. Valle et al. [68] compared linear frequency and amplitude coding of grasping force using electrical stimulation delivered via an intraneural feedback interface. The experiments demonstrated that amplitude coding resulted in better force modulation and less adaptation. The parameter modulation can be continuous or discrete. When using discrete encoding, the force range is divided into several levels (e.g., low, medium, and high), and the stimulation transmits the current level of the measured force. This simplifies the interpretation of the feedback, as the user needs to recognize few discrete levels instead of continuous modulation [90, 91]. If multiple channels are available, the information can be conveyed using spatial coding, in which the current level of grasping force is indicated by activating a specific stimulator within the array [92]. For example, in [31], ten-coin vibrators placed along the forearm were used to transmit ten levels of force during prosthesis grasping. The spatial coding can be combined with parameter modulation leading to mixed coding [27]. Here, each stimulator can be activated at several intensities and/or frequencies, and the force level is indicated by a combination of activated stimulator and selected stimulation intensity/frequency. This approach allows communicating many force levels clearly and with minimal training, because it is easier to discriminate a combination of two qualitatively different parameters (N locations × M frequencies/intensities) compared to $N \times M$ levels of a single parameter (intensity, frequency, or location). Mixed coding has been used with Michelangelo Hand to provide high-resolution feedback that was still easy to interpret (>95% success rate in recognizing 15 force levels) [27] (Fig. 15.5). Finally, a phenomenon of tactile illusion can be exploited to provide feedback using electro- and vibrotactile stimulation [41]. To produce the illusion, two stimulators are activated simultaneously at specific intensities to elicit a perception of tactile sensation located between the two stimulators. The location of the phantom sensation can be modulated by adjusting the ratio between the two intensities to produce a moving tactile stimulus.

A popular method to transmit prosthesis grasping force is to use modality-matched interfaces such as mechanotactile stimulators in the form of linear force applicators or squeezing braces and cuffs. In one of the early studies [55], pressure feedback delivered through a cuff was compared to vibration feedback using a force-matching task, and the tests demonstrated that the modality-matched interface (pressure) resulted in less error. In [93, 94], linear pushers (servo motors) were embedded into a prosthetic socket to apply low-level forces to the residual limb. The tactile stimulation was delivered to the skin locations that elicited sensations in the lost fingers (phantom map), resulting in both modality-matched and somatotopically congruent feedback. In a later study [46], this interface outperformed vibration feedback in a multi-site sensory discrimination task.

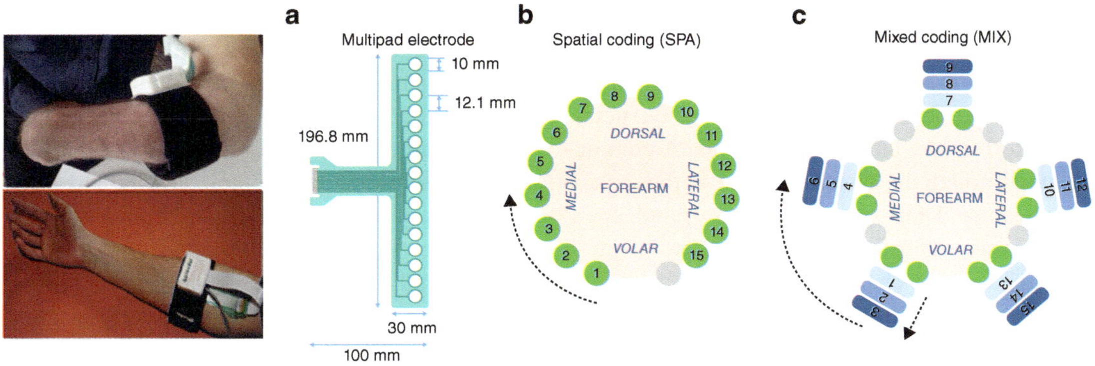

Fig. 15.5 A flexible electrode for multichannel electrotactile stimulation. (**a**) The electrode dimensions and placement; (**b**) spatial coding of grasping force, where each electrode pad when activated represents 1 level of force (15 in total); (**c**) mixed coding of grasping force, in which 5 pairs of pads are activated at 3 different frequencies, resulting in 15 force levels (5 pairs × 3 frequencies). The mixed coding allows better success rate in recognizing the force levels. (Adapted from [27, 39])

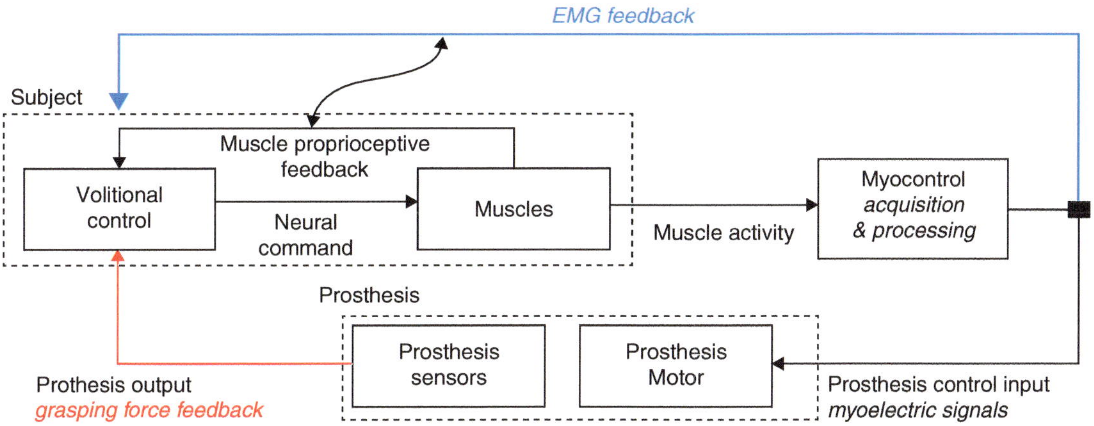

Fig. 15.6 A novel approach to feedback in prosthetics. Instead of feeding back the prosthesis output (sensor data), such as generated grasping force, the idea is to use the feedback to transmit the input into the prosthesis, that is, the myoelectric control signal generated by the user. As the prosthesis operates proportionally, the generated EMG anticipates the grasping force and thereby allows predictive control. This also augments the natural proprioceptive feedback from user muscles. (Adapted from [98])

More recently, it was proposed that force feedback might not be an optimal choice for facilitating force control during routine grasping with a prosthesis. This apparently counterintuitive conclusion stems from the fact that grasping is a fast and routine action, in which the grasping force is anticipated even before the object has been contacted [95]. Once the prosthesis contacts a rigid object, the force increases fast and the feedback comes too late to make corrections. Therefore, to support the routine grasping, the feedback should transmit variables that allow for predictive control of force, such as prosthesis velocity [96]. Due to the nature of prosthesis operation, the closing velocity is proportional to the grasping force, and therefore, the velocity anticipates the force, providing more time for corrections. Following similar reasoning, a novel approach to feedback was presented by [97]; instead of feeding back the prosthesis output, such as measured force or closing velocity, it was proposed to transmit the command input, i.e., a myoelectric signal (EMG feedback, Fig. 15.6). And indeed, the tests have demonstrated [98] that EMG feedback transmitted using electrotactile stimulation with spatial coding (4 locations × 2 frequencies) outperformed the classic force feedback communicated through the same interface. In these studies, a single DOF of a prosthesis (hand open/close) was controlled proportionally using a smoothed single-channel EMG, and this signal was also transmitted back to the user via EMG feedback. Following similar reasoning, the group of Jonathon Sensinger used auditory feedback to convey the output of more advanced myoelectric controllers based on pattern classification and regression [99, 100]. The feedback communicated proportional and simultaneous activation of two DOFs using four distinct frequencies.

Recently, the group of Christian Cipriani [101, 102] proposed using vibrotactile bursts to transmit only discrete events, such as object contact and release, instead of continuous information (e.g., magnitude of grasping force). This approach is inspired by the insights from the human motor control proposing that the grasping motion unfolds in phases, which are demarcated by specific events and controlled by phase-specific controllers [103, 104]. The role of discrete vibrotactile feedback is therefore to facilitate the control by communicating the timing of these events and phases of the grasping action. In a recent study [105], where the feedback communicated both discrete events and continuous myoelectric information, the results demonstrated that the discrete vibrotactile stimulation dominated the continuous acoustic feedback. In addition to contact and release events, discrete feedback has been used to convey the slippage of the object from the grasp [101, 106].

Compared to force feedback, there was substantially less work in providing artificial proprioception. In one of the early studies, the elbow angle of a prosthetic arm was transmitted through vibrotactile stimulation using the tactile illusion approach [28]. The joint angle was mapped to a vibrotactile stimulus moving along the forearm between the two stimulators. In more recent studies [107, 108], the aperture of a virtual and real prosthesis was transmitted using spatial coding with ten vibrators placed on the forearm. The results demonstrated that transversal and longitudinal placement of vibrators resulted in a similar performance. In [109], the position of a virtual wrist was conveyed using vibrotactile stimulation with mixed coding. Two interfaces were tested, three custom-made voice-coil tactors and three coin motors to deliver vibrations perpendicular and tangential to the skin, respectively. Each motor could vibrate at 4 frequencies resulting in 11 discrete levels of feedback. The tests showed that there was no significant difference in performance between the two interfaces. In a recent study [39], dynamic stimulation patterns were proposed to intuitively transmit several prosthesis variables, including wrist rotation and hand aperture, with the possibility to combine the patterns and thereby communicate several variables simultaneously. The stimulation electrodes were positioned circumferentially along the forearm. The hand aperture was transmitted by generating two tactile stimuli that move towards or away from each other to indicate hand closing and opening, respectively. The wrist rotation was coded by rotating the two stimuli around the forearm. The patterns were tested using a novel system for electrotactile stimulation, including an easy to apply, flexible array electrode that could be integrated into a prosthesis socket. In [110], multichannel electrotactile stimulation was used to communicate finger movement during simultaneous and proportional myoelectric control of a simulated and real dexterous robotic hand. Mixed encoding was employed, with one electrode for each controlled degree of freedom (DoF) and frequency modulation to transmit the amount of flexion in that DoF. Four voice-coil vibrators placed around the waist and activated at three levels were used to transmit a configuration of a virtual hand (4 DoFs) [111]. Two coding schemes were tested, in which the vibrators communicated individual DoF angles or activation levels of the four kinematic synergies (principal components). In [48], a wearable device delivering skin stretch was used to communicate the position of a simulated arm. The elbow angle was mapped to the amount of rotation of the device end-effector. In another study, the skin stretch produced through a rocker mechanism [50, 112] was employed to transmit hand aperture while controlling a robotic hand. In particular, the study by Battaglia et al. [50] has demonstrated the challenges of providing feedback from a compliant underactuated hand, in which the fingers can move independently and therefore the aperture cannot be described by a single number. Finally, an interesting approach to provide modality-matched proprioceptive feedback is to evoke a kinesthetic illusion [23]. This can be done by vibrating the tendons to activate muscle spindles, which creates an illusory joint motion. However, to produce the illusion, the vibration intensity needs to be high, and this requires a bulky device. In a recent work [113], this approach has been used to restore the kinesthetic sense in amputees for closed-loop myoelectric control of a dexterous prosthetic hand. The perception of complex grasp movements was successfully created by vibrating the muscles.

Finally, several studies developed and tested stimulation interfaces capable of providing more feedback variables simultaneously, most often a combination of hand aperture and grasping force. Arakeri et al. [114] used two channels of electrotactile stimulation placed on the neck to communicate force and aperture. D'Anna et al. [65] used several intraneural electrodes with multiple contacts implanted into median and ulnar nerves to elicit tactile sensations in two nonoverlapping locations in the phantom limb. The two spatially distinct percepts communicated touch and joint angle using sensory substitution. A similar approach using extraneural electrodes (FINE) was presented in [115], where a proprioceptive percept (finger flexion) could be elicited to communicate aperture information through modality-matched feedback.

Impact of Feedback

Performance

Considering that sensory feedback is essential for motor control in able-bodied subjects [103], it is surprising to find that the results reported in literature regarding the benefits of feedback in prosthetics are actually contradictory. For example, an early study [116] demonstrated that electrotactile feedback improved the control of simulated grasping even in the presence of abundant visual cues (grasping a compliant object). On the other side, in a recent study [117], the artificial feedback failed to improve the performance, even in the context of a dual task which drew visual attention away from the prosthesis.

The only studies in which the benefits of artificial feedback are clearly evident and consistently reported are the ones in which the subjects were fully isolated from all other feedback sources [19, 62, 107]. Typically, the subjects wore noise cancelling headphones and blinds so that the hearing and vision were completely blocked [19, 65] or substantially impaired [50]. Therefore, the subjects could not see or hear the prosthesis movements. These studies demonstrated that artificial electro- and vibrotactile feedback improved prosthesis operation when compared to the condition of full sensory deprivation. For example, the feedback resulted in better control of force [19] and aperture [107], stiffness recognition [118], object identification [114, 115], and positioning of wrist [109] and elbow [28]. However, this is hardly a surprising outcome, as any feedback is likely to be better than no feedback at all. These studies are still important as they demonstrate the effectiveness of the proposed feedback interfaces; however, they do not say much about the utility of the artificial feedback in a daily life. Similar comment holds for the studies in which some elements of the prosthesis control loop are simulated, for example, when using vibrotactile feedback to communicate the interaction forces between a haptic robot and a virtual object [119, 120].

The results reported in the studies performed using a more realistic setup are often openly contradictive. In one of the early studies [121], it was demonstrated that modality-matched force feedback provided by a linear force applicator increased the success rate when manipulating a brittle object. However, in a later study [85], force feedback transmitted via a single vibrator (coin motor) using intensity modulation did not improve the success rate in grasping daily-life objects using a prosthesis. A coin type vibrator has also been used in [87], where subjects grasped four objects of different weight. When the force feedback was provided, the subjects decreased their grasping forces, improving thereby the economy of the grasp. Force feedback has been tested in [84] using an interactive force matching task during which the subjects were asked to grasp an object and produce low, medium, and high force. Force information was coded by modulating the duration and frequency of 200-Hz vibration bursts. The force feedback decreased the error only for medium forces and in the group of subjects experienced with myoelectric control. In a recent study [47], the grasping force feedback delivered through a torque applicator failed to improve the performance, despite using a custom-made low-impedance gripper providing a precise and continuous force modulation. As demonstrated in [31], the vibrotactile force feedback might not be beneficial if the task is simple and prosthesis operation predictable. The feedback improved grasping performance only when an uncertainty was introduced in feedforward control. In a longitudinal study [102], the subjects used a prosthesis equipped with a simple contact feedback for daily-life tasks, and they were tested weekly using a virtual egg test to mimic grasping of delicate brittle objects. The performance improved across testing sessions only in the feedback condition, and the subjects broke fewer blocks in the sessions in which the feedback was provided. Another longitudinal study [122] demonstrated that electrotactile feedback on grasping force was useful, but the effectiveness of feedback decreased substantially due to short- (within session) and long-term (across session) learning. In [62] the performance of SHAP test, which is a standard clinical test for the assessment of prosthesis control in functional tasks, was similar with and without somatosensory feedback. Markovic et al. [124] employed multichannel vibrotactile feedback to communicate contact, active function, and grasping force in a multifunctional prosthesis. The experiments in amputees showed that the effectiveness of feedback depended on the task as well as the level of training. In several recent studies, the feedback improved performance in tasks that included grasping delicate objects, such as restacking of soft cups [18, 124] and replacing virtual eggs [67, 99].

The few studies investigating the artificial proprioception failed to demonstrate a significant benefit of feedback if the visual input was not concurrently blocked. A most illustrative example was a study [125] in which the subjects used myoelectric interface to control a cursor shown on the computer screen, while the feedback on the cursor position was transmitted by moving the contralateral arm using a robotic manipulandum, hence providing supplemental feedback using natural proprioception. Even in this ideal case, the additional proprioceptive input did not improve the performance compared to visual feedback alone. Similar results were reported in [126] where supplemental vibrotactile feedback did not improve learning to control a simulated prosthesis.

Embodiment

The artificial somatosensory feedback can have a substantial psychological as well as therapeutic impact on prosthesis user. Sensory information is instrumental for the sense of embodiment, which is a subjective feeling that something belongs to someone's body. The feedback can be used to manipulate the feeling of body ownership in able-bodied subjects to such an extent that an artificial object can be integrated into the body scheme. For example, in a rubber hand illusion (RHI), an experimenter simultaneously strokes a silicone hand in full view and the subject hand, which is hidden under a table or behind the screen. After only a few minutes of stimulation, the subject develops an illusion that the silicone hand is a part of his/her own body; in other words, the subject embodies the silicone hand and, at the same time, disembodies his/her real hand. The integration of multiple sensory inputs that are congruent in time and space is critical for the illusion to take place. Practically, this means that the stimulation has to be delivered synchronously and to the same location on the silicone and real hand. Importantly for prosthetics, the RHI can also be induced in amputees [127], especially in those that have a sensory map of the phantom hand on the residual limb, as this allows the delivery of spatially congruent stimulation. However, the illusion was weaker compared to stimulating the contralateral hand (classic RHI setup). The RHI could be evoked even when using an artificial hand with a robotic appearance and while the hand has been actively operated via myoelectric control [128]. Even closer to a real-life scenario, in [129] the illusion was produced in two patients with targeted muscle reinnervation using an automatic feedback loop comprising a sensorized prosthesis and a tactor. The tactor stimulated the phantom map on the residual limb proportionally to the measured grasping force. Furthermore, the RHI could be successfully induced in able-bodied and amputee subjects using vibration motors [130]. The touch on the rubber hand was transmitted by delivering vibration to the forearm using a feedback interface convenient for practical applications (coin motor). This study therefore demonstrated that the RHI can be produced even when the stimulation was not matched in modality. In [131], able-bodied subjects received a 4-week training in prosthesis grasping with electrotactile touch feedback. Although the stimulation was incongruent in modality and location, the training elicited a visuo-tactile cross modal congruency effect (CCE) in the peripersonal space of the prosthesis. The CCE characterizes normal limbs, and it is positively associated with the feeling of embodiment. Recent studies demonstrated that somatosensory feedback delivered through different stimulation modalities (e.g., vibration, electrical, and mechanical) seems to improve the embodiment [73, 74], especially after a longer time use of a sensate prosthesis [75].

As demonstrated in [132], sensory stimulation per se can decrease phantom limb pain (PLP). Sensory discrimination training can reverse the maladaptive plastic changes in the brain, which is believed to be one of the main reasons behind PLP. Importantly, a prosthesis equipped with sensory feedback can have a therapeutic effect [133, 134]. A group of amputees undergone a training over several sessions with a single DOF prosthesis (gripper) providing an electrotactile feedback on the prosthesis grasping force using a set of electrodes placed circumferentially around the forearm. The training substantially decreased the intensity and frequency of the PLP episodes.

Discussion

Providing an effective sensory feedback to prosthesis users is a long-standing challenge in prosthetics. The first prototypes of prostheses equipped with artificial tactile feedback have been developed in the 1970s [135, 136], and a first review providing historical perspective was published in the 1980s [22]. Nowadays, the technology has immensely advanced; modern prostheses are dexterous robotic systems with individually controllable fingers [83], and there are miniature surface and implantable stimulators capable of delivering flexible stimulation profiles over multiple channels. Nevertheless, an effective commercially available feedback interface with established clinical utility is still missing.

Therefore, it seems that the technology is not a real bottleneck. Instead, we propose that what is missing is a clear understanding about the role of feedback in closed-loop prosthesis control and that this question needs to be addressed through basic research investigating the very nature of feedback in prosthetics, as also voiced in a recent review [25]. There are some recent studies that indeed recognize this point [29, 137, 138]. The main assumption is that the feedback needs to be regarded within a broader context of human motor control (Fig. 15.1). Humans acquire dynamic models of the environment and use those internal models to anticipate and control predictively [139]. Furthermore, when estimating the state of the environment, humans can integrate information from multiple feedback sources as well as previous experience (expectation) to obtain an optimal estimate [140]. Some recent studies have demonstrated that similar principles operate in amputee and able-bodied subjects when using myoelectric control [29, 30]. Nevertheless, a systematic and deep investigation of the motor control strategies that are employed by prosthesis users is still to be undertaken. We need models of computational motor control, akin to those used in able-bodied subjects [141, 142], which can predict and explain the behavior of amputees when controlling myoelectric prostheses during functional tasks. Such models can inform the design of feedback interfaces that will make a difference, not only in the controlled conditions of research labs but also in daily-life prosthesis use.

Importantly, these principles when considered can indeed lead to an effective feedback interface. For example, as shown in [97, 98], the feedback that is designed to facilitate predictive control by feeding back the user's own EMG outperformed the classic approach based on transmitting the information on the grasping force. Audible feedback of the myoelectric signals can improve performance and facilitate the development of internal models in prosthesis control [99, 100]. In addition, as demonstrated in [96], operating a prosthesis even without artificial sensory feedback is still closed-loop control, because the user receives information from the prosthesis and his/her own body. For example, the subjects were able to estimate the prosthesis grasping force visually from its velocity of closing. Since most myoelectric prostheses respond proportionally to a command signal, the subjects can also rely on the natural muscle proprioception (sense of effort) to produce desired grasping forces [123]. To improve the control performance, the artificial sensory feedback needs to provide more than what is already intrinsically available in the prosthesis. Importantly, as shown in a recent study [143], the incidental sources of feedback can provide high-fidelity information about the prosthesis state (closing velocity).

Therefore, on one side, the artificial feedback needs to be designed to mimic the role of natural sensory feedback within the human motor control loop. On the other side, however, the nature of the biological feedback is very different from its supposed artificial replacement. Biological somatosensory feedback includes multiple modalities and spatially distributed sensing, whereas prosthesis feedback has been implemented using discrete stimulation components. A typical feedback interface comprises a single sensor measuring the total grasping force and a single stimulator (vibration motor) transmitting this information back to the user [82]. Nevertheless, there are some promising developments towards decreasing this gap in the future. There are advanced sensors mimicking their biological counterparts, for example, sensorized compliant fingers [144] that register vibrations, distributed touch, and temperature, or artificial skins that can completely cover the prosthetic hand to provide high-density distributed tactile sensing [145]. These sensors are originally developed for applications in robotics, but it has been increasingly recognized that they can be usefully applied in prosthetics. In a recent study [146], a system was presented to transmit tactile information detected by an artificial skin to a human subject by delivering electrical stimulation through a matrix of electrodes placed on the forearm (Fig. 15.7). The experimental evaluation demonstrated that the subjects could interpret spatially distributed stimulation and successfully recognize shapes of different complexity (lines, geometries, letters) that were applied to the artificial skin by an experimenter. Osborn et al. [147] presented a multilayered artificial skin and electrical stimulation system that are able to capture and elicit sensation of touch as well as pain arising when interacting with a sharp object.

Finally, as the feedback is an integral part of the motor control loop (Fig. 15.1), there is a strong interaction with the other components of that loop, especially with the feedforward (command) interface [27, 36]. In general, the feedback cannot have a strong impact on performance, even when transmitted through an ideal visual interface, if the prosthesis control is unreliable. Using good feedback, the user can indeed receive information on the prosthesis state and then decide on the proper command to send to the prosthesis. However, this is still futile if she/he cannot reliably generate that command by activating her/his muscles. Unfortunately, unreliable control is characteristic to most of the con-

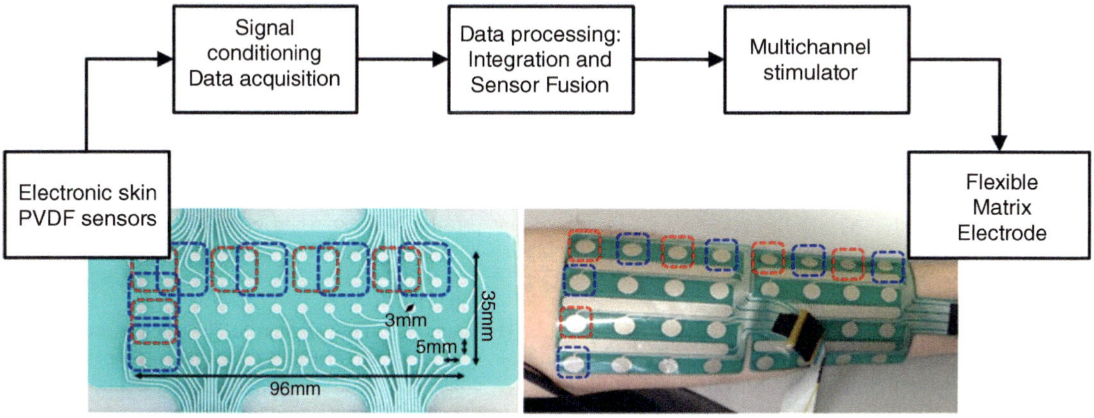

Fig. 15.7 Artificial electrotactile feedback from an electronic skin. The skin embeds a matrix of sensitive tactile sensors (taxels). The tactile signals are recorded, processed, and translated into stimulation parameters. The stimulation is delivered to the subject using a matrix electrode placed around the forearm. Therefore, the subject can "feel" a tactile stimulus applied to the skin. (Adapted from [146])

temporary prosthetic devices, as they rely on surface EMG which leads to noisy command signals that are less and less reliable as the contraction strength increases [3]. Therefore, in order for the user to be able to exploit the benefit of feedback, better control interfaces as well as mechatronic solutions are required. As described in other chapters, there are some promising developments in this direction, including implantable solutions for EMG recording [148] and osseointegrated interfacing [149] providing more stable control signals, as well as the prototypes of low-impedance prosthetic devices [47] that enable precise and continuous modulation of grasping force.

References

1. Beckerle P, et al. Feel-good robotics: requirements on touch for embodiment in assistive robotics. Front Neurorobot. 2018;12:84. https://doi.org/10.3389/fnbot.2018.00084.
2. Lederman SJ, Klatzky RL. Haptic perception: a tutorial. Attent Percept Psychophys. 2009;71(7):1439–59. https://doi.org/10.3758/APP.71.7.1439.
3. Parker P, Englehart K, Hudgins B. Myoelectric signal processing for control of powered limb prostheses. J Electromyogr Kinesiol. 2006;16(6):541–8. https://doi.org/10.1016/j.jelekin.2006.08.006.
4. Vincent Evolution 2 Hand. 2019. http://vincent-systems.de/en/prosthetics/vincent-evolution-2/. Accessed 9 Oct 2014.
5. Ability Hand. 2020. https://www.psyonic.co/ability-hand. Accessed 5 Nov 2020.
6. Luke Arm. 2020. https://www.mobiusbionics.com/luke-arm/. Accessed 5 Nov 2020.
7. Biddiss E, Beaton D, Chau T. Consumer design priorities for upper limb prosthetics. Disabil Rehabil Assist Technol. 2007;2(6):346–57. https://doi.org/10.1080/17483100701714733.
8. Cordella F, et al. Literature review on needs of upper limb prosthesis users. Front Neurosci. 2016;10:209. https://doi.org/10.3389/fnins.2016.00209.
9. Lewis S, et al. User demands for sensory feedback in upper extremity prostheses. In: 2012 IEEE international symposium on medical measurements and applications proceedings, IEEE; 2012. p. 1–4. https://doi.org/10.1109/MeMeA.2012.6226669.
10. Lewis S, et al. Satisfaction of prosthesis users with electrical hand prostheses and their suggested improvements. Biomed Eng. 2013. https://doi.org/10.1515/bmt-2013-4385.
11. Bach-y-Rita P, Kercel SW. Sensory substitution and the human–machine interface. Trends Cognit Sci. 2003;7(12):541–6. https://doi.org/10.1016/j.tics.2003.10.013.
12. Kaczmarek KA, et al. Electrotactile and vibrotactile displays for sensory substitution systems. IEEE Trans Biomed Eng. 1991;38(1):1–16. https://doi.org/10.1109/10.68204.
13. Szeto AY, Saunders FA. Electrocutaneous stimulation for sensory communication in rehabilitation engineering. IEEE Trans Biomed Eng. 1982;29(4):300–8. http://www.ncbi.nlm.nih.gov/pubmed/7068167. Accessed 12 Apr 2013.

14. Gomez-Tames J, Gonzalez J, Yu W. A simulation study on the dominance of the tissues' conductivity in the muscle recruitment. J Med Imaging Health Inform. 2013;3(1):72–8. https://doi.org/10.1166/jmihi.2013.1139.
15. Gonzalez J, et al. Auditory display as a prosthetic hand biofeedback. J Med Imaging Health Inform. 2011;1(4):325–33. https://doi.org/10.1166/jmihi.2011.1051.
16. Markovic M, et al. Stereovision and augmented reality for closed loop control of grasping in hand prostheses. J Neural Eng. 2014;11(4):046001.
17. Markovic M, et al. GLIMPSE: Google Glass interface for sensory feedback in myoelectric hand prostheses. J Neural Eng. 2017;14(3):036007. https://doi.org/10.1088/1741-2552/aa620a.
18. Clemente F, et al. Intraneural sensory feedback restores grip force control and motor coordination while using a prosthetic hand. J Neural Eng. 2019;16(2):026034. https://doi.org/10.1088/1741-2552/ab059b.
19. Raspopovic S, et al. Restoring natural sensory feedback in real-time bidirectional hand prostheses. Sci Transl Med. 2014;6(222):222ra19. https://doi.org/10.1126/scitranslmed.3006820.
20. Tabot GA, et al. Restoring the sense of touch with a prosthetic hand through a brain interface. Proc Natl Acad Sci U S A. 2013;110(45):18279–84. https://doi.org/10.1073/pnas.1221113110.
21. Engdahl SM, et al. Surveying the interest of individuals with upper limb loss in novel prosthetic control techniques. J Neuroeng Rehabil. 2015;12(1):53. https://doi.org/10.1186/s12984-015-0044-2.
22. Childress DS. Closed-loop control in prosthetic systems: historical perspective. Ann Biomed Eng. 1980;8(4–6):293–303. https://doi.org/10.1007/BF02363433.
23. Schofield JS, et al. Applications of sensory feedback in motorized upper extremity prosthesis: a review. Expert Rev Med Devices. 2014;11(5):499–511. https://doi.org/10.1586/17434440.2014.929496.
24. Svensson P, et al. A review of invasive and non-invasive sensory feedback in upper limb prostheses. Expert Rev Med Devices. 2017;14(6):439–47. https://doi.org/10.1080/17434440.2017.1332989.
25. Sensinger JW, Dosen S. A Review of Sensory Feedback in Upper-Limb Prostheses From the Perspective of Human Motor Control. Front Neurosci. 2020;14:345.
26. Harris CM, Wolpert DM. Signal-dependent noise determines motor planning. Nature. 1998;394(6695):780–4. https://doi.org/10.1038/29528.
27. Dosen S, et al. Multichannel electrotactile feedback with spatial and mixed coding for closed-loop control of grasping force in hand prostheses. IEEE Trans Neural Syst Rehabil Eng. 2017;25(3):183–95. https://doi.org/10.1109/TNSRE.2016.2550864.
28. Mann RW. Kinesthetic sensing for the EMG controlled "Boston Arm". IEEE Trans Man-Mach Syst. 1970;11(1):110–5. https://doi.org/10.1109/TMMS.1970.299971.
29. Dosen S, Markovic M, Wille N, et al. Building an internal model of a myoelectric prosthesis via closed-loop control for consistent and routine grasping. Exp Brain Res. 2015;233(6):1855–65. https://doi.org/10.1007/s00221-015-4257-1.
30. Lum PS, et al. Internal models of upper limb prosthesis users when grasping and lifting a fragile object with their prosthetic limb. Exp Brain Res. 2014;232:3785–95. https://doi.org/10.1007/s00221-014-4071-1.
31. Saunders I, Vijayakumar S. The role of feed-forward and feedback processes for closed-loop prosthesis control. J Neuroeng Rehabil. 2011;8(1):60. https://doi.org/10.1186/1743-0003-8-60.
32. D'Alonzo M, et al. HyVE: hybrid vibro-electrotactile stimulation for sensory feedback and substitution in rehabilitation. IEEE Trans Neural Syst Rehabil Eng. 2014;22(2):290–301. https://doi.org/10.1109/TNSRE.2013.2266482.
33. D'Alonzo M, et al. HyVE—hybrid vibro-electrotactile stimulation—is an efficient approach to multi-channel sensory feedback. IEEE Trans Haptics. 2014;7(2):181–90. https://doi.org/10.1109/TOH.2013.52.
34. Geng B, et al. Evaluation of sensation evoked by electrocutaneous stimulation on forearm in nondisabled subjects. J Rehabil Res Dev. 2012;49(2):297. https://doi.org/10.1682/JRRD.2010.09.0187.
35. Paredes LP, et al. The impact of the stimulation frequency on closed-loop control with electrotactile feedback. J Neuroeng Rehabil. 2015;12(1):35. https://doi.org/10.1186/s12984-015-0022-8.
36. Dosen S, Markovic M, Hartmann C, et al. Sensory feedback in prosthetics: a standardized test bench for closed-loop control. IEEE Trans Neural Syst Rehabil Eng. 2015;23(2):267–76. https://doi.org/10.1109/TNSRE.2014.2371238.
37. D'Anna E, et al. A somatotopic bidirectional hand prosthesis with transcutaneous electrical nerve stimulation based sensory feedback. Sci Rep. 2017;7(1):10930. https://doi.org/10.1038/s41598-017-11306-w.
38. Shin H, et al. Evoked haptic sensations in the hand via non-invasive proximal nerve stimulation. J Neural Eng. 2018;15(4):046005. https://doi.org/10.1088/1741-2552/aabd5d.
39. Štrbac M, et al. Integrated and flexible multichannel interface for electrotactile stimulation. J Neural Eng. 2016;13(4):046014. https://doi.org/10.1088/1741-2560/13/4/046014.
40. Bolanowski SJ, et al. Four channels mediate the mechanical aspects of touch. J Acoust Soc Am. 1988;84(5):1680–94. http://www.ncbi.nlm.nih.gov/pubmed/3209773. Accessed 19 May 2014.
41. Choi S, Kuchenbecker KJ. Vibrotactile display: perception, technology, and applications. Proc IEEE. 2013;101(9):2093–104. https://doi.org/10.1109/JPROC.2012.2221071.

42. Jones LA, Sarter NB. Tactile displays: guidance for their design and application. Hum Factors. 2008;50(1):90–111. http://www.ncbi.nlm.nih.gov/pubmed/18354974. Accessed 19 May 2014.
43. Dosen S, et al. A novel method to generate amplitude-frequency modulated vibrotactile stimulation. IEEE Trans Haptics. 2016;9(1):3–12. https://doi.org/10.1109/TOH.2015.2497229.
44. Cipriani C, D'Alonzo M, Carrozza MC. A miniature vibrotactile sensory substitution device for multifingered hand prosthetics. IEEE Trans Biomed Eng. 2012;59(2):400–8. https://doi.org/10.1109/TBME.2011.2173342.
45. Azadi M, Jones LA. Vibrotactile actuators: effect of load and body site on performance. In: 2014 IEEE haptics symposium (HAPTICS), IEEE; 2014. p. 351–6. https://doi.org/10.1109/HAPTICS.2014.6775480.
46. Antfolk C, D'Alonzo M, Controzzi M, et al. Artificial redirection of sensation from prosthetic fingers to the phantom hand map on transradial amputees: vibrotactile versus mechanotactile sensory feedback. IEEE Trans Neural Syst Rehabil Eng. 2013;21(1):112–20. https://doi.org/10.1109/TNSRE.2012.2217989.
47. Brown JD, et al. An exploration of grip force regulation with a low-impedance myoelectric prosthesis featuring referred haptic feedback. J Neuroeng Rehabil. 2015;12:104. https://doi.org/10.1186/s12984-015-0098-1.
48. Wheeler J, et al. Investigation of rotational skin stretch for proprioceptive feedback with application to myoelectric systems. IEEE Trans Neural Syst Rehabil Eng. 2010;18(1):58–66. https://doi.org/10.1109/TNSRE.2009.2039602.
49. Schoepp KR, et al. Design and integration of an inexpensive wearable mechanotactile feedback system for myoelectric prostheses. IEEE J Transl Eng Health Med. 2018;6:1–11. https://doi.org/10.1109/JTEHM.2018.2866105.
50. Battaglia E, et al. Skin stretch haptic feedback to convey closure information in anthropomorphic, under-actuated upper limb soft prostheses. IEEE Trans Haptics. 2019;12(4):508–20. https://doi.org/10.1109/TOH.2019.2915075.
51. Rossi M, et al. HapPro: a wearable haptic device for proprioceptive feedback. IEEE Trans Biomed Eng. 2019;66(1):138–49. https://doi.org/10.1109/TBME.2018.2836672.
52. Bark K, et al. Rotational skin stretch feedback: a wearable haptic display for motion. IEEE Trans Haptics. 2010;3(3):166–76. https://doi.org/10.1109/TOH.2010.21.
53. Akhtar A, et al. Passive mechanical skin stretch for multiple degree-of-freedom proprioception in a hand prosthesis; 2014. p. 120–8. https://doi.org/10.1007/978-3-662-44196-1_16.
54. Prattichizzo D, et al. Towards wearability in fingertip haptics: a 3-DoF wearable device for cutaneous force feedback. IEEE Trans Haptics. 2013;6(4):506–16. https://doi.org/10.1109/TOH.2013.53.
55. Patterson PE, Katz JA. Design and evaluation of a sensory feedback system that provides grasping pressure in a myoelectric hand. J Rehabil Res Dev. 1992;29(1):1–8. http://www.ncbi.nlm.nih.gov/pubmed/1740774. Accessed 28 June 2013.
56. Casini S, et al. Design and realization of the CUFF—clenching upper-limb force feedback wearable device for distributed mechano-tactile stimulation of normal and tangential skin forces. In: 2015 IEEE/RSJ international conference on intelligent robots and systems (IROS), IEEE; 2015. p. 1186–93. https://doi.org/10.1109/IROS.2015.7353520.
57. Dong J. On the use of subdermal electrical stimulation for restoration of sensory feedback. Aalborg University Press, PhD thesis; 2020.
58. Tan DW, et al. A neural interface provides long-term stable natural touch perception. Sci Transl Med. 2014;6(257):257ra138. https://doi.org/10.1126/scitranslmed.3008669.
59. Gaunt RA, Hokanson JA, Weber DJ. Microstimulation of primary afferent neurons in the L7 dorsal root ganglia using multielectrode arrays in anesthetized cats: thresholds and recruitment properties. J Neural Eng. 2009;6(5):055009. https://doi.org/10.1088/1741-2560/6/5/055009.
60. Loutit AJ, et al. Characterisation and functional mapping of surface potentials in the rat dorsal column nuclei. J Physiol. 2017;595(13):4507–24. https://doi.org/10.1113/JP273759.
61. Flesher SN, et al. Intracortical microstimulation of human somatosensory cortex. Sci Transl Med. 2016;8(361):361ra141. https://doi.org/10.1126/scitranslmed.aaf8083.
62. Schiefer M, et al. Sensory feedback by peripheral nerve stimulation improves task performance in individuals with upper limb loss using a myoelectric prosthesis. J Neural Eng. 2016;13(1):016001. https://doi.org/10.1088/1741-2560/13/1/016001.
63. Schiefer MA, et al. 'Artificial tactile and proprioceptive feedback improves performance and confidence on object identification tasks. PLoS One. 2018;13(12):e0207659. https://doi.org/10.1371/journal.pone.0207659. Edited by M. Sakakibara.
64. Tan DW, et al. Stability and selectivity of a chronic, multi-contact cuff electrode for sensory stimulation in human amputees. J Neural Eng. 2015;12(2):026002. https://doi.org/10.1088/1741-2560/12/2/026002.
65. D'Anna E, et al. A closed-loop hand prosthesis with simultaneous intraneural tactile and position feedback'. Sci Robot. 2019;4(27):eaau8892. https://doi.org/10.1126/scirobotics.aau8892.
66. Horch K, et al. Object discrimination with an artificial hand using electrical stimulation of peripheral tactile and proprioceptive pathways with intrafascicular electrodes. IEEE Trans Neural Syst Rehabil Eng. 2011;19(5):483–9. https://doi.org/10.1109/TNSRE.2011.2162635.
67. Valle G, et al. Biomimetic intraneural sensory feedback enhances sensation naturalness, tactile sensitivity, and manual dexterity in a bidirectional

prosthesis. Neuron. 2018;100(1):37–45.e7. https://doi.org/10.1016/j.neuron.2018.08.033.
68. Valle G, et al. Comparison of linear frequency and amplitude modulation for intraneural sensory feedback in bidirectional hand prostheses. Sci Rep. 2018;8(1):16666. https://doi.org/10.1038/s41598-018-34910-w.
69. George JA, et al. Biomimetic sensory feedback through peripheral nerve stimulation improves dexterous use of a bionic hand. Sci Robot. 2019;4(32):eaax2352. https://doi.org/10.1126/scirobotics.aax2352.
70. Wendelken S, et al. Restoration of motor control and proprioceptive and cutaneous sensation in humans with prior upper-limb amputation via multiple Utah Slanted Electrode Arrays (USEAs) implanted in residual peripheral arm nerves. J Neuroeng Rehabil. 2017;14(1):121. https://doi.org/10.1186/s12984-017-0320-4.
71. Saal HP, Bensmaia SJ. Biomimetic approaches to bionic touch through a peripheral nerve interface. Neuropsychologia. 2015;79:344–53. https://doi.org/10.1016/j.neuropsychologia.2015.06.010.
72. Strauss I, et al. Characterization of multi-channel intraneural stimulation in transradial amputees. Sci Rep. 2019;9(1):19258. https://doi.org/10.1038/s41598-019-55591-z.
73. Cuberovic I, et al. Learning of artificial sensation through long-term home use of a sensory-enabled prosthesis. Front Neurosci. 2019;13:1–24. https://doi.org/10.3389/fnins.2019.00853.
74. Graczyk EL, et al. Home use of a neural-connected sensory prosthesis provides the functional and psychosocial experience of having a hand again. Sci Rep. 2018;8(1):9866. https://doi.org/10.1038/s41598-018-26952-x.
75. Page DM, et al. Motor control and sensory feedback enhance prosthesis embodiment and reduce phantom pain after long-term hand amputation. Front Hum Neurosci. 2018;12:352. https://doi.org/10.3389/fnhum.2018.00352.
76. Petrini FM, et al. Six-month assessment of a hand prosthesis with intraneural tactile feedback. Ann Neurol. 2019;85(1):137–54. https://doi.org/10.1002/ana.25384.
77. Dong J, et al. The variability of psychophysical parameters following surface and subdermal stimulation: a multiday study in amputees. IEEE Trans Neural Syst Rehabil Eng. 2019;28(1):174–80. https://doi.org/10.1109/TNSRE.2019.2956836.
78. Dong J, et al. The short-term repeatability of subdermal electrical stimulation for sensory feedback. IEEE Access. 2020;8:63983–92. https://doi.org/10.1109/ACCESS.2020.2984534.
79. Clemente F, Cipriani C. A novel device for multimodal sensory feedback in hand prosthetics: design and preliminary prototype. In: 2014 IEEE haptics symposium (HAPTICS), IEEE; 2014. p. 569–73. https://doi.org/10.1109/HAPTICS.2014.6775518.
80. Huang H, et al. Multi-modal Sensory feedback system for upper limb amputees. In: 2017 new generation of CAS (NGCAS), IEEE; 2017. p. 193–6. https://doi.org/10.1109/NGCAS.2017.62.
81. Keehoon K, et al. On the design of miniature haptic devices for upper extremity prosthetics. IEEE/ASME Trans Mechatron. 2010;15(1):27–39. https://doi.org/10.1109/TMECH.2009.2013944.
82. Antfolk C, D'Alonzo M, Rosén B, et al. Sensory feedback in upper limb prosthetics. Expert Rev Med Devices. 2013;10(1):45–54. https://doi.org/10.1586/erd.12.68.
83. Belter JT, et al. Mechanical design and performance specifications of anthropomorphic prosthetic hands: a review. J Rehabil Res Dev. 2013;50(5):599–618. http://www.ncbi.nlm.nih.gov/pubmed/24013909. Accessed 9 Jul 2014.
84. Chatterjee A, et al. Testing a prosthetic haptic feedback simulator with an interactive force matching task. JPO J Prosthet Orthot. 2008;20(2):27–34. https://doi.org/10.1097/01.JPO.0000311041.61628.be.
85. Cipriani C, et al. On the shared control of an EMG-controlled prosthetic hand: analysis of user–prosthesis interaction. IEEE Trans Robot. 2008;24(1):170–84. https://doi.org/10.1109/TRO.2007.910708.
86. Jorgovanovic N, et al. Virtual grasping: closed-loop force control using electrotactile feedback. Comput Math Methods Med. 2014;2014:120357. https://doi.org/10.1155/2014/120357.
87. Pylatiuk C, Kargov A, Schulz S. Design and evaluation of a low-cost force feedback system for myoelectric prosthetic hands. JPO J Prosthet Orthot. 2006;18(2):57–61. https://doi.org/10.1097/00008526-200604000-00007.
88. Szeto AY, Farrenkopf GR. Optimization of single electrode tactile codes. Ann Biomed Eng. 1992;20(6):647–65. http://www.ncbi.nlm.nih.gov/pubmed/1449232. Accessed 18 Nov 2013.
89. Szeto AYJ, Lyman J. Comparison of codes for sensory feedback using electrocutaneous tracking. Ann Biomed Eng. 1977;5(4):367–83. https://doi.org/10.1007/BF02367316.
90. Anani AB, Ikeda K, Körner LM. Human ability to discriminate various parameters in afferent electrical nerve stimulation with particular reference to prostheses sensory feedback. Med Biol Eng Comput. 1977;15(4):363–73.
91. Anani AB, Körner LM. Afferent electrical nerve stimulation: human tracking performance relevant to prosthesis sensory feedback. Med Biol Eng Comput. 1979;17(4):425–34. https://doi.org/10.1007/BF02447053.
92. Guemann M, et al. Effect of vibration characteristics and vibror arrangement on the tactile perception of the upper arm in healthy subjects and upper limb amputees. J Neuroeng Rehabil. 2019;16(1):138. https://doi.org/10.1186/s12984-019-0597-6.

93. Antfolk C. A tactile display system for hand prostheses to discriminate pressure and individual finger localization. J Med Biol Eng. 2010;30(6):355. https://doi.org/10.5405/jmbe.766.
94. Cipriani C, et al. A novel concept for a prosthetic hand with a bidirectional interface: a feasibility study. IEEE Trans Biomed Eng. 2009;56(11):2739–43. https://doi.org/10.1109/TBME.2009.2031242.
95. Johansson RS, Cole KJ. Sensory-motor coordination during grasping and manipulative actions. Curr Opin Neurobiol. 1992;2(6):815–23. http://www.ncbi.nlm.nih.gov/pubmed/1477545. Accessed 17 Sept 2013.
96. Ninu A, et al. Closed-loop control of grasping with a myoelectric hand prosthesis: which are the relevant feedback variables for force control? IEEE Trans Neural Syst Rehabil Eng. 2014;22(5):1041–52. https://doi.org/10.1109/TNSRE.2014.2318431.
97. Dosen S, Markovic M, Somer K, et al. EMG biofeedback for online predictive control of grasping force in a myoelectric prosthesis. J Neuroeng Rehabil. 2015;12(1):55. https://doi.org/10.1186/s12984-015-0047-z.
98. Schweisfurth MA, et al. Electrotactile EMG feedback improves the control of prosthesis grasping force. J Neural Eng. 2016;13(5):056010. https://doi.org/10.1088/1741-2560/13/5/056010.
99. Shehata AW, et al. Improving internal model strength and performance of prosthetic hands using augmented feedback. J Neuroeng Rehabil. 2018;15(1):70. https://doi.org/10.1186/s12984-018-0417-4.
100. Shehata AW, Scheme EJ, Sensinger JW. Audible feedback improves internal model strength and performance of myoelectric prosthesis control. Sci Rep. 2018;8(1):8541. https://doi.org/10.1038/s41598-018-26810-w.
101. Aboseria M, et al. Discrete vibro-tactile feedback prevents object slippage in hand prostheses more intuitively than other modalities. IEEE Trans Neural Syst Rehabil Eng. 2018;26(8):1577–84. https://doi.org/10.1109/TNSRE.2018.2851617.
102. Clemente F, et al. Non-invasive, temporally discrete feedback of object contact and release improves grasp control of closed-loop myoelectric transradial prostheses. IEEE Trans Neural Syst Rehabil Eng. 2016;24(12):1314–22. https://doi.org/10.1109/TNSRE.2015.2500586.
103. Flanagan JR, Bowman MC, Johansson RS. Control strategies in object manipulation tasks. Curr Opin Neurobiol. 2006;16(6):650–9. https://doi.org/10.1016/j.conb.2006.10.005.
104. Johansson RS, Flanagan JR. Tactile sensory control of object manipulation in humans. In: The senses: a comprehensive reference. Elsevier; 2008. p. 67–86. https://doi.org/10.1016/B978-012370880-9.00346-7.
105. Engels LF, et al. When less is more—discrete tactile feedback dominates continuous audio biofeedback in the integrated percept while controlling a myoelectric prosthetic hand. Front Neurosci. 2019;13:578. https://doi.org/10.3389/fnins.2019.00578.
106. Zollo L, et al. Restoring tactile sensations via neural interfaces for real-time force-and-slippage closed-loop control of bionic hands. Sci Robot. 2019;4(27):1–12. https://doi.org/10.1126/scirobotics.aau9924.
107. Witteveen HJB, et al. Vibro- and electrotactile user feedback on hand opening for myoelectric forearm prostheses. IEEE Trans Biomed Eng. 2012;59(8):2219–26. https://doi.org/10.1109/TBME.2012.2200678.
108. Witteveen HJ, Rietman HS, Veltink PH. Vibrotactile grasping force and hand aperture feedback for myoelectric forearm prosthesis users. Prosthet Orthot Int. 2015;39(3):204–12. https://doi.org/10.1177/0309364614522260.
109. Erwin A, Sup FC. A haptic feedback scheme to accurately position a virtual wrist prosthesis using a three-node tactor array. PLoS One. 2015;10(8):e0134095. https://doi.org/10.1371/journal.pone.0134095.
110. Patel GK, et al. Multichannel electrotactile feedback for simultaneous and proportional myoelectric control. J Neural Eng. 2016;13(5):056015. https://doi.org/10.1088/1741-2560/13/5/056015.
111. Cheng A, et al. Conveying the configuration of a virtual human hand using vibrotactile feedback. In: 2012 IEEE haptics symposium (HAPTICS), IEEE; 2012. p. 155–62. https://doi.org/10.1109/HAPTIC.2012.6183784.
112. Battaglia E, et al. The rice haptic rocker: skin stretch haptic feedback with the Pisa/IIT SoftHand. In: 2017 IEEE world haptics conference (WHC), IEEE; 2017. p. 7–12. https://doi.org/10.1109/WHC.2017.7989848.
113. Marasco PD, et al. Illusory movement perception improves motor control for prosthetic hands. Sci Transl Med. 2018;10(432):1–13. https://doi.org/10.1126/scitranslmed.aao6990.
114. Arakeri TJ, Hasse BA, Fuglevand AJ. Object discrimination using electrotactile feedback. J Neural Eng. 2018;15(4):046007. https://doi.org/10.1088/1741-2552/aabc9a.
115. Schiefer MA, et al. Artificial tactile and proprioceptive feedback improves performance and confidence on object identification tasks. PLoS One. 2018;13(12):1–18. https://doi.org/10.1371/journal.pone.0207659.
116. Zafar M, Van Doren CL. Effectiveness of supplemental grasp-force feedback in the presence of vision. Med Biol Eng Comput. 2000;38(3):267–74. https://doi.org/10.1007/BF02347046.
117. Raveh E, Friedman J, Portnoy S. Visuomotor behaviors and performance in a dual-task paradigm with and without vibrotactile feedback when using a myoelectric controlled hand. Assist Technol. 2017;30(8):1–7. https://doi.org/10.1080/10400435.2017.1323809.
118. Witteveen HJB, et al. Stiffness feedback for myoelectric forearm prostheses using vibrotactile stimulation. IEEE Trans Neural Syst Rehabil Eng. 2014;22(1):53–61. https://doi.org/10.1109/TNSRE.2013.2267394.

119. Rombokas E, et al. Vibrotactile sensory substitution for electromyographic control of object manipulation. IEEE Trans Biomed Eng. 2013;60(8):2226–32. https://doi.org/10.1109/TBME.2013.2252174.
120. Stepp CE, An Q, Matsuoka Y. Repeated training with augmentative vibrotactile feedback increases object manipulation performance. PLoS One. 2012;7(2):e32743. https://doi.org/10.1371/journal.pone.0032743.
121. Meek SG, Jacobsen SC, Goulding PP. Extended physiologic taction: design and evaluation of a proportional force feedback system. J Rehabil Res Dev. 1989;26(3):53–62. http://www.ncbi.nlm.nih.gov/pubmed/2666644. Accessed 28 June 2013.
122. Strbac M, et al. Short- and long-term learning of feedforward control of a myoelectric prosthesis with sensory feedback by amputees. IEEE Trans Neural Syst Rehabil Eng. 2017;25(11):2133–45. https://doi.org/10.1109/TNSRE.2017.2712287.
123. Markovic M, Schweisfurth MA, et al. Myocontrol is closed-loop control: incidental feedback is sufficient for scaling the prosthesis force in routine grasping. J Neuroeng Rehabil. 2018;15(1):81. https://doi.org/10.1186/s12984-018-0422-7.
124. Markovic M, Schweisfurth MA, et al. The clinical relevance of advanced artificial feedback in the control of a multi-functional myoelectric prosthesis. J Neuroeng Rehabil. 2018;15(1):28. https://doi.org/10.1186/s12984-018-0371-1.
125. Pistohl T, et al. Artificial proprioceptive feedback for myoelectric control. IEEE Trans Neural Syst Rehabil Eng. 2015;23(3):498–507. https://doi.org/10.1109/TNSRE.2014.2355856.
126. Hasson CJ, Manczurowsky J. Effects of kinematic vibrotactile feedback on learning to control a virtual prosthetic arm. J Neuroeng Rehabil. 2015;12(1):31. https://doi.org/10.1186/s12984-015-0025-5.
127. Ehrsson HH, et al. Upper limb amputees can be induced to experience a rubber hand as their own. Brain. 2008;131(12):3443–52. https://doi.org/10.1093/brain/awn297.
128. Rosén B, et al. Referral of sensation to an advanced humanoid robotic hand prosthesis. Scand J Plastic Reconstruct Surg Hand Surg. 2009;43(5):260–6. https://doi.org/10.3109/02844310903113107.
129. Marasco PD, et al. Robotic touch shifts perception of embodiment to a prosthesis in targeted reinnervation amputees. Brain. 2011;134(Pt 3):747–58. https://doi.org/10.1093/brain/awq361.
130. D'Alonzo M, Clemente F, Cipriani C. Vibrotactile stimulation promotes embodiment of an alien hand in amputees with phantom sensations. IEEE Trans Neural Syst Rehabil Eng. 2015;23(3):450–7. https://doi.org/10.1109/TNSRE.2014.2337952.
131. Marini F, et al. Crossmodal representation of a functional robotic hand arises after extensive training in healthy participants. Neuropsychologia. 2014;53:178–86. https://doi.org/10.1016/j.neuropsychologia.2013.11.017.
132. Flor H, et al. Effect of sensory discrimination training on cortical reorganisation and phantom limb pain. Lancet. 2001;357(9270):1763–4. https://doi.org/10.1016/S0140-6736(00)04890-X.
133. Dietrich C, et al. Sensory feedback prosthesis reduces phantom limb pain: proof of a principle. Neurosci Lett. 2012;507(2):97–100. https://doi.org/10.1016/j.neulet.2011.10.068.
134. Dietrich C, et al. Leg prosthesis with somatosensory feedback reduces phantom limb pain and increases functionality. Front Neurol. 2018;9:270. https://doi.org/10.3389/fneur.2018.00270.
135. Beeker TW, During J, Den Hertog A. Artificial touch in a hand-prosthesis. Med Biol Eng. 1967;5(1):47–9. http://www.ncbi.nlm.nih.gov/pubmed/6037625.
136. Prior RE, et al. Supplemental sensory feedback for the VA/NU myoelectric hand. Background and preliminary designs. Bull Prosthet Res. 1976;170–91.
137. Johnson RE, et al. Does EMG control lead to distinct motor adaptation? Front Neurosci. 2014;8:302. https://doi.org/10.3389/fnins.2014.00302.
138. Johnson RE, et al. EMG versus torque control of human-machine systems: equalizing control signal variability does not equalize error or uncertainty. IEEE Trans Neural Syst Rehabil Eng. 2017;25(6):660–7. https://doi.org/10.1109/TNSRE.2016.2598095.
139. Davidson PR, Wolpert DM. Widespread access to predictive models in the motor system: a short review. J Neural Eng. 2005;2(3):S313–9. https://doi.org/10.1088/1741-2560/2/3/S11.
140. Berniker M, Kording K. Bayesian approaches to sensory integration for motor control. Wiley Interdiscip Rev Cognit Sci. 2011;2(4):419–28. https://doi.org/10.1002/wcs.125.
141. Shadmehr R, Krakauer JW. A computational neuroanatomy for motor control. Exp Brain Res. 2008;185(3):359–81. https://doi.org/10.1007/s00221-008-1280-5.
142. Todorov E, Jordan MI. Optimal feedback control as a theory of motor coordination. Nat Neurosci. 2002;5(11):1226–35. https://doi.org/10.1038/nn963.
143. Wilke MA, et al. Psychometric characterization of incidental feedback sources during grasping with a hand prosthesis. J Neuroeng Rehabil. 2019;16(1):155. https://doi.org/10.1186/s12984-019-0622-9.
144. SynTouch. BioTac senzorized finger. 2014. http://www.syntouchllc.com/Media/Blog/2013-08-27-S.
145. Kim J, et al. Stretchable silicon nanoribbon electronics for skin prosthesis. Nat Commun. 2014;5:5747. https://doi.org/10.1038/ncomms6747.
146. Franceschi M, et al. A system for electrotactile feedback using electronic skin and flexible matrix electrodes: experimental evaluation. IEEE Trans Haptics. 2017;10(2):162–72. https://doi.org/10.1109/TOH.2016.2618377.
147. Osborn LE, et al. Prosthesis with neuromorphic multilayered e-dermis perceives touch and

pain. Sci Robot. 2018;3(19):eaat3818. https://doi.org/10.1126/scirobotics.aat3818.

148. Pasquina PF, et al. First-in-man demonstration of a fully implanted myoelectric sensors system to control an advanced electromechanical prosthetic hand. J Neurosci Methods. 2015;244:85–93. https://doi.org/10.1016/j.jneumeth.2014.07.016.

149. Mastinu E, et al. Grip control and motor coordination with implanted and surface electrodes while grasping with an osseointegrated prosthetic hand. J Neuroeng Rehabil. 2019;16(1):49. https://doi.org/10.1186/s12984-019-0511-2.

Part VI

Prosthetic Replacement in Patients with Inner Amputations

Deafferentation Pain Following Brachial Plexus Avulsion Injuries

16

Laura A. Hruby

Brachial plexus avulsion injuries represent one of the most severe nerve injuries resulting in debilitating sensory and motor deficits. Besides functional impairments, most affected patients suffer from a chronic pain syndrome, referred to as deafferentation pain. Its origin is thought to arise due to scar formation in the dorsal horn after traumatic nerve root avulsion. At a later stage, central mechanisms including cortical reorganization maintain the pain. Drug therapy often fails to improve pain with interest growing in behavioral treatment modalities. Most importantly, it has been shown that successful nerve reconstruction results both in functional recovery and reduction of deafferentation pain as the central nervous system regains sensory feedback from the previously denervated limb.

Background

A chronic pain syndrome, referred to as avulsion pain or deafferentation pain, frequently develops after nerve roots are traumatically torn from the spinal cord [38]. One of the first descriptions was published by Frazier and Skillern in 1911 and included the vivid exemplification of a patient's pain experience, who had suffered a closed brachial plexus avulsion injury after a man had fallen down on his left shoulder and head from the fourth floor [13]:

> *The pain is continuous; it does not stop a minute either day or night. It is either burning or compressing (like a vise) or dragging (a sense of weight) in character, or a combination of all these at the same time. Every few minutes this pain is intensified in a paroxysm lasting from a few seconds to a minute or longer. In addition there is, every few minutes, a jerking sensation similar to that obtained by touching, in the proper way, a Leyden jar, namely, a jerk, or a succession of jerks, either in the hand only or running the whole length of the arm and at times very severe. I have a graphic picture of it in my mind; it is like zigzag made in the skies by a stroke of lightning. The pain is felt sometimes in the very arm itself, but most of the time away from the arm, in what I got into the habit of calling an imaginary arm. The upper part of the arm is mostly free from pain; the lower part from a little above the elbow to the tips of the fingers, never. Pain of every character, burning, compression, dragging or jerking is increased by walking, standing or even sitting up, so that the recumbent position is a necessity for most of the day.*

Despite the fact that this exemplification dates back more than 100 years ago, patients with multiple nerve root avulsions today still portray their pain using a similar vocabulary. Deafferentation pain usually occurs soon after the traumatic

L. A. Hruby (✉)
Department of Orthopedics and Trauma Surgery, Medical University of Vienna, Vienna, Austria

Clinical Laboratory for Bionic Extremity Reconstruction, Division of Plastic and Reconstructive Surgery, Department of Surgery, Medical University of Vienna, Vienna, Austria
e-mail: laura.hruby@meduniwien.ac.at

O. C. Aszmann, D. Farina (eds.), *Bionic Limb Reconstruction*,
https://doi.org/10.1007/978-3-030-60746-3_16

event, which resulted in brachial plexus avulsion [38]. In lower root avulsions, persistent pain in the deafferented hand area is experienced by 70–90% of patients [26].

Deafferentation as a medical term includes the total deprivation of sensory input from the periphery. In brachial plexus avulsion injuries, all input from afferent axons of a peripheral nerve is interrupted when the dorsal rootlets are torn from the dorsal horn. Why this deafferentation process frequently entails the development of a chronic, intractable pain syndrome has been the subject of numerous studies and reports. In the cat spinal cord, Loeser et al. showed abnormal neuronal firing patterns after selective rhizotomy [19] similar to the pathological changes seen in an epileptic focus in the brain [3]. These findings were confirmed in a human spinal cord specimen, showing spontaneous hyperactivity of neurons in the dorsal horn, which are thought to lose their "normal" graded response to peripheral stimuli but instead fire with abnormally prolonged bursts to any stimulus above threshold [20]. Despite total loss of sensation, affected patients frequently report of allodynia in the deafferented hand, which may be explained by the segmental neuronal hyperactivity of afferents located in the injured, scarred dorsal horn.

Neuronal hyperactivity and spontaneous ectopic discharges in the spinal cord are key mechanisms resulting in deafferentation pain after brachial plexus avulsion injuries. The prolonged and therapy-resistant nature of this pain syndrome, however, may be explained by its complex distribution throughout supraspinal somatosensory pathways. Recordings of thalamic nuclei, where sensory input from the spinal cord is relayed, have revealed spontaneously hyperactive neurons in patients with chronic deafferentation pain making it an all the more "central pain" syndrome [14, 15, 31]. Furthermore, adaptive changes in the cortex have also been described.

An fMRI (functional magnetic resonance imaging) study on cortical activation following unimanual finger movement has shown that in the healthy subject, the primary sensorimotor cortex contralateral to the hand movement is activated, while concomitantly the ipsilateral sensorimotor cortex is deactivated [2]. The motor control of unimanual tasks therefore entails interhemispheric interaction to inhibit interference from the opposite brain hemisphere. Interestingly, it has been found that interhemispheric inhibition is discontinued in brachial plexus avulsion patients [16]. The authors reasoned that following brachial plexus avulsion injuries, there is no need to inhibit an already disabled arm while moving the opposite, intact extremity. Interhemispheric functional connectivity is also weakened between the two primary motor areas in patients with brachial plexus injury implying a desynchronization of brain areas due to the loss of neurological connection to the affected arm and hand [18].

As oppose to this inhibition of interhemispheric connectivity, the amount of reorganization of the primary somatosensory cortex contralateral to the deafferented hand directly correlates with the intensity of deafferentation pain [10, 17]. As early as 6 months after nerve root avulsion, the interruption of afferent, sensory input from the hand is accompanied by the expansion of inputs from the face into the former hand cortical territory, and eventually, touch on the face can activate the hand area of the primary somatosensory cortex [12, 17]. In other words, the cortical hand area disappears as a result of complete brachial plexus injury, which may be referred to an inner amputation of the hand, and is invaded by adjacent cortical areas, particularly the face. Although the term "phantom limb" usually applies to conventional amputees, it may also be used in association with the inner amputation resulting from global brachial plexus avulsion, with many patients experiencing similar movement sensations and painful symptoms in the affected limb despite its denervation [23]. As in amputees, phantom limb phenomena are associated with the mentioned reorganization of cortical structures [12, 37].

The spontaneous nature of deafferentation pain is attributed to bursts of neuronal hyperactivity at the spinal cord as well as at supraspinal areas specifically the thalamus and cortex. Patients typically describe "electric, shock-like" paroxysmal pain attacks [9], which last several seconds to minutes and can occur up to 20 times per day. Added to this, most patients report of a "crushing or dull background pain," which is not as intense, however, continuously felt throughout the day and night. In many patients, weather changes with sudden temperature drops or increases, wind, emotional stress, as well as physical stress can aggravate pain intensity and frequency.

Treatment Strategies

The treatment of chronic deafferentation pain remains a challenge. Drug therapy including narcotics and opioids and more invasive procedures such as stellate ganglion blockade, sympathectomy, limb amputation, transcutaneous electrical stimulation (TENS), and dorsal column stimulation all share one characteristic: their efficacy is non-predictable and mostly insufficient [39]. Furthermore, adverse side effects of opioid painkillers limit their tolerance and lead to high abandonment rates. Additionally, frequent and intense pain attacks may cause feelings of stress, nausea, anxiety, and social withdrawal [10], which can also negatively impact on occupational activities and necessitate extended sickness absence rates.

Dorsal Root Entry Zone (DREZ) Lesioning

In 1952, Rexed divided the superficial dorsal horn of the cat spinal cord into a series of parallel laminae [30], which has since been applied to other species including humans [40]. Pain from the arm and hand is transmitted by primary nociceptive afferents, which enter the spinal cord mainly via Rexed laminae I and II (= substantia gelatinosa of Rolando) [30] (Fig. 16.1). In traumatic dorsal root avulsion, a scar is formed at the spinal cord-rootlet junction, and consequent neuronal hyperexcitability with spontaneous discharge of neurons in Rexed laminae I and II mimics burning pain being transmitted from the affected hand [40]. Additionally, the inhibitory effect of large, myelinated afferents on nociceptive C fibers, which usually occur in the substantia gelatinosa, is diminished following deafferentation [41]. The dorsal root entry zone (DREZ) has therefore been postulated for surgical

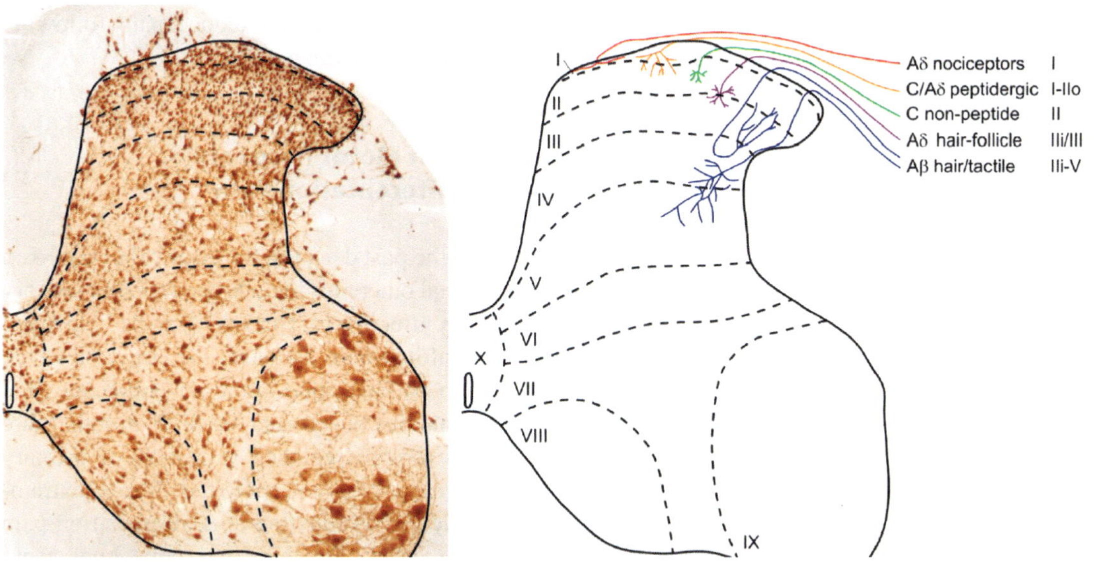

Fig. 16.1 Organization of the laminae in the cat midlumbar spinal cord as originally described by Rexed (left). Schematic drawing of the laminar organization of the dorsal horn and primary afferent inputs (right). (Taken from [40] with permission)

ablation to reduce deafferentation pain by suppression of the nociceptive pathways [33, 36]. The technique aims at destroying the deafferented hyperactive neurons in the scarred dorsal horn and is referred to as microsurgical DREZotomy (MDT). Either microcoagulation using radiofrequency thermocoagulation or microsurgical bipolar coagulation may be performed to ablate the small-caliber nociceptive fibers in the DREZ and the medial part of Lissauer's tract, within which nociceptive afferents travel to adjacent spinal segments to be relayed [27, 32, 34, 35]. This invasive procedure is recommended only until at least 12 months after injury due to the natural tendency of spontaneous improvement to tolerable pain levels and because most patients accustom themselves to the burden of pain with time [38]. Success rates differ between single centers, and recurrence of pain occurs in a considerable number of patients [1, 22]. Complications associated with this procedure include long tract impairment resulting in motor weakness and sensory ataxia due to damage of adjacent pathways in the spinal cord [1]. Permanent adverse sequelae occur in 10% of patients [38]. The invasive nature of the surgery, which includes a cervical laminectomy to adequately expose the spinal cord, is also to be considered. As outlined before, deafferentation pain following avulsion injury is not restricted to pathological changes in the dorsal horn itself, which is why selective procedures such as DREZ ablation frequently fail to resolve pain in deafferentation states in a relatively high number of patients [31]. Bearing in mind the risk of major complications due to iatrogenic damage of adjacent structures and pathways in the spinal cord, we do not recommend this procedure, especially not in otherwise healthy young individuals.

Behavioral Treatment Approaches

Due to the therapy-refractory nature of deafferentation pain, psychological approaches in pain management including relaxation techniques, mirror therapy, graded motor imagery, cognitive behavioral therapy, and body awareness training have been proposed [21]. In other neuropathic pain syndromes, such as spinal cord injury (SCI), phantom limb pain in conventional amputees, and complex regional pain syndrome, these treatment approaches are widely distributed. Since neuropathic pain is thought to be maintained by related cortical reorganization mechanisms, i.e., shrinkage of the cortical area of the deafferented limb and its invasion by adjacent cortical areas, the primary goal of therapy is to activate cortical areas corresponding to the affected limb, which ultimately leads to pain reduction [24]. Visual illusions, which trick the patient's brain of moving a "healthy," functional limb, have been shown to correct the mismatch between motor output and sensory feedback and thereby significantly reduce neuropathic pain [25, 29].

Interestingly, most patients report that distracting activities such as gardening, manual labor, or driving a car have the potential to temporarily reduce deafferentation pain. Patients who are absorbed in their work or hobbies can thereby gain considerable periods of relief of pain, making distraction by far the most potent analgesic [28]. Almost all patients, however, feel the pain instantly return once they stop that particular activity [28]. Occupational therapies, which aim at reintegrating patients into working life, can also efficiently reduce pain following brachial plexus avulsion [8].

Effects of Motor Recovery on Deafferentation Pain

During the past decades, the body of literature on functional outcomes including motor and sensory recovery after brachial plexus repair has expanded to a multitudinous amount. Recently, however, researchers have also come to appreciate that functional improvements in limb mobility are associated with regained activation of primary sensorimotor areas related to the affected extremity as well as a reduction in deafferentation pain [11]. In 1988, Bruxelle and colleagues were among the first to report positive effects of surgical reconstruction in patients with brachial plexus avulsion injuries on deafferentation pain in a large patient cohort [8]. He showed that 3 years after the surgery, deafferentation pain had decreased in 63% of patients who had some sort of functional recovery. The positive effect of functional restoration on avulsion pain was confirmed in a number of following studies [4, 5, 7]. Interestingly, it has been recognized that patients who experienced no pain improvement did not show evidence of returning motor function [5].

In complete brachial plexus injuries, nerve transfers are performed to partially restore motor function in the arm and hand [6]. Treatment success is dependent on the regeneration of axons from a donor nerve into the target muscle. Besides the regeneration of motor axons, which will ultimately initiate muscle activation, sensory axons originating in the re-innervated muscle transmit proprioceptive feedback from this muscle to the central nervous system. This phenomenon may be termed as *reafferentation,* since the brain regains sensory feedback from the previously denervated limb. Consequently, deafferentation pain is reduced. In summary, successful reconstructive nerve repair results in functional recovery and sensory reafferentation of the central nervous system and can thereby reduce deafferentation pain [8].

References

1. Aichaoui F, Mertens P, Sindou M. Dorsal root entry zone lesioning for pain after brachial plexus avulsion: results with special emphasis on differential effects on the paroxysmal versus the continuous components. A prospective study in a 29-patient consecutive series. Pain. 2011;152(8):1923–30. https://doi.org/10.1016/j.pain.2011.03.037.
2. Allison JD, Meador KJ, Loring DW, Figueroa RE, Wright JC. Functional MRI cerebral activation and deactivation during finger movement. Neurology. 2000;54(1):135–42.
3. Atkinson JR, Ward AA Jr. Intracellular studies of cortical neurons in chronic epileptogenic foci in the monkey. Exp Neurol. 1964;10:285–95.
4. Berman J, Anand P, Chen L, Taggart M, Birch R. Pain relief from preganglionic injury to the brachial plexus by late intercostal nerve transfer. J Bone Joint Surg. 1996;78(5):759–60.
5. Berman JS, Birch R, Anand P. Pain following human brachial plexus injury with spinal cord root avulsion and the effect of surgery. Pain. 1998;75(2–3):199–207.
6. Bertelli JA, Ghizoni MF. Reconstruction of complete palsies of the adult brachial plexus by root grafting using long grafts and nerve transfers to target nerves. J Hand Surg Am. 2010;35(10):1640–6. https://doi.org/10.1016/j.jhsa.2010.06.019.
7. Bertelli JA, Ghizoni MF. Results and current approach for brachial plexus reconstruction. J Brachial Plex Peripher Nerve Inj. 2011;6(1):2. https://doi.org/10.1186/1749-7221-6-2.
8. Bruxelle J, Travers V, Thiebaut JB. Occurrence and treatment of pain after brachial plexus injury. Clin Orthop Relat Res. 1988;237:87–95.
9. Chew DJ, Murrell K, Carlstedt T, Shortland PJ. Segmental spinal root avulsion in the adult rat: a model to study avulsion injury pain. J Neurotrauma. 2013;30(3):160–72. https://doi.org/10.1089/neu.2012.2481.
10. De Ridder D, De Mulder G, Verstraeten E, Sunaert S, Moller A. Somatosensory cortex stimulation for deafferentation pain. Acta Neurochir Suppl. 2007;97(Pt 2):67–74.
11. Dimou S, Biggs M, Tonkin M, Hickie IB, Lagopoulos J. Motor cortex neuroplasticity following brachial plexus transfer. Front Hum Neurosci. 2013;7:500. https://doi.org/10.3389/fnhum.2013.00500.
12. Flor H, Elbert T, Muhlnickel W, Pantev C, Wienbruch C, Taub E. Cortical reorganization and phantom phenomena in congenital and traumatic upper-extremity amputees. Exp Brain Res. 1998;119(2):205–12.
13. Frazier CH, Skillern PG. Supraclavicular subcutaneous lesions of the brachial plexus not associated with skeletal injuries. J Am Med Assoc. 1911;57(25):1957–63.
14. Gucer G, Niedermeyer E, Long DM. Thalamic EEG recordings in patients with chronic pain. J Neurol. 1978;219(1):47–61.
15. Hirayama T, Dostrovsky JO, Gorecki J, Tasker RR, Lenz FA. Recordings of abnormal activity in patients with deafferentation and central pain. Stereotact Funct Neurosurg. 1989;52(2–4):120–6.
16. Hsieh JC, Cheng H, Hsieh HM, Liao KK, Wu YT, Yeh TC, et al. Loss of interhemispheric inhibition on the ipsilateral primary sensorimotor cortex in patients with brachial plexus injury: fMRI study. Ann Neurol. 2002;51(3):381–5.
17. Jain N, Florence SL, Kaas JH. Reorganization of somatosensory cortex after nerve and spinal cord injury. News Physiol Sci. 1998;13:143–9.
18. Liu B, Li T, Tang WJ, Zhang JH, Sun HP, Xu WD, et al. Changes of inter-hemispheric functional connectivity between motor cortices after brachial plexuses injury: a resting-state fMRI study. Neuroscience. 2013;243:33–9. https://doi.org/10.1016/j.neuroscience.2013.03.048.
19. Loeser JD, Ward AA Jr. Some effects of deafferentation on neurons of the cat spinal cord. Arch Neurol. 1967;17(6):629–36.
20. Loeser JD, Ward AA Jr, White LE Jr. Chronic deafferentation of human spinal cord neurons. J Neurosurg.

1968;29(1):48–50. https://doi.org/10.3171/jns.1968.29.1.0048.
21. Mehta S, Orenczuk K, McIntyre A, Willems G, Wolfe DL, Hsieh JT, et al. Neuropathic pain post spinal cord injury part 1: systematic review of physical and behavioral treatment. Top Spinal Cord Inj Rehabil. 2013;19(1):61–77. https://doi.org/10.1310/sci1901-61.
22. Mehta S, Orenczuk K, McIntyre A, Willems G, Wolfe DL, Hsieh JT, et al. Neuropathic pain post spinal cord injury part 2: systematic review of dorsal root entry zone procedure. Top Spinal Cord Inj Rehabil. 2013;19(1):78–86. https://doi.org/10.1310/sci1901-78.
23. Melzack R. Phantom limbs. Sci Am. 1992;266(4):120–6.
24. Moseley GL. Graded motor imagery is effective for long-standing complex regional pain syndrome: a randomised controlled trial. Pain. 2004;108(1–2):192–8. https://doi.org/10.1016/j.pain.2004.01.006.
25. Moseley GL. Using visual illusion to reduce at-level neuropathic pain in paraplegia. Pain. 2007;130(3):294–8. https://doi.org/10.1016/j.pain.2007.01.007.
26. Narakas AO. The treatment of brachial plexus injuries. Int Orthop. 1985;9(1):29–36.
27. Nashold BS Jr, Ostdahl RH. Dorsal root entry zone lesions for pain relief. J Neurosurg. 1979;51(1):59–69. https://doi.org/10.3171/jns.1979.51.1.0059.
28. Parry CB. Pain in avulsion lesions of the brachial plexus. Pain. 1980;9(1):41–53.
29. Ramachandran VS, Rogers-Ramachandran D, Cobb S. Touching the phantom limb. Nature. 1995;377(6549):489–90. https://doi.org/10.1038/377489a0.
30. Rexed B. The cytoarchitectonic organization of the spinal cord in the cat. J Comp Neurol. 1952;96(3):414–95.
31. Rinaldi PC, Young RF, Albe-Fessard D, Chodakiewitz J. Spontaneous neuronal hyperactivity in the medial and intralaminar thalamic nuclei of patients with deafferentation pain. J Neurosurg. 1991;74(3):415–21. https://doi.org/10.3171/jns.1991.74.3.0415.
32. Sindou M. Microsurgical DREZotomy (MDT) for pain, spasticity, and hyperactive bladder: a 20-year experience. Acta Neurochir. 1995;137(1–2):1–5.
33. Sindou M, Fischer G, Goutelle A, Mansuy L. [Selective surgery of posterior nerve roots. First results of surgery for pain]. Neuro-Chirurgie. 1974;20(5):391–408.
34. Sindou M, Mertens P, Wael M. Microsurgical DREZotomy for pain due to spinal cord and/or cauda equina injuries: long-term results in a series of 44 patients. Pain. 2001;92(1–2):159–71.
35. Sindou M, Mifsud JJ, Boisson D, Goutelle A. Selective posterior rhizotomy in the dorsal root entry zone for treatment of hyperspasticity and pain in the hemiplegic upper limb. Neurosurgery. 1986;18(5):587–95.
36. Sindou M, Quoex C, Baleydier C. Fiber organization at the posterior spinal cord-rootlet junction in man. J Comp Neurol. 1974;153(1):15–26. https://doi.org/10.1002/cne.901530103.
37. Teixeira MJ, da Paz MG, Bina MT, Santos SN, Raicher I, Galhardoni R, et al. Neuropathic pain after brachial plexus avulsion—central and peripheral mechanisms. BMC Neurol. 2015;15:73. https://doi.org/10.1186/s12883-015-0329-x.
38. Thomas DG. Brachial plexus injury: deafferentation pain and dorsal root entry zone (DREZ) coagulation. Clin Neurol Neurosurg. 1993;95 Suppl:S48–9.
39. Thomas DG, Sheehy JP. Dorsal root entry zone lesions (Nashold's procedure) for pain relief following brachial plexus avulsion. J Neurol Neurosurg Psychiatry. 1983;46(10):924–8.
40. Todd AJ. Neuronal circuitry for pain processing in the dorsal horn. Nat Rev Neurosci. 2010;11(12):823–36. https://doi.org/10.1038/nrn2947.
41. Woolf CJ, Wall PD. Chronic peripheral nerve section diminishes the primary afferent A-fibre mediated inhibition of rat dorsal horn neurones. Brain Res. 1982;242(1):77–85.

Treatment Algorithm for Bionic Hand Reconstruction in Patients with Global Brachial Plexopathies

17

Laura A. Hruby and Agnes Sturma

Injuries to the brachial plexus represent one of the most severe nerve injuries with permanent and devastating effects on a patient's physical and psychological well-being. Despite the fact that modern reconstructive surgeries yield objectively good motor results for the shoulder and elbow, hand function remains negligible in a considerable subset of patients. Prosthetic hand replacement has recently been described for patients in whom primary and secondary reconstructions have failed to improve hand function. Here we present the treatment algorithm for the concept of bionic hand reconstruction in patients with global brachial plexopathies.

L. A. Hruby (✉)
Department of Orthopedics and Trauma Surgery, Medical University of Vienna, Vienna, Austria

Clinical Laboratory for Bionic Extremity Reconstruction, Division of Plastic and Reconstructive Surgery, Department of Surgery, Medical University of Vienna, Vienna, Austria
e-mail: laura.hruby@meduniwien.ac.at

A. Sturma
Clinical Laboratory for Bionic Extremity Reconstruction, Division of Plastic and Reconstructive Surgery, Department of Surgery, Medical University of Vienna, Vienna, Austria

Department of Bioengineering, Imperial College London, London, UK

Global Brachial Plexopathies Beyond Biological Rehabilitation

Following brachial plexus injury, primary nerve reconstruction should be initiated early as denervated muscle cells undergo degenerative fibrosis resulting in loss of motor end plates as soon as the nerval input is discontinued [9, 17]. The progressive atrophy of muscles is accompanied by stiffness of joints leading to contractures in the hand.

Primary reconstructions include direct neural repair, nerve grafts, as well nerve transfers [2, 3, 14]. In longstanding lesions with irreversible atrophy of muscles or delayed surgical treatment, secondary reconstructions are performed, which include free functional muscle transfers (FFMT), tendon transfers, tenodesis, and arthrodesis [5, 16, 17]. Functional outcomes after primary and secondary reconstructions of complete avulsion injuries have been significantly improved by advances in microsurgical reconstructive techniques, and shoulder stability and elbow function are successfully restored in the majority of patients. Some patients, however, remain severely handicapped by a paralyzed and insensate hand [4].

For such patients, bionic hand reconstruction represents a game-changing treatment option [1]. This procedure prepares patients for future prosthetic hand use by surgically improving the biotechnological interface. Bionic reconstruction ranges from selective nerve and muscle transfers to establish electromyographic (EMG) signal

O. C. Aszmann, D. Farina (eds.), *Bionic Limb Reconstruction*,
https://doi.org/10.1007/978-3-030-60746-3_17

sites in the arm of the patient, which—in the future—will drive a prosthetic device, to the elective amputation of the functionless hand and final fitting with a prosthetic hand. The clinical sequence of all relevant steps has been outlined in a comprehensive treatment algorithm [8].

Algorithm for Bionic Hand Reconstruction

Initial Consultation

Upon initial consultation, a thorough case history is obtained, which summarizes circumstances, which led to brachial plexus injury, primary care after the injury, medical records on primary and secondary reconstructions performed abroad, as well as expectations and/or reasons for consultation. Clinical examinations include a detailed musculoskeletal assessment of hand and arm function, electrodiagnostic and imaging studies using high-resolution ultrasound, and/or magnetic resonance imaging (MRI), which are performed to assess the state of affected muscles and nerves.

Biological treatment alternatives are discussed with the patient with an interdisciplinary team of reconstructive surgeons, a physiotherapist, a clinical psychologist, physiatrists, and orthopedic technicians specialized in prosthetics. If hand function is deemed futile because of dismal function, absent sensation, and lack of treatment alternatives, the next step in the algorithm is identification of surface EMG signals detected in muscles of the forearm and upper arm that would be available for control of a prosthetic hand.

Identification of Surface EMG Signals

Global brachial plexopathies including multiple root avulsions may cause complete paralysis of the hand even if timely surgical reconstruction is performed [15]. Despite negligible hand function, some muscles present contractile activity, which—although without clinical significance (M1 according to the British Medical Research Council scale)—can be made visible with transcutaneous electrodes. During initial review, positions on the patient's skin with (relatively) high amplitudes are detected, which mark sEMG hotspots available for future prosthetic control. The patient's muscular activity yielding reliable EMG signals is visualized for the patient using various biofeedback methods. EMG waves appear on a computer screen in different colors representing different signal sites. Additionally, these signals are simultaneously translated into prosthetic function with a table-top prosthesis mounted next to the patient. This arrangement allows to predict prosthetic hand function, which can be expected after bionic reconstruction (see Fig. 17.1).

Reliable prosthetic control necessitates at least two separable surface EMG signals [13]. These linearly control the prosthesis, while

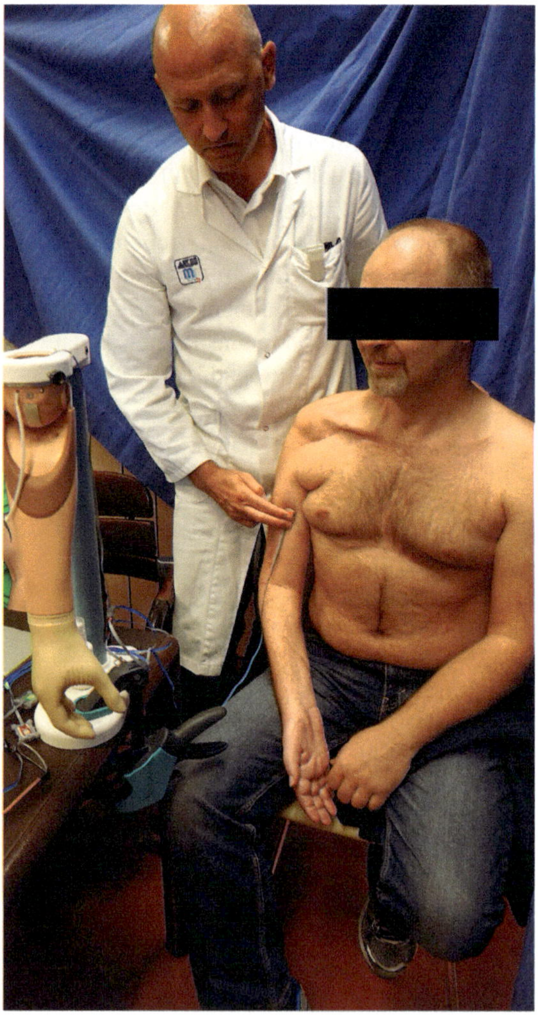

Fig. 17.1 Set-up of a table-top prosthesis. Transcutaneous electrodes are placed over muscles, where they sense and record EMG activity. (© Laura Hruby)

co-contraction of both muscle signals is used to switch between different functional levels of the prosthetic device (the elbow, wrist, and hand in above-elbow amputees; the wrist and hand in below-elbow amputees). In some patients, two signals are detected upon initial consultation, who are then offered further training to optimize signal consistency and separation before elective amputation and prosthetic hand replacement. Other patients present with only one sEMG signal. These selective nerve transfers in combination with free functional muscle transfers (FFMT) are performed to establish additional myoactivity for future prosthetic control.

Nerve Transfers

During initial consultation, the clinical examination includes the assessment of a positive Tinel Hoffmann's sign, which implies the presence of viable axons within nerves of the forearm and upper arm. During surgical exploration, which is oftentimes indispensable to evaluate residual nerve function, these findings are confirmed using intraoperative nerve stimulation. In some patients, however, long denervation intervals lead to muscle atrophy, and muscles no longer show adequate response to nerve stimulation. For that purpose, an intra-operative fast staining method may be used, which screens nerves for acetylcholine-positive motor axons to be used for re-innervating a new target muscle. FFMTs need to be performed in a subset of patients due to longstanding lesions and atrophied residual muscles in the arm. Hence, the muscle transfer is done a priori to create a signal site for future prosthetic control. Although the newly established muscular activity is without clinical significance, it is still detectable with EMG sensors and sufficient to drive a prosthetic hand.

EMG-Guided Rehabilitation and Training

Following surgery, regular follow-up visits take place in a 3-month interval. As with classical surgical interventions, wound and pain management is initiated in the early stage. It is important to inform patients that deafferentation pain can even increase subsequent to the surgical manipulation of nerves during nerve transfer surgery. Upon nerve regeneration, specific rehabilitation regimes targeting muscle re-innervation are applied starting at 3–6 months post surgery. Rehabilitative approaches include strengthening of weak muscles of the adjacent joints (e.g., the biceps muscle for a planned transradial amputation), motor imagery of hand movements, and cognitive training using EMG biofeedback methods [7, 11, 12]. During EMG signal training, the patients learn how to separately activate their individual muscle signals and how to modulate the amplitude. Subsequent to nerve transfer surgery, this process may take some time, since nerves are rerouted during surgery and provide neural input to new muscle targets, which can represent a major cognitive challenge for some patients. This cognitive relearning is promoted with home training tools, which visualize EMG activity and feedback signal intensity to the patient, allowing sufficient training units without the need to see a therapist on a regular basis. The ultimate goal of signal training is to ascertain that a patient will be able to reliably control his future prosthetic hand.

Fitting of a Hybrid Hand

The fitting of a hybrid hand is performed for training purposes as well as to document prospective prosthetic function, which can be expected after elective amputation. The hybrid hand is a splint-like construction with a myoelectric hand prosthesis mounted onto or below the impaired plexus hand. It is fitted individually to each patient as soon as reliable signal control has been achieved after extensive training units. Electrodes sensing the patient's myoelectric activity are integrated in the shaft and record EMG activity, which is then translated into hand function by the prosthetic's mechanics. Using this setup, patients perform objective hand function tests, which allow documentation of future prosthetic function, e.g., for insurance companies that will pay for the prosthetic fitting. Some patients also use the hybrid hand as a home training tool, which improves handling during activities of daily living

(ADL). Residual upper limb function including shoulder stability and range of motion in adjacent joints can be assessed with the hybrid hand setup, which may also be used to improve these functions during training. Finally, only patients with a clear functional benefit based on the hybrid hand are considered for amputation (evaluated with video documentation comparing functions of the plexus hand with the hybrid hand; scores of objective hand function tests with plexus hand and hybrid hand).

Psychological Assessment

Before elective amputation, each patient has to pass a psychosocial assessment procedure consisting of a semi-structured interview (see Chap. 19). This interview addresses five different domains: psychosocial adjustment following the accident, which led to brachial plexus injury; affective relationship to the plexus hand; psychosocial status of the patient; motivational aspects for an amputation; and expectations of prosthetic function after successful bionic reconstruction. Most importantly, it has to be evaluated whether the patient is cognitively able to estimate the consequences of an elective amputation or to be fully aware of the irreversible decision associated with it. For that reason, several psychological disorders represent contraindications for the process of bionic reconstruction. These include posttraumatic stress disorders, alcohol and/or substance abuse, insufficient emotional coping, and expected noncompliance associated with prosthetic fitting and maintenance. For patients who do not qualify for bionic reconstruction due to any of the above reasons, psychological support is provided.

Amputation of the Functionless Hand/Arm

Depending on where detectable myoactivity including newly generated EMG signal sites is located, the elective amputation of the flail limb is performed either at a distal level (transradial amputation) or at a more proximal level (short transhumeral amputation). Transradial amputations are performed 17 cm from the lateral epicondyle when at least two separable surface EMG signals are detectable in the forearm (e.g., one at the volar aspect, mostly the pronator teres muscle, and one at the dorsal aspect at the extensor compartment). This stump length allows enough room for the various components of the future prosthetic hand but also maintains the best possible leverage. In all cases, the most sensitive skin surface was used to cover the stump in order to obtain a fully sensate stump, which improves prosthetic handling due to biological sensory feedback [1].

In some patients with longstanding, more severe, and complete brachial plexopathies, residual myoactivity may not be found in the forearm and is not possible to be established by means of selective nerve and/or muscle transfers due to the extensive neurological damage caused by multiple root avulsions. Furthermore, muscle activity is so faint that biological elbow function is absent, which needs to be reconstructed as well by prosthetic means. In such patients, surface EMG signals may be detectable more proximally, e.g., at the supra- and infraspinatus muscles and the pectoralis major muscle [6]. Short transhumeral or glenohumeral amputations then allow optimal fitting of a prosthetic arm including a prosthetic elbow.

Final Fitting of a Prosthetic Hand or Arm

Since in patients with global brachial plexopathies atrophy of the affected limb has already taken place, prosthetic fitting can be initiated as soon as 4–6 weeks after amputation which is in contrast to the time frame for conventional amputations [10]. Regular prosthetic usage increases dexterity. Training sessions with an instructional therapist, however, are recommended to further improve prosthetic handling in everyday life.

Whereas contracting one muscle will initiate a specific prosthetic movement, co-contraction of the two EMG signals will achieve switching between functional levels of the prosthesis. In transradial amputees, these are rotation of the hand (pronation and supination) and opening and closing of the hand. In transhumeral amputees, the functional levels include the elbow (flexion and extension), hand rotation (pronation and supination), and hand opening and closing.

Functional assessments are performed at least 3 months after prosthetic fitting and regular prosthetic usage (see Chap. 18).

All consequent steps of the treatment algorithm for bionic hand reconstruction in patients with global brachial plexopathies are schematically outlined in Fig. 17.2.

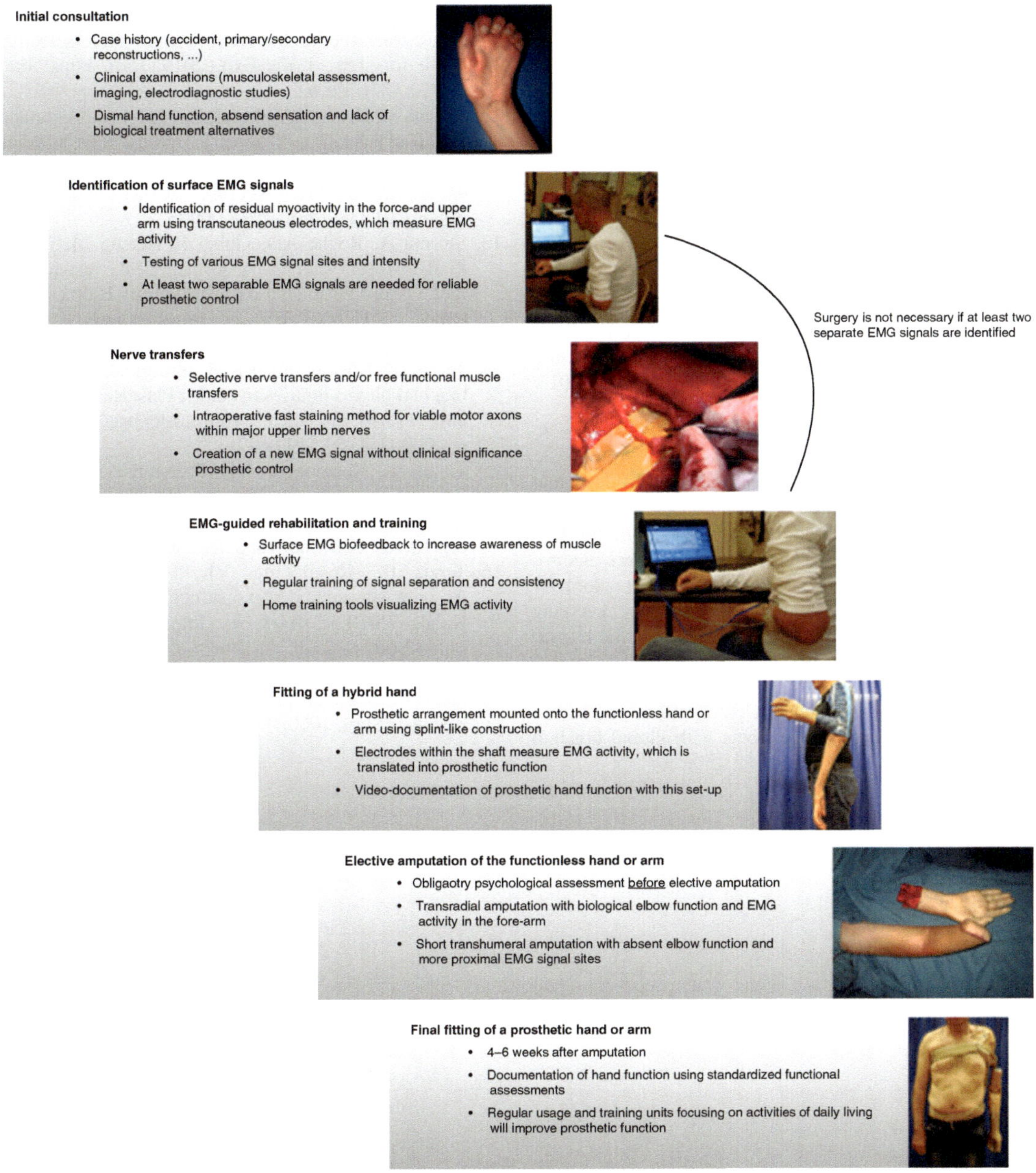

Fig. 17.2 Treatment algorithm for bionic hand reconstruction in patients with global brachial plexus avulsion injuries. (© Laura Hruby)

References

1. Aszmann OC, Roche AD, Salminger S, Paternostro-Sluga T, Herceg M, Sturma A, et al. Bionic reconstruction to restore hand function after brachial plexus injury: a case series of three patients. Lancet. 2015;385:2183–9. https://doi.org/10.1016/S0140-6736(14)61776-1.
2. Bertelli JA, Ghizoni MF. Reconstruction of complete palsies of the adult brachial plexus by root grafting using long grafts and nerve transfers to target nerves. J Hand Surg Am. 2010;35(10):1640–6. https://doi.org/10.1016/j.jhsa.2010.06.019.
3. Bertelli JA, Ghizoni MF. Results and current approach for brachial plexus reconstruction. J Brachial Plex Peripher Nerve Inj. 2011;6(1):2. https://doi.org/10.1186/1749-7221-6-2.
4. Doi K, Sakai K, Fuchigami Y, Kawai S. Reconstruction of irreparable brachial plexus injuries with reinnervated free-muscle transfer. Case report. J Neurosurg. 1996;85(1):174–7. https://doi.org/10.3171/jns.1996.85.1.0174.
5. Doi K, Sakai K, Kuwata N, Ihara K, Kawai S. Double free-muscle transfer to restore prehension following complete brachial plexus avulsion. J Hand Surg Am. 1995;20(3):408–14. https://doi.org/10.1016/S0363-5023(05)80097-8.
6. Hruby LA, Gstoettner C, Sturma A, Salminger S, Mayer JA, Aszmann OC. Bionic upper limb reconstruction: a valuable alternative in global brachial plexus avulsion injuries—a case series. J Clin Med. 2019;9(1):23. https://doi.org/10.3390/jcm9010023.
7. Hruby LA, Sturma A, Aszmann OC. Surface electromyographic biofeedback as a rehabilitation tool for patients with global brachial plexus injury receiving bionic reconstruction. J Vis Exp. 2019;(151). https://doi.org/10.3791/59839.
8. Hruby LA, Sturma A, Mayer JA, Pittermann A, Salminger S, Aszmann OC. Algorithm for bionic hand reconstruction in patients with global brachial plexopathies. J Neurosurg. 2017;127(5):1–9. https://doi.org/10.3171/2016.6.JNS16154.
9. Moran SL, Steinmann SP, Shin AY. Adult brachial plexus injuries: mechanism, patterns of injury, and physical diagnosis. Hand Clin. 2005;21(1):13–24. https://doi.org/10.1016/j.hcl.2004.09.004.
10. Salminger S, Roche AD, Hruby LA, Sturma A, Riedl O, Bergmeister KD, et al. Prosthetic reconstruction to restore function in transcarpal amputees. J Plast Reconstr Aesthet Surg. 2016;69(3):305–10. https://doi.org/10.1016/j.bjps.2015.10.029.
11. Sturma A, Hruby LA, Farina D, Aszmann OC. Structured motor rehabilitation after selective nerve transfers. J Vis Exp. 2019;(150). https://doi.org/10.3791/59840.
12. Sturma A, Hruby LA, Prahm C, Mayer JA, Aszmann OC. Rehabilitation of upper extremity nerve injuries using surface EMG biofeedback: protocols for clinical application. Front Neurosci. 2018;12:906. https://doi.org/10.3389/fnins.2018.00906.
13. Sturma A, Roche AD, Gobel P, Herceg M, Ge N, Fialka-Moser V, et al. A surface EMG test tool to measure proportional prosthetic control. Biomed Tech (Berl). 2015;60(3):207–13. https://doi.org/10.1515/bmt-2014-0022.
14. Terzis JK, Kostopoulos VK. The surgical treatment of brachial plexus injuries in adults. Plast Reconstr Surg. 2007;119(4):73e–92e. https://doi.org/10.1097/01.prs.0000254859.51903.97.
15. Terzis JK, Kostopoulos VK. Free muscle transfer in posttraumatic plexopathies: part III. The hand. Plast Reconstr Surg. 2009;124(4):1225–36. https://doi.org/10.1097/PRS.0b013e3181b5a322.
16. Terzis JK, Kostopoulos VK. Free muscle transfer in posttraumatic plexopathies part II: the elbow. Hand. 2010;5(2):160–70. https://doi.org/10.1007/s11552-009-9223-6.
17. Terzis JK, Papakonstantinou KC. The surgical treatment of brachial plexus injuries in adults. Plast Reconstr Surg. 2000;106(5):1097–122; quiz 123–4.

Functional and Psychosocial Outcomes of Bionic Reconstruction and Impact on Quality of Life, Body Image Perception and Deafferentation Pain

18

Laura A. Hruby, Agnes Sturma, and Anna Pittermann

Global brachial plexopathies represent one of the most severe nerve injuries and usually affect young individuals at the prime of their life. Restoration of motor function in the shoulder and elbow do not automatically translate into improved functionality in daily life and/or psychological well-being. A vast majority of brachial plexus patients report to feel functionally limited to a great extent due to non-recovery or elbow recovery only. As with traditional primary and secondary reconstructions, it is thus important to highlight functional as well as psychosocial outcome variables following bionic reconstruction in patients with complete brachial plexus injuries. Here we describe effects of bionic hand reconstruction on various aspects of life including functionality, overall quality of life, body image, and deafferentation pain.

L. A. Hruby (✉)
Department of Orthopedics and Trauma Surgery, Medical University of Vienna, Vienna, Austria

Clinical Laboratory for Bionic Extremity Reconstruction, Department of Surgery, Medical University of Vienna, Vienna, Austria
e-mail: laura.hruby@meduniwien.ac.at

A. Sturma
Clinical Laboratory for Bionic Extremity Reconstruction, Department of Surgery, Medical University of Vienna, Vienna, Austria

Department of Bioengineering, Imperial College London, London, UK

A. Pittermann
Clinical Laboratory for Bionic Extremity Reconstruction, Department of Surgery, Medical University of Vienna, Vienna, Austria

Department of Clinical Psychology, General Hospital of Vienna, Vienna, Austria

Background

Global brachial plexopathies including multiple nerve root avulsions have permanent and devastating effects on a patient's physical, psychological, and socioeconomic well-being [23]. Typically, global brachial plexopathies affect young individuals at the prime of their life and in otherwise healthy condition [27]. Advances in microsurgical techniques as well as increasing numbers of civilian brachial plexus injuries (BPI) have promoted tremendous progress in brachial plexus repair [4]. Although stabilization of the shoulder joint and restoration of elbow function is achieved in the majority of patients, some still suffer from various physically and psychologically debilitating sequelae, which can further hinder psychosocial adjustment following the accident [11]. A "good" motor result, which may

O. C. Aszmann, D. Farina (eds.), *Bionic Limb Reconstruction*,
https://doi.org/10.1007/978-3-030-60746-3_18

satisfy the reconstructive surgeon, does not always meet patients' expectations as isolated functional gain might not improve the overall functionality in daily life and therefore does not automatically translate into psychological well-being [12, 19].

As with classic primary and secondary reconstructions, it is thus important to evaluate not only functional outcome measures of bionic reconstruction but also psychosocial outcome variables, which focus on patient-centered, subjective data highlighting benefits and merits of prosthetic hand replacement in patients with BPIs [15]. This chapter is dedicated to summarize effects of bionic reconstruction on various aspects of life including functionality in daily life, overall quality of life, body image, and deafferentation pain. Single steps of the procedure ranging from identification of eligible patients to final prosthetic fitting can be found in Chap. 17.

Functional Outcomes

In all patients with brachial plexus injuries, qualifying for bionic reconstruction global arm and hand function is assessed before elective amputation and after final prosthetic fitting with two-objective hand function tests (ARAT [31] and SHAP [21]) and a patient-reported questionnaire addressing subjective disability (DASH [18]). For detailed information on functional testing see Chap. 7.

As can be expected, all patients interested in the procedure of bionic reconstruction mainly report to feel functionally limited to a great extent due to either non-recovery or elbow flexion recovery only following various primary and secondary reconstructions performed elsewhere. Several national and international media reports have promoted interest in bionic reconstruction with the majority of patients explicitly wishing for functional improvement upon initial consultation.

It has been shown in various studies that bionic reconstruction enables prosthetic hand use after elective transradial amputation of the impaired plexus hand [1, 16]. A stable shoulder joint and good elbow function (>M4) to control the prosthetic hand in three-dimensional space have previously been defined as prerequisites for prosthetic hand replacement [16].

The concept of bionic reconstruction, however, is also applicable for patients without sufficient elbow function. If residual myoactivity can be detected at a more proximal level with identification of two or more separable EMG signals, which will later translate into antagonistic prosthetic functions, a short transhumeral or glenohumeral amputation of the functionless arm is performed to allow fitting of a prosthetic arm including a prosthetic elbow [14]. Widespread research in the field of prosthetics has promoted tremendous progress, particularly bringing forth improvements to replicate motor function and control [9]. Logically, the number of neuromuscular units that can be established as well as their function (i.e., signal consistency and amplitude) is related to enhanced control of a myoelectric prosthetic device. As a consequence of extensive neurological damage following a brachial plexus avulsion injury, however, residual myoactivity in these patients is often faint with absent, weak, or cognitively confusing signals [1]. Therefore, intense cognitive training programs are essential to realize good prosthetic hand use after elective amputation [26]. Patients in whom incompliance is expected thus do not qualify for bionic reconstruction. It is also important to stress that a myoelectric prosthesis by no means compares with a biological hand in terms of functionality [13]. However, given the futile hand function in patients with long-standing inveterate brachial plexopathies, excellent prospective prosthetic hand use justifies elective amputation and prosthetic hand replacement [16]. Patients need to be thoroughly informed that the bionic hand or arm will always remain an "assist" extremity, which will still considerably expand manual capacity during daily life activities due to regained bimanual dexterity [16].

Improved functionality after prosthetic replacement of the impaired plexus hand is not only reflected by significant increases in objective hand function scores, which have been reported in the literature [1, 16]. Likely even more important, statements of patients who have undergone bionic reconstruction highlight its efficacy to regain hand function. After having lived without any function in their limb for years or even decades, most patients are overwhelmed, for example, when they realize for the first time that they are able to prepare a meal all by themselves using two hands again. Patients can go back to pursuing hobbies, are reintegrated into working life, and report of increased self-confidence due to regained functionality.

Effects of Bionic Reconstruction on Quality of Life

Even if some function can be restored, patients with global brachial plexopathies still have to cope with several stressors including reduced self-sufficiency, dependence on others, occupational retraining following the accident, financial instability due to unemployment, as well as dissatisfaction with aesthetic appearance of the withered arm and hand [11, 19]. Symptoms of depression and/or anxiety appear in nearly one third of patients with complete BPI [19].

As an integral part of our assessments, quality of life is evaluated with the SF-36 Health Survey [30]. The questionnaire addresses eight independent subscales: physical functioning, physical role functioning, bodily pain, general health, vitality, social role functioning, emotional role functioning, and mental health. As is in agreement with previous studies, upon initial consultation, most of our patients have inferior results in regard to physical functioning and mental health as compared to healthy norm samples. Vitality and social role functioning are usually limited to a great extent.

Patients who have undergone bionic reconstruction stress that prosthetic hand replacement has changed their life in many ways. Not only do they report of improved physical functioning but also increased vitality and social and emotional functioning due to regained bimanual interaction with their environment. Returning to former social activities is usually accompanied by appreciable improvements in patient satisfaction [25]. Increased self-sufficiency and reduced reliance on others are rated as one of the most important benefits resulting from bionic reconstruction.

Effects of Bionic Reconstruction on Body Image

Clinicians familiar with brachial plexus injuries know that the affected limb will eventually become atrophic and cold with a distal bluish discoloration due to autonomic paralysis [5]. Since our hands represent the instruments with which we interact with our environment daily, this apparent physical deficit may cause psychological distress and potential social pressure [20]. Psychological sequelae include a disturbed body image and negative self-evaluation [20].

We regularly assess body image perception with the Body Image Questionnaire (FKB-20) before and after bionic reconstruction. This questionnaire is widely used for the diagnosis of body image disorders and evaluation of subjective body awareness [7]. Two scales can be identified: the scale "negative body evaluation" allows conclusions about physical appearance, body image, and associated subjective well-being with a person's body image [8]; the scale "vital body dynamics" addresses bodily strength, fitness, and health, which are subjectively rated by the patient [8].

By evaluating pre- and post-bionic reconstruction scale scores we have found that prosthetic hand replacement restores an intact body image by resolution of the negative body evaluation present in the majority of brachial plexus patients [15]. During follow-up visits patients, regularly report of a high grade of embodiment of the prosthetic hand, stating that the prosthesis had become an integral component of their self-image using phrases like "For me this is not a mechatronic device. This is my new hand. I put it on right after waking up and mostly fall asleep at night having forgotten to take it off" [15].

Effects of Bionic Reconstruction on Deafferentation Pain

Deafferentation pain following brachial plexus avulsion injury is a severe chronic pain syndrome, which affects 70–90% of patients [2, 24, 28]. Mechanisms for its development are described in Chap. 16.

The approach of elective amputation has already been described for severe cases of complete BPI without the primary intention to replace the limb with a prosthesis [22]. Despite the fact that none of these used functional prostheses, afterward all were still satisfied with the decision to have the impaired hand amputated, as it served to relief patients of the burden of a flail, insensate arm [6, 12]. Importantly, however, chronic deafferentation pain did not improve by sole amputation of the deafferented hand. This is not surprising since the pain's origin is thought to arise in the dorsal horn of the spinal cord [29]. Additionally, supraspinal central mechanisms are then thought to maintain the pain [10]. Following the "inner amputation," which occurs after brachial plexus avulsion, the central nervous system (CNS) is deprived of its neurological connection to an extremely relevant piece of anatomy—the hand—which in fact entertains most of the primary motor and sensory cortex related to movement [23]. By replacing the functionless, insensate appendage with a prosthetic hand, the CNS again receives afferent input from the previously deafferented hand. Firstly, the patient receives visual feedback from his functioning hand, which he starts to use again on a daily basis, thereby slowly expelling the deafferented phantom hand from his awareness.

Motor recovery following classic reconstructive surgery, may it be very limited, has been associated with reduced deafferentation in various studies [2, 3, 17]. The fact that bionic reconstruction—successfully restoring hand function in patients who have lived without any function for a long time—effectively reduces deafferentation pain is thus not surprising. We refer to this phenomenon as "functional re-afferentation."

References

1. Aszmann OC, Roche AD, Salminger S, Paternostro-Sluga T, Herceg M, Sturma A, et al. Bionic reconstruction to restore hand function after brachial plexus injury: a case series of three patients. Lancet. 2015;385:2183–9. https://doi.org/10.1016/S0140-6736(14)61776-1.
2. Berman J, Anand P, Chen L, Taggart M, Birch R. Pain relief from preganglionic injury to the brachial plexus by late intercostal nerve transfer. J Bone Joint Surg. 1996;78(5):759–60.
3. Berman JS, Birch R, Anand P. Pain following human brachial plexus injury with spinal cord root avulsion and the effect of surgery. Pain. 1998;75(2–3):199–207.
4. Bertelli JA, Ghizoni MF. Results and current approach for brachial plexus reconstruction. J Brachial Plex Peripher Nerve Inj. 2011;6(1):2. https://doi.org/10.1186/1749-7221-6-2.
5. Carlstedt T. Central nerve plexus injury. London/Hackensack, NJ: Imperial College Press/Distributed by World Scientific Pub; 2007.
6. Chuang DC. Adult brachial plexus reconstruction with the level of injury: review and personal experience. Plast Reconstr Surg. 2009;124(6 Suppl):e359–69. https://doi.org/10.1097/PRS.0b013e3181bcf16c.
7. Clement U, Lowe B. [Validation of the FKB-20 as scale for the detection of body image distortions in psychosomatic patients]. Psychother Psychosom Med Psychol. 1996;46(7):254–9.
8. Clement U, Der Löwe B. "Fragebogen zum Körperbild (FKB-20)": Literaturüberblick, Beschreibung und Prüfung eines Meßinstruments. Diagnostica. 1996;42:352–76.
9. Farina D, Aszmann O. Bionic limbs: clinical reality and academic promises. Sci Transl Med. 2014;6(257):257ps12. https://doi.org/10.1126/scitranslmed.3010453.
10. Flor H, Diers M, Andoh J. The neural basis of phantom limb pain. Trends Cogn Sci. 2013;17(7):307–8. https://doi.org/10.1016/j.tics.2013.04.007.
11. Franzblau L, Chung KC. Psychosocial outcomes and coping after complete avulsion traumatic brachial plexus injury. Disabil Rehabil. 2015;37(2):135–43. https://doi.org/10.3109/09638288.2014.911971.
12. Franzblau LE, Shauver MJ, Chung KC. Patient satisfaction and self-reported outcomes after complete brachial plexus avulsion injury. J Hand Surg Am. 2014;39(5):948–55.e4. https://doi.org/10.1016/j.jhsa.2014.01.022.
13. Graham B, Adkins P, Tsai TM, Firrell J, Breidenbach WC. Major replantation versus revision amputation and prosthetic fitting in the upper extremity: a late functional outcomes study. J Hand Surg Am. 1998;23(5):783–91.
14. Hruby LA, Gstoettner C, Sturma A, Salminger S, Mayer JA, Aszmann OC. Bionic upper limb reconstruction: a valuable alternative in global brachial

plexus avulsion injuries—a case series. J Clin Med. 2019;9(1):23. https://doi.org/10.3390/jcm9010023.
15. Hruby LA, Pittermann A, Sturma A, Aszmann OC. The Vienna psychosocial assessment procedure for bionic reconstruction in patients with global brachial plexus injuries. PLoS One. 2018;13(1):e0189592. https://doi.org/10.1371/journal.pone.0189592.
16. Hruby LA, Sturma A, Mayer JA, Pittermann A, Salminger S, Aszmann OC. Algorithm for bionic hand reconstruction in patients with global brachial plexopathies. J Neurosurg. 2017;127(5):1–9. https://doi.org/10.3171/2016.6.JNS16154.
17. Htut M, Misra P, Anand P, Birch R, Carlstedt T. Pain phenomena and sensory recovery following brachial plexus avulsion injury and surgical repairs. J Hand Surg. 2006;31(6):596–605. https://doi.org/10.1016/j.jhsb.2006.04.027.
18. Hudak PL, Amadio PC, Bombardier C. Development of an upper extremity outcome measure: the DASH (disabilities of the arm, shoulder and hand) [corrected]. The Upper Extremity Collaborative Group (UECG). Am J Ind Med. 1996;29(6):602–8. https://doi.org/10.1002/(SICI)1097-0274(199606)29:6<602::AID-AJIM4>3.0.CO;2-L.
19. Kretschmer T, Ihle S, Antoniadis G, Seidel JA, Heinen C, Borm W, et al. Patient satisfaction and disability after brachial plexus surgery. Neurosurgery. 2009;65(4 Suppl):A189–96. https://doi.org/10.1227/01.NEU.0000335646.31980.33.
20. Kumnig M. The psychological assessment of candidates for reconstructive hand transplantation. Transpl Int. 2012;25(5):573–85.
21. Light CM, Chappell PH, Kyberd PJ. Establishing a standardized clinical assessment tool of pathologic and prosthetic hand function: normative data, reliability, and validity. Arch Phys Med Rehabil. 2002;83(6):776–83.
22. Maldonado AA, Kircher MF, Spinner RJ, Bishop AT, Shin AY. The role of elective amputation in patients with traumatic brachial plexus injury. J Plast Reconstr Aesthet Surg. 2016;69(3):311–7. https://doi.org/10.1016/j.bjps.2015.10.019.
23. Nadi M, Midha R. Editorial: myoelectric functional hand prosthesis for total brachial plexus injury. J Neurosurg. 2017;127(5):1–3. https://doi.org/10.3171/2016.7.JNS161501.
24. Parry CB. Pain in avulsion of the brachial plexus. Neurosurgery. 1984;15(6):960–5.
25. Salminger S, Roche AD, Hruby LA, Sturma A, Riedl O, Bergmeister KD, et al. Prosthetic reconstruction to restore function in transcarpal amputees. J Plast Reconstr Aesthet Surg. 2016;69(3):305–10. https://doi.org/10.1016/j.bjps.2015.10.029.
26. Sturma A, Göbel, P., Herceg, M., Gee, N., Roche, A.D., Fialka-Moser, V., Aszmann, O.C. Advanced rehabilitation for amputees after selective nerve transfers: EMG-guided training and testing. In: Winnie Jensen, Andersen OK, Metin Akay, editor. Replace, repair, restore, relieve—bridging clinical and engineering solutions in neurorehabilitation. 2014. p. 169–177.
27. Terzis JK, Papakonstantinou KC. The surgical treatment of brachial plexus injuries in adults. Plast Reconstr Surg. 2000;106(5):1097–122; quiz 123–4.
28. Thomas DG. Brachial plexus injury: deafferentation pain and dorsal root entry zone (DREZ) coagulation. Clin Neurol Neurosurg. 1993;95 Suppl:S48–9.
29. Todd AJ. Neuronal circuitry for pain processing in the dorsal horn. Nat Rev Neurosci. 2010;11(12):823–36. https://doi.org/10.1038/nrn2947.
30. Ware JE Jr, Gandek B. Overview of the SF-36 health survey and the International Quality of Life Assessment (IQOLA) project. J Clin Epidemiol. 1998;51(11):903–12.
31. Yozbatiran N, Der-Yeghiaian L, Cramer SC. A standardized approach to performing the action research arm test. Neurorehabil Neural Repair. 2008;22(1):78–90. https://doi.org/10.1177/1545968307305353.

19

Ethical Considerations and Psychological Evaluation in Elective Amputation for Brachial Plexus Injuries

Anna Pittermann, Laura A. Hruby, Agnes Sturma, and Oskar C. Aszmann

Elective amputation and prosthetic reconstruction of a nonfunctional limb provide a treatment strategy in cases where other medical interventions failed to achieve functional improvements. Since this is an irreversible, life-changing event, the process of decision-making before the procedure is a key element in ethical action. The intervention is only considered if the medical and psychological status of the patient indicates it and the patient himself expresses the desire to undergo bionic reconstruction. To allow an informed decision of the patient, clinicians need to clearly communicate the current clinical situation and give realistic expectations of different treatment options. Additional psychological evaluation and a waiting period allow the patient to reflect his considerations for undergoing amputation and provide the psychologist with valuable information about possible contraindication. While psychological counseling usually ends with a final session after the elective amputation, it can be restarted in case of psychosocial problems.

A. Pittermann (✉)
Department of Clinical Psychology, General Hospital of Vienna, Vienna, Austria

Clinical Laboratory for Bionic Extremity Reconstruction, Department of Surgery, Medical University of Vienna, Vienna, Austria

Division of Plastic and Reconstructive Surgery, Department of Surgery, Medical University of Vienna, Vienna, Austria
e-mail: anna.pittermann@akhwien.at

L. A. Hruby
Clinical Laboratory for Bionic Extremity Reconstruction, Department of Surgery, Medical University of Vienna, Vienna, Austria

A. Sturma
Clinical Laboratory for Bionic Extremity Reconstruction, Department of Surgery, Medical University of Vienna, Vienna, Austria

Department of Bioengineering, Imperial College London, London, UK

O. C. Aszmann
Clinical Laboratory for Bionic Extremity Reconstruction, Department of Surgery, Medical University of Vienna, Vienna, Austria

Division of Plastic and Reconstructive Surgery, Department of Surgery, Medical University of Vienna, Vienna, Austria

Ethical Thoughts on Elective Amputation in Patients with Brachial Plexus Injuries (BPI)

The topic of elective amputation remains controversial and very often provokes refusal as a first reaction. For that reason it is all the more important to thoroughly go through ethical and psychosocial concerns regarding this issue.

Brown [2] very extensively described the cases of 15 patients with elective amputation in unsatisfactory hands and the considerations to be taken into account before such an amputation. Many of these are also true for elective amputation in patients with brachial plexus injuries (BPI).

O. C. Aszmann, D. Farina (eds.), *Bionic Limb Reconstruction*,
https://doi.org/10.1007/978-3-030-60746-3_19

The amputation in the context of bionic reconstruction represents an irreversible, life-changing procedure. Even if the concerned limb is without functional use to the patient and even if a prosthesis can be integrated well into one's body schema [9, 10], it does make a psychological and social difference to be a person with an existing limb that cannot be used or to be an amputee wearing a prosthesis.

Most of the patients suitable for elective amputation and bionic reconstruction experience a long period of time where grief, hope, disappointment, and anger are the most prominent emotions related to their hand. They often undergo a high number of surgical interventions and many other attempts to get any function and/or feeling back into their injured hand, until they might come to a point where they accept the situation as it is and decide not to focus on any possible future improvements but concentrate on the present situation and how they can make the most out of it. Even patients with the best surgical outcomes can result to be highly unsatisfied and functionally limited due to pain and loss of dexterity [8] and in the end wish for an amputation. Making the decision to have an elective amputation might take a long time in which the individual has to think through all the aspects of his decision and make many experiences and process a lot of information.

Even though the amputation always represents a great loss to the patient, they very often describe that the loss they feel has already happened on the day they had their accident that led to the BPI. Realizing this might be a key element in the decision for elective amputation.

The process of accepting the irreversible damage to their hand can be seen as integrating the loss into one's identity, something that makes it possible to adapt to the situation and make some kind of peace with it. The duration of this process varies individually and may be cyclic with several setbacks.

Actually not only the affected patient has to deal with the disappointment that goes along with the awareness that no more functional improvement will be achieved and that an elective amputation might be the best option. Also the involved surgeon has to accept that in some patients no more benefit can result from additional interventions. This evaluation might be difficult for a professional trained and used to cure or at least to improve but not to amputate as a solution. Recognizing at what point persistent attempts are not realistic but more of a burden for the patient, to put the own ego as a surgeon on hold and only focus on the patient's benefit might turn out to be a crucial step in the whole treatment process.

The surgeon's main task in patients suitable for elective amputation and bionic reconstruction is to clearly communicate to the patient the clinical situation and the realistic expectations of different treatment options. At this stage it might be important for some patients to also gather information from other experts in this field, which is something that should be encouraged. The more information and insight a patient has, the easier it will be for him to make a decision he can relate to for the rest of his life. Once the patient has a full understanding of the extent of the damage and surgical possibilities, the patient can compare the options he has and make an informed consent.

In the case of bionic reconstruction, it is also crucial for the patient to be aware of the intense rehabilitation program essential after potential nerve surgery as well as for optimal future prosthetic control [5]. The patient has to be informed about the appearance and use of a prosthesis. Ideally he should have the opportunity to talk to and to observe a patient who already went through the process of bionic reconstruction. Limitations as well as pros and cons of a prosthesis have to be discussed with the patient.

There are many factors that contribute to a patient's decision for or against elective amputation. Some patients might want to get rid of pain, and others wish for an improvement of hand function or have the feeling a prosthesis will improve their appearance. Patients might decide to live with their functionless limb because of religious thoughts, and others base their decision on body image reasons. In any case the decision for elective amputation is a very personal and a very individual one and should be treated as such. The patient's social and cultural background, his occupation, personality, and coping mechanisms have to be explored and show that patient selection is crucial in the context of elective amputation and bionic reconstruction [12]. Psychological evaluation prior to elective amputation should be performed in any case.

Psychological Evaluation Prior to Elective Amputation and Bionic Hand Substitution

While evidence for the functional benefits of bionic reconstruction in patients with brachial plexus injuries is available [1, 5], the procedure still remains irreversible and should thus only be performed in cases where patients are fully aware of its consequences.

A standardized way of evaluating patients prior to elective amputation and bionic hand substitution is the "Vienna psychosocial assessment procedure for bionic reconstruction (VPAP)" [4]. The VPAP follows a primarily qualitative approach, augmented by three validated, patient report questionnaires. The main element of this assessment procedure is a semi-structured interview conducted by a clinical psychologist (Anna Pittermann). The three administered questionnaires are the SF-36 Health Survey, addressing overall quality of life, the FKB-20 Body Image Questionnaire on body image issues and the VAS (visual analogue scale) assessing deafferentation pain.

The aim of the semi-structured interview is to explore patients' psychological, social, and emotional situation after BPI, their psychosocial status, and their expectations and motivations for elective amputation and prosthetic replacement of the functionless limb. This evaluation should go deep and be very patient-centered, something that can be achieved best by a semi-structured interview.

In the absence of an existing concept for standardized evaluation before elective amputation and bionic reconstruction, the VPAP is based on the fundamental questions that have to be asked before elective amputation. Factors that have to be taken into consideration, also described by Brown [2], are for example:

- Pain: Consistent, heavy pain might be a reason for the patient to wish for an amputation (see also [11]). In the case of BPI, pain very often is based on a more complex reflex phenomenon, and the patient has to understand that his pain might not be eased by amputation (alone).
- Function: Will daily living activities be better accomplished with a prosthesis? Are the patients' environmental and social circumstances organized in a way that they will allow the patient to make use of his prosthesis?
- Appearance: Is the patient suffering from the appearance of his functionless limb, and did he have enough time to get used to his altered appearance? Is the patient capable of imaging himself as an amputee? Are there any severe body image concerns?
- Psychosocial Circumstances: The personal situation of a patient, existing psychiatric illnesses, coping mechanisms, cognitive abilities, and social support can have an immense influence on the decision of a patient and have to be explored.

All the above described factors together with conclusions derived from the model of adaptation process to severe hand injuries by Grob et al. [3] and the evaluation procedures for hand transplantation described and developed by Kumnig et al. [6, 7] formed the basis for the development of the VPAP.

The semi-structured interview of the VPAP is divided into sections of questions relating to five different categories, A to E (see Table 19.1). The patients respond freely to these questions, and their answers are transcribed verbatim by the interviewer. After the interview the answers are analyzed on the basis of a pre-defined scoring system. The duration of the interview is 90–120 min.

Table 19.1 Structure of the VPAP interview for the assessment of candidates for elective amputation and bionic reconstruction

Scale	Category	Item examples	Scale score (range)
A	Psychosocial adjustment after BPI	Symptoms of post-traumatic stress disorder Resources and stress management Debilitating circumstances related to the accident	0–6
B	Self-perception of the injured arm	Experience of functional disability Pain Difficulties regarding physical appearance Neglect Body image issues Difficulties regarding social reactions	0–5
C	Global psychosocial status	Education Employment status Social resources Substance abuse Psychiatric history	0–9
D	Motivational aspects related to elective amputation	Decision-making process Sufficient information level (irreversibility of decision; awareness of the fact that deafferentation pain will not be cured by amputation etc.) Outcome expectations	0–5
E	Prosthesis	Information level about prosthesis Awareness of functional limitations of a prosthetic hand Adherence level regarding difficulties with the prosthetic hand Level of compliance regarding instructions in handling the prosthetic device and training demands Social reactions to prosthetic device	0–5

Each of the five interview sections (A–E) consists of five to nine different items of potential conflicts/symptomatology, and each item is assigned a value of 0 or 1 with 0 indicating no problem in the addressed domain and 1 indicating the presence of risk factors. The minimum total score is 0 and the maximum total score 30. This approach of data analysis allows intraindividual comparison of results as well as interindividual comparison of scores among different patients.

Table 19.2 represents a graphical scheme of patient flow in patients with BPI. If during the first appointment psychosocial problems like heavy pain, social difficulties, body image issues, symptoms of a post-traumatic stress disorder, and/or other psychiatric disorders are observed, psychological counseling should be provided for the patient and/or his family. The same is true for problems occurring during a following appointment.

For patients not suitable for elective amputation and bionic reconstruction, the evaluation process ends at this point with subsequent psychological support if necessary. For patients who are potential candidates for elective amputation, the options are being discussed, and a waiting period where the patient can think through his options is being planned. If after this period (usually a couple of weeks) the patient still wishes for elective amputation, an appointment for psychosocial assessment is being scheduled.

In case there appear severe difficulties or any contraindications for elective amputation (e.g., severe personality disorders, alcohol abuse, acute post-traumatic stress disorder, insufficient information level about elective amputation or bionic reconstruction), the process of bionic reconstruction is not initiated, and psychological counseling continues. In case no contraindications can be found, the elective amputation is being performed.

Psychological counseling ends after a final session after the elective amputation but can be restarted if psychosocial problems become observable during the rehabilitation phase.

Table 19.2 Graphical scheme of patient flow

Treatment plan	Variable	Action
First appointment	Psychosocial problems – Heavy pain – Post-traumatic stress disorder (PTSD) – Social difficulties – Adaptation disorders	Psychological counseling
Following appointments	Psychosocial problems	Psychological counseling
Endpoint		No elective amputation
Or Patient suitable for elective amputation and bionic reconstruction		Possibility of elective amputation is being discussed
Waiting period		
Decision for elective amputation		Psychosocial assessment/evaluation process (VPAP)
Elective amputation		
Postoperative		Psychological counseling
Outpatient clinic/rehabilitation phase	Psychosocial problems	Psychological counseling

References

1. Aszmann OC, Roche AD, Salminger S, Paternostro-Sluga T, Herceg M, Sturma A, Hofer C, Farina D. Bionic reconstruction to restore hand function after brachial plexus injury: a case series of three patients. Lancet. 2015;385(9983):2183–9.
2. Brown PW. Sacrifice of the unsatisfactory hand. J Hand Surg. 1979;4(5):417–23.
3. Grob M, Papadopulos NA, Zimmermann A, Biemer E, Kovacs L. The Psychological Impact of Severe Hand Injuries. J Hand Surge. 2008;33(3):358–62.
4. Hruby LA, Pittermann A, Sturma A, Aszmann OC. The Vienna psychosocial assessment procedure for bionic reconstruction in patients with global brachial plexus injuries. PLoS One. 2018;13(1):e0189592. https://doi.org/10.1371/0189592.
5. Hruby LA, Sturma A, Mayer JA, Pittermann A, Salminger S, Aszmann OC. Algorithm for bionic hand reconstruction in patients with global brachial plexopathies. J Neurosurg. 2017;27(5):1163–71.
6. Kumnig M, Jowsey SG, Rumpold G, Weissenbacher A, Hautz T, Engelhardt TO, Brandacher G, Gabl M, Ninkovic M, Rieger M, Zelger B, Blautz M, Margreiter R, Pierer G, Pratschke J, Schneeberger. The psychological assessment of candidates for reconstructive hand transplantation. Transpl Int. 2012;25(5):573–85.
7. Kumnig M, Jowsey SG, Moreno E, Brandacher G, Azari K, Rumpold G. An overview of psychosocial assessment procedures in reconstructive hand transplantation. Transpl Int. 2014;27(5):417–27.
8. Liu Y, Lao J, Gao K, Gu Y, Xin Z. Outcome of nerve transfers for traumatic complete brachial plexus avulsion: results of 28 patients by DASH and NRS questionnaires. J Hand Surg. 2012;37(5):413–21.
9. Mayer A, Kudar K, Bretz K, Tihanyi J. Body schema and body awareness of amputees. Prosthet Orthot Int. 2008;32(3):363–82.
10. Murray CD. The social meanings of prosthesis use. J Health Psychol. 2005;10(3):425–41.
11. Quan DL, Dudek NL, Marks M, Boutet M, Varpio L. A qualitative study of factors influencing the decision to have an elective amputation. J Bone Joint Surg Am. 2011;93:2087–92.
12. Smith JK, Myers KP, Holloway RG, Landau ME. Ethical considerations in elective amputation after traumatic peripheral nerve injuries. Neurol Clin Pract. 2014;4(4):280–6.

Part VII

Rehabilitation in Upper Limb Prosthetics

Principles of Occupational and Physical Therapy in Upper Limb Amputations

20

Diane J. Atkins and Agnes Sturma

The unique background, education, and training of occupational and physical therapists enable them to have a profound influence on an individual with upper limb loss. Understanding the many phases of amputee prosthetic rehabilitation is required in order to meet the functional needs of individuals fit with body-powered and myoelectric prostheses. Thoughtfully designed protocols include interventions before surgical treatment, after surgery, pre-prosthetic interventions, and prosthetic training with the aim to enable the amputee to achieve an optimal functional outcome. These protocols are designed to support the individual with upper limb loss in regaining their independence and are modified dependent on the individual life situations and the prosthetic device used. In view of the emerging advances in upper limb surgery and technology, a collaborative relationship with the entire team of surgeons, prosthetists, and rehabilitation professionals is more important than ever as we prepare each patient to reach their maximum potential.

D. J. Atkins (✉)
Department of Physical Medicine and Rehabilitation, Baylor College of Medicine, Houston, TX, USA

A. Sturma
Clinical Laboratory for Bionic Extremity Reconstruction, Medical University of Vienna, Vienna, Austria

Department of Bioengineering, Imperial College London, London, UK

Introduction

We are enriched by the personal aura and spirit of persons who manage life with disabilities, each in his individual and special way.—E. Marquardt [4].

The impact of losing one or both arms cannot be overstated. The hand functions in prehensile activities as a sensory organ and as a means of communication. Any loss interferes with an individual's productivity and feeling of completeness, as well as alters their interaction with the environment [4]. It immediately creates a psychological crisis for the individual and a major life event that this person must face and cope throughout their life (see Chap. 2). The individual experiences amputation through his or her own unique perceptual filter, making it a different experience for each person, one that is dependent on life history, personality style, stage of life, social support network, and other subjective factors. The coping responses may include emotional reactions, psychological defense strategies, behavioral reactions, attitude changes, and alterations in interpersonal relationships. The loss of any body part is a serious threat to the individual's core identity, as it shakes the foundation upon which this identity is built [4]. Thus, amputation is not only associated with the actual loss of body structure and function [12] but needs to be treated by the medical team as the life-changing event it is.

O. C. Aszmann, D. Farina (eds.), *Bionic Limb Reconstruction*,
https://doi.org/10.1007/978-3-030-60746-3_20

While psychological problems associated with the amputation need to be treated by qualified psychologists and psychotherapists, occupational and physical therapists support individuals to regain autonomy, social role functioning, and participation in activities of their daily life. Their unique background, education and, training enable them to be instrumental in treating the individual with upper limb loss. Their unique skill sets allow them to prepare patients to meet their many challenges. Today, the most effective surgeons, engineers, and rehabilitation clinicians blend their unique skills and bring a collaborative spirit and relationship to their work with occupational and physical therapists. Therapists can assist in enabling the amputee to be in control of his or her life, provide a safe atmosphere of emotional expression, and provide an environment that will insure acceptance and protection of their dignity.

This chapter is designed to first give an overview about general considerations and steps within rehabilitation to then describe specific interventions for more advanced technologies.

Evaluation

During a therapist's first encounter with an upper limb amputee, a comprehensive evaluation should occur. In an ideal setting, this evaluation is done by the multidisciplinary team together, including therapists, medical doctors, psychologists, and prosthetists. Assessment includes many domains: etiology and onset, date of birth, dominance, concomitant medical issues, range of motion and muscle strength, shape and integrity of the residual limb, status of the opposite extremity, pain (phantom/residual limb/in other body parts), viable muscle sites for myoelectric control, background education, vocational goals, home environment, and family support. In view of the advances in surgical and technological treatment options, the combined expertise of surgeons, rehabilitation physicians, prosthetists, and therapists has become even more essential in the successful treatment of the upper limb amputee. Additionally, it is highly recommended to involve the full clinical team in the decision-making process, should any far-reaching interventions (as TMR or osseointegration) be considered.

Prehension evaluation for the sound limb can be performed as appropriate. Although no standardized evaluation for prehension deficits is identified for this population, it can be necessary to determine baseline function of the remaining limb, as the intact limb will be responsible for conducting all fine motor and dexterity tasks for vocational, leisure, and occupational purposes.

Early Post-amputation Therapy Programs

Awareness of successful management of the upper limb amputee, who has just sustained traumatic limb loss, is crucial. During this first phase of care, the individual has little if any control over what is happening, and they must depend upon the healthcare team to provide the best treatment possible.

The post-amputation phase varies based upon the severity of the patient's injury but generally lasts 1–3 weeks. The therapist is integrally involved with the primary goals of providing reassurance and sharing the basic guidelines of what will follow. Support and appropriate information sharing is imperative, as this is a time of grieving and uncertainty.

The components of this post-amputation phase include:

1. Comprehensive evaluation
2. Promote wound healing
3. Control incisional and phantom limb pain
4. Control edema
5. Maintain joint range of motion
6. Explore the patient's and family's feelings regarding change in body
7. Provide psychological and emotional support (in collaboration with a psychologist)
8. Provide basic information regarding prosthetic alternatives
9. Orientation to the basic self-care activities of daily living (ADL) by utilizing adaptive equipment and techniques to promote independence

10. Share information relating to grieving and loss of limb (in collaboration with a psychologist)

Pre-prosthetic Preparation

Research already performed in the 1980s has demonstrated that individuals fit with a prosthesis within 30 days of amputation exhibited a 93% rehabilitation success rate with a 100% return to work rate within 4 months of injury and those fit beyond 30 days exhibited a 42% rehabilitation success rate with a 15% return to work rate within 6–24 months [40]. This 30-day period is termed the "golden window" [25] and indicates the importance of timely prosthetic training and fitting. The 30-day window, however, is very challenging based on wound healing, insurance approval process, etc. Therefore, more recent research focused on the added value of a fitting within a 6-month window and found that it increases the acceptance of a prosthesis by 16 times [7].

Ideally, the pre-prosthetic program should begin when the sutures are removed and the patient is medically stable. This is dependent upon the extent of the injury and associated medical conditions. The occupational and/or physical therapist will be managing and monitoring this extremely important phase of preparing the limb for a prosthesis.

The objectives of the pre-prosthetic program are:

1. Promote residual limb shrinkage and shaping
2. Promote residual limb desensitization
3. Provide pain management strategies
4. Maintain normal joint range of motion
5. Maintain skin mobility
6. Increase muscle strength
7. Provide instruction in the proper hygiene of the limb
8. Instruction in change of dominance activity, if indicated
9. Maximize self-reliance in the performance of ADL tasks
10. Myo-test for potential site selection for myoelectric control and provide instructions in increasing their muscular strength, separation, and endurance
11. Provide comprehensive information regarding prosthetic alternatives (best together with a prosthetist)
12. Explore and document patient functional needs and goals regarding the future

It is imperative that psychological support is offered and modified to appropriately respond to the rapidly changing needs of the individual as they advance through this process. Support is further enhanced by the addition of a peer support group or peer visitor, which provides a significant aid in coping for the amputee [41].

Proper shaping and shrinking of the residual limb is a critical component of the pre-prosthetic phase. It is accomplished by figure-of-eight compression wrapping from an elastic bandage or with a tubular elastic bandage, such as tubigrip or compressogrip (both Mölnlycke Health Care group, Gothenburg, Sweden) as shown in Fig. 20.1.

When establishing an effective program, maintenance of full and active range of motion

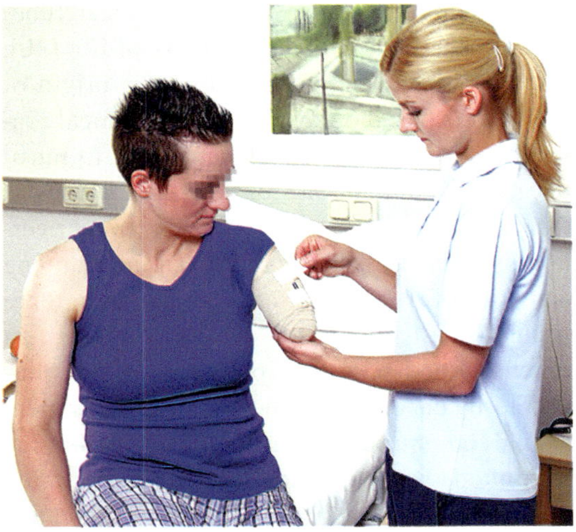

Fig. 20.1 Figure-of-eight compression wrapping ensures proper shaping and shrinking. (Used with permission from Ottobock)

of the scapula, glenohumeral, elbow, and forearm is crucial in order to aid in the prosthetic control motions and maximize the functional potential of the prosthesis. Forearm range of motion may begin to decrease in as little as 2–3 weeks, if this motion is not encouraged. This is especially important for transradial amputees with forearms longer than 50% where this motion can be transferred to the prosthetic socket. Forearm pronation and supination is a valuable voluntary motion required in virtually all bilateral tasks. Without pronation and supination of the forearm, the person may use shoulder abduction which can cause overuse injuries at a later time [30].

Pain management strategies may be necessary throughout the whole rehabilitation process for both acute post-surgery pain and chronic amputation-specific pain. The latter can be divided in residual limb pain (located in the amputation stump) and phantom limb pain. As described in the designated chapters about phantom limb pain, different treatment strategies are available, including drug therapies [17, 29], psychological interventions [5], behavioral treatment strategies and newer treatment methods as mirror therapy [32], Graded Motor Imagery [27], sensory discrimination training [47], and virtual reality [28].

Additionally, overuse complaints are relatively common and may affect physical function, prosthesis use, and the quality of life [30]. Their treatment might depend on the origin of pain and is based on the therapist's clinical reasoning. In general, it is recommended to introduce the ergonomic use of assistive devices at this early stage of rehabilitation to prevent overuse syndromes. Therefore, when introducing activities of daily living after amputation, there should be a focus on the quality of movement execution and posture. Appropriate techniques should be encouraged, and adaptive equipment should be introduced in order to encourage maximum independence while minimizing compensatory movements. However, gadget tolerance and reliance on these devices will decline over time.

Myoelectric Site Testing and Training

With the advancing emphasis on a variety of myoelectric prosthetic control options, myo-site testing and training are major goals of the pre-prosthetic therapy program. The objectives of this intervention are to identify, instruct, and train the patient to independently, correctly, and efficiently use specific limb musculature to activate and perform basic myoelectric prosthetic functions. In systems with direct prosthetic control (i.e., activity measured from one surface electrode refers to one prosthetic movement), site selection is important. It requires the specialized skills of trained therapists and prosthetists to identify the best possible electrode placement and the most efficient control scheme for each individual's abilities and needs. Optimal sites are selected based upon what is most intuitive for the patient and what is optimal within the socket interface design. Close collaboration between the therapist and prosthetist is critical and will dramatically impact a successful outcome.

Electrode site identification, which is critical for direct myoelectric control, takes place in therapy with the use of surface electrodes and with software such as the MyoBoy software (Otto Bock, Minneapolis, MN, and Duderstadt, Germany), the MyoLabII (Motion Control, Inc. Salt Lake City, UT), virtu-limb (Össur Touch Solutions, Dublin, OH, and Scotland, as shown in Fig. 20.2), or therapeutic biofeedback units.

Once ideal sites are identified, motor training begins. Motor training can take place utilizing a variety of computer based software where the focus is independent activation of each muscle to be incorporated in the prosthetic design. When separation of control is achieved, proportional control is introduced, and the elicited strength of a selected muscle contraction will control the speed and grip force of the terminal device. A more detailed explanation of the myo-site training protocol can be found in Smurr et al. (2008) [41].

If this pre-prosthetic phase is managed well by experienced therapists, this sets the stage for a

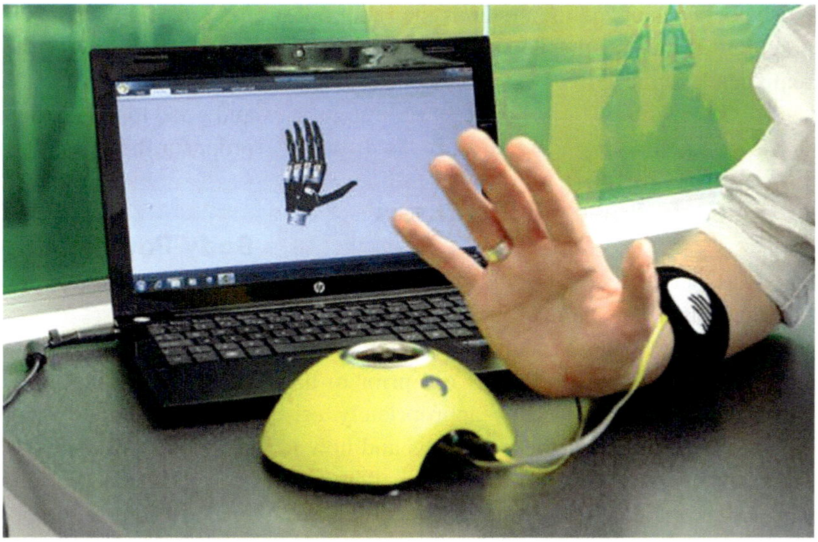

Fig. 20.2 Myoelectric site testing and training with virtu-limb from Össur Touch Solutions. (Used with permission from Össur Touch Solutions)

sense of satisfaction, motivation, and optimism for the next critically important step of actual training with the prosthesis.

Body-Powered Prosthetic Training

Although a great deal of attention has been placed upon the state-of-the-art advances in electric prosthetic deign and componentry, there remains an ever-present need for body-powered prostheses designed for the more robust, heavy duty wearer and user. These prostheses are well-suited for environments that include water, dirt, machinery, extremes of hot and cold, as well as vocations involving farming and ranching [38]. Users of body-powered prostheses report an increased kinesthetic and proprioceptive feedback through their cables and harness, enabling them to work with their prosthesis without a need to constantly visually attend what activity they are performing. While some individuals with an amputation prefer the lighter weight and less need for maintenance and repair with a body-powered device, others need and desire for a more advanced electric prosthesis continues to exist for many. There is no clear advantage in terms of general functional performance between body-powered and externally powered prosthetic devices [10]. However, due to different use cases for the devices and different ideas of the individuals in regard to device preferences and body image, giving them a choice is highly recommended. Selecting the right prosthesis for the individual based on their goals and lifestyle is critical to be done with the multidisciplinary team to ensure the pros and cons of the various options are taken into consideration for each individual. Often it is the recommendation of the physician to provide both an electric and body-powered prosthesis for back-up and for use when an electric prosthesis is being modified or repaired.

The prosthetic training skills described in this section are focused upon the body-powered user, but the majority of training principles can apply to the myoelectric prosthetic user as well.

Before initiating a program of upper extremity prosthetic training, one must realistically orient the patient to what the prosthesis can and cannot do. If the individual has an unrealistic expectation about the usefulness of the prosthesis as a replacement arm, he may be dissatisfied with the ultimate functioning of the prosthesis and may reject it altogether. On the other hand, if the expectations of the amputee are more realistic at the beginning of training, the ultimate acceptance will be based upon the ability of the prosthesis to improve the individual's performance. It is imperative, then, that the therapist be honest and positive about the function of the prosthesis. If he/she believes in and understands the functional potential of the prosthesis, success can be more realistically achieved.

Following Delivery of the Body-Powered Prosthesis

The time between casting and the delivery of a prosthesis is often characterized by eager anticipation and hope that this artificial arm will enable the individual to function as he did prior to the amputation. If the patient is appropriately oriented to the realities of the prosthesis and how it looks and operates, he/she will be in a better position to accept the limitations of the prosthesis when it is delivered to him. With this said, however, if another successful user, with a similar level of amputation and type of prosthesis, can be invited to an early therapy session, the advantages of seeing what is possible from a successful wearer and user, hope can be restored, and the prosthetic training process is a more positive and goal-oriented experience.

The day the prosthesis is delivered is ideally the day that therapy with the device should begin. A quiet, non-distracting environment is suggested for the initial visits, as this is often a time of self-consciousness and awkwardness when learning to don, doff, and control the device.

The following goals should be addressed during the first several visits after fitting:

- Orientation to prosthetic component terminology
- Independence in donning and doffing the prosthesis
- Orientation to a wearing schedule (not to exceed 15–30-min periods initially, gradually increased over time, with frequent skin checks throughout the day)
- Care of the residual limb and prosthesis

Now that the body-powered prosthesis is ready to be worn and used, orientation and practice of the following body-motions are necessary: scapular abduction, chest expansion, shoulder depression, extension and abduction (necessary for body-powered elbow operation), humeral flexion, elbow flexion/extension, and forearm pronation and supination. Full range of motion, and good to normal muscle strength, is desired in order for the prosthesis to optimally operate.

Body-Powered Prosthetic Controls Training

Prior to initiating any unilateral or bilateral tasks, it is essential that the individual with upper limb loss is completely comfortable with how each component of his/her prosthesis operates with the required body motion. Repeated drills are helpful, positioning the prosthesis at various heights and ranges in order to increase confidence in the arm's operation is recommended. This avoids unnecessary frustration when activity training begins. When operation of the components of the prosthesis becomes second nature, grasp and release activities of different size, shape, and textured objects can begin. The important concepts of pre-positioning and tension control of the terminal device are important to understand and reinforce. Drills that include grasp and release in various planes are suggested as well.

Close attention must be paid to the individual's awkward or compensatory body motions when approaching an object. Often the amputee will adjust his body to accommodate, rather than adjust or reposition the elbow and wrist position, if possible. A mirror can be effective in assisting the amputee to see the way his body is positioned and subsequently self-correct. It can also be helpful to encourage the patient to consciously think how his/her own arm would need to be positioned to accomplish a task, from shoulder to hand and pre-position his prosthesis accordingly. Reminders to maintain an upright posture and avoid extraneous movement are also important.

Controls training for the bilateral upper limb amputee will take an extended period of time. Learning to separate the control motions of prostheses is a complex and coordinated motor process that will require frequent repetition. Passing an object back and forth, in different planes, such as a ruler, may help in reinforcing this pattern.

Functional Use Training for Body-Powered as Well as Myoelectric Prostheses

Feedback from individuals with upper limb amputation, when interviewed at least 1 year after amputation, indicated that functional prosthetic training should be provided as soon as possible after a prosthesis is provided [48]. Durance and O-Shea reinforced this principle when stating that 90% of persons who received training were able to functionally use their prosthesis, compared to only 50% of those who did not receive any training [14].

The sequence of training an individual with a body-powered and myoelectric prosthesis mirrors each other. It includes:

- Controls practice (grasp and release) as shown in Figs. 20.3a, b and 20.4
- Unilateral functional skills
- Bilateral functional skills
- Advanced bilateral skills encouraging community reintegration

For most prosthetic fitting options, functional use training is the most difficult and prolonged stage of the prosthetic training process. The

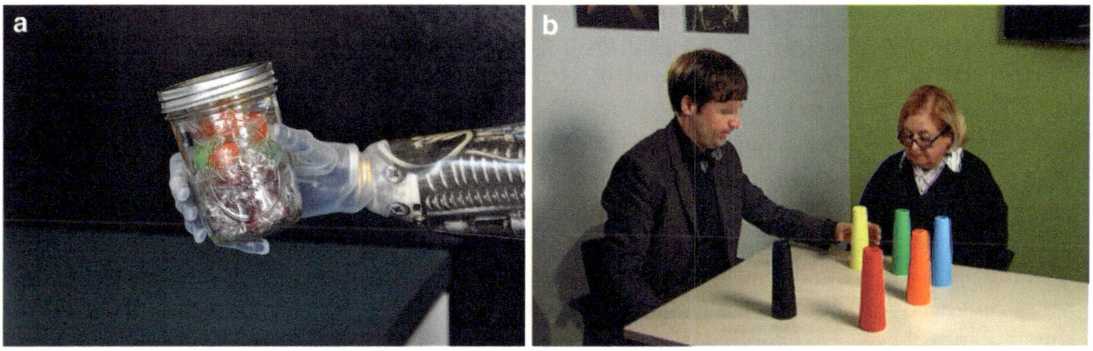

Fig. 20.3 (**a** and **b**) Grasp and release training utilizing different sizes and shapes of objects. (Used with permission from Össur Touch Solutions)

Fig. 20.4 Jenga is a game that reinforces grasp and release at different levels. (Used with permission from Ottobock)

length of time is dependent on a number of factors that include length of residual limb, complexity of prosthetic controls, muscle strength, and range of motion. The motor learning, coordination abilities, and frustration factor of the patient are not to be underestimated. All of these can lead to one feeling emotionally and physically drained and fatigued. When a therapist senses one's emotional and physical tolerance has been met, it is time to stop for the day.

It is important to reinforce to the unilateral amputee that his prosthesis will usually play a non-dominant functional role and one that is most useful for gross prehension as in holding and stabilizing objects, while the sound limb performs fine motor prehension activities. With this said, however, the newer hand technology that offers compliant grasp and multi-grip abilities allow the user to utilize the hand more like a normal hand in contrast to previous myoelectric hands. Unilateral patterns of independence occur quickly in the amputee who has lost an arm or hand. It is therefore essential, if possible, to fit the unilateral amputee within 6 months of the amputation. These individuals will show a greater propensity of wearing and successfully using with prosthesis. This applies all amputees fitted with body-powered or electric components [3, 7].

It is appropriate to practice simple bilateral activities of daily living that are useful and purposeful. These activities could include but are not restricted to:

- Cutting food as shown in Fig. 20.5a, b
- Using scissors
- Dressing activities (see Fig. 20.6)
- Tying shoelaces (see Fig. 20.7)
- Toileting independence
- Opening a jar or bottle
- Washing dishes
- Simple meal preparation, as preparing a coffee (see Fig. 20.8)

The importance of pre-positioning prior to approaching these tasks cannot be overemphasized. The amputee should be instructed to orient the components of the prosthesis in space to a position that resembles that of a normal limb engaged in the same task. Mastering ADL skills that are ergonomically correct is important in order to avoid overuse syndromes. Conditions such as tenosynovitis, epicondylitis, carpal tunnel syndrome, shoulder impingement, and diffuse repetitive overuse injuries are not uncommon in the individual with upper limb loss [21]. These overuse complaints may affect physical function, prosthesis use, and the quality of life [30]. Some of them can be explained by the limited degrees of freedom current prosthesis on the market offer, especially the lack of wrist movements [9]. They need to be treated dependent on the type of pain and the origin of the overuse syndrome. Generally, promoting the use of a prosthetic/assistive device while encouraging physiological movement patterns is considered beneficial. Therefore, when introducing

Fig. 20.5 (**a** and **b**) Cutting food is a bilateral task that reinforces the value of a multi-articulating hand. (Used with permission from Össur Touch Solutions)

Fig. 20.6 A multi-articulated hand used for dressing activities. (Used with permission from Össur Touch Solutions)

Fig. 20.7 Tying shoelaces requires bilateral fine motor skills. (Used with permission form Ottobock)

activities of daily living at this phase of rehabilitation, there should be a focus on the quality of movement execution and posture.

Advanced bilateral skills are introduced when prosthetic proficiency is demonstrated. These activities are specific to the patient's interests, hobbies, family, and work goals. These tasks could include:

- Housekeeping (see Fig. 20.9)
- Yard work
- Home repair and using light tools
- Grocery shopping (as shown in Fig. 20.10a–c)
- Advanced meal preparation
- Childcare
- Exercise pursuits (see Fig. 20.11)

Fig. 20.8 Preparing a coffee is an easy task for the beginning of ADL training. (Used with permission from Ottobock)

Fig. 20.9 Housekeeping tasks reinforce bilateral function. (Used with permission of Handspring Clinical Services)

Fig. 20.10 (**a**–**c**) Reaching for objects at the grocery store requires proficiency at different levels. (Used with permission from Össur Touch Solutions)

Fig. 20.11 Exercise pursuits are enhanced with improved body symmetry. (Used with permission from Ottobock)

Training Principles of Electric Multi-articulating Hands

Since the launch of the Touch Bionics (now part of Össur) i-Limb hand and digits in 2007, electric multi-articulating hands have continued to gain momentum in their design, ease of use, increasing functionality, and patient desire to utilize them in their daily activities. Each of the most commonly prescribed electric hands exhibit specific characteristics and a training protocol that is unique to each. Current research suggests that spending more time on learning fine control aspects of the prosthetic hand will result in better performance and acquisition of prosthetic skills [8].

In the following, specific modifications in prosthetic training to make best use of specific major components will be highlighted. The following points are general training principles that will assist the therapist as he/she endeavors to effectively train the individual with a myoelectric prosthesis:

- Independence in donning and doffing the prosthesis is essential at the outset.

- Functional training should reinforce proper posture and minimize compensatory movements.
- Muscles utilized in myoelectric control fatigue easily and sufficient rest periods are required, especially at the beginning of training.
- The comfort and fit of the prosthetic socket interface, and properly programmed components, are critically important and require close interaction between the therapist and prosthetist. Without this, rejection of the prosthesis is inevitable.
- Emphasis should be placed on the value of the prosthesis as being utilized as a functional assist in activities such as cutting food, taking money from a wallet, tying shoes, using a zipper, and buckling a belt.
- Functional training should progress from simple to more complex tasks with short- and long-term goals incorporated in this process. In multi-articulate hands, begin with mastery of one or two grip patterns before introducing others.
- Tasks should focus on the client's lifestyle and employment history as skills and confidence increase.
- Carefully listen to your patient throughout the training process, and encourage their ongoing communication and feedback.

With an increasing number of devices entering the market, this chapter cannot provide specific protocols for therapy with each of them. It is highly recommended to look up specifications and training tips on the manufacturer's website, if a patient is fitted with a device unknown to the therapist. Additionally, many companies offer practical trainings for their devices. Acknowledging that this list is not exhaustive, some of the specific points to consider of major hands are included below:

Training Considerations with i-Limb Hands

- There are several versions of i-Limb hands that offer different features.
- All versions of the i-Limb hands have similar out of the box features where training should begin:
  - Grasp and release of various sized and shaped objects training on the compliant grip of the hand where having a motor in each digit allows it to conform to the object being held.
  - Stalling digits can be accomplished by putting pressure against a digit stopping it from moving to achieve functional positions such as stopping the index digit to type on a computer.
  - Proportional control (allowing for grip on delicate objects) and vari-grip (allowing for increased grip strength on items as needed) should be introduced.
- The thumb of the i-Limb hands is unique in that it can be manually moved along the full range of rotation from lateral pinch into a palmar position as well as stopped anywhere along that rotation such as for cylindrical grip on a handle. Some versions also have powered rotation to pre-position the thumb when a grip is entered for increased dexterity.
- Depending on the version of the hand, between 12 and 24 pre-programmed grips are available to enhance the clients' ability to complete their goals. The grips selected for the user should be based on their goals.
- A therapist and prosthetist can guide the patient to the most optimal control option to select a desired grip. The alternatives include gesture control (moving the hand in space to activate a grip), app control (touching a grip on screen), grip chips (Bluetooth chip to change grip), and triggers through muscle control (such as co-contraction).
- Whatever method for switching between the grip is chosen, this needs to be trained in order to gain reliable control. Otherwise, the patient will never make use of the majority of grip patterns.
- If the flexion wrist is used as part of the fitting, orientation to its five lockable positions is important in almost all unilateral and bilateral activities to decrease compensation. An unlocked position can also be selected

enabling free movement through its range of motion.
- A recommended training protocol is available at: www.touchbionics.com/training.

Training Considerations with Michelangelo Hand

- The primary features of the Michelangelo hand include an active thumb, index and middle finger with the ring, and little finger passively following the other fingers.
- All fingers are driven together, but it can change the relative motion of fingers and thumb to give two distinct prehensile patterns, tip and lateral opposition.
- The wrist joint is flexible and includes a lock button which enables the user to utilize in the flexible or rigid mode.
- Training is based upon various activities utilizing the lateral, opposition mode, neutral, and rotation.
- Training can be enhanced by the use of Axon Software enabling the user to learn how to precisely control and efficiently use the Michelangelo hand.

Training Considerations with the Bebionic Hand

- The Bebionic hand can provide 14 different grip patterns and hand positions using proportional myoelectric control. The thumb can be passively moved to two different positions allowing for opposition and lateral grips.
- If the Bebionic hand is used with conventional two-site myoelectric control, switching between the grips and positions can be done by the use of co-contraction of both myosignals or with an "open" signal, when the hand is already open. Additionally, a button on the back side of the hand gives another option to switch between degrees of freedom or between grip options.
- Again, the chosen switching methods need to be trained in order to gain reliable control and ensure that problems with this do not impede use of the different grips.
- Training can be supported with sEMG biofeedback that can be provided by the associated software.
- The Bebionic hand can also be used in combination with a pattern recognition system. This allows to access different grip patterns in a relatively easy and intuitive way without the need of switching.
- If this control method is chosen, training needs to focus first on exploring possible muscle activation pattern. Later these patterns should be refined and trained to ensure that each of them is distinctive and that they are consistent over time. For this, adequate visual feedback of the muscular activity and the pattern is required.
- As soon as a reliable control with the chosen control method is possible, training should focus on the use of the available grip pattern in activities of daily living.

Training Considerations for the Partial Hand Amputee Fit with i-Digits

- Challenges may exist when working with an individual with partial hand loss, as research indicates that partial hand patients experience a higher level of negative emotional response than other levels of upper limb amputation [26].
- Depending on the remaining intact fingers and residual limb, one unique aspect of fitting the partial hand prosthesis is that remaining fingers can provide sensory feedback to the user enhancing their ability to feel the grip they have on objects.
- If wrist and forearm range of motion remains, this can also enhance the individual's ability to utilize the prosthesis with less compensation at more proximal joints.
- Externally powered multi-articulating digits, with a manually rotatable thumb and compliant grasping abilities, are available for the partial hand amputee.

Fig. 20.12 (**a**, **b**) Partial hand amputees are able to accomplish fine motor tasks with great ease. (Used with permission from Össur Touch Solutions)

- The stages of training from simple to complex mirror that of those with the i-Limb hands, stressing the utilization of these devices in bilateral tasks to enhance overall functionality. An example for fine motor tasks is given in Fig. 20.12a, b.
- As in other prosthetic fitting and training protocols, optimal outcomes for the partial hand client are achieved when the prosthetist and therapist work closely together. This is especially true for the partial hand as each configuration is unique and customization of each device is essential. Working as a multidisciplinary team to ensure the alignment of the digits is functional, and the myo-sites (often utilizing remaining intrinsic muscles of the hand) work well without interfering with remaining finger, wrist, or forearm motion greatly enhances the function of the end-user.
- For additional information on the recommended fitting and training process, see www.touchbionics.com/training.

Training Protocol After Targeted Muscle Re-innervation in High Upper Limb Amputees

As described in a previous chapter, targeted muscle re-innervation (TMR) is meant to improve the neuromuscular interface for prosthetic control by the use of selective nerve transfers [23, 24]. Thus, TMR allows to control a multi-degree of freedom prosthesis by using muscular activity [36]. In contrast to conventional myo-electric fittings, more than two electrode sites are used for prosthetic control. The principles of physical and occupational therapy, as described before, are still applicable for patients who undergo TMR surgery. Nevertheless, the use of more myo-signals significantly increases the complexity of control, and the physiological nerve regeneration after surgery changes the motor commands and the timeline for fitting [45]. Thus, here the specific interventions needed to ensure good prosthetic control after TMR surgery are described. While teamwork is considered as a success factor for all prosthetic fittings, its importance has been emphasized in the rehabilitation of the TMR patient [11, 35, 42].

Interventions Before TMR Surgery

It is recommended to see a patient who is planning to undergo TMR surgery within a multidisciplinary team. This allows the medical professionals to discuss the whole procedure with the patient in order to assess medical, psychological status, and requirements for surgery and subsequent fitting. Information should include details about the surgery and the process, the patient's responsibilities within the process (time at hospital, home program, traveling, costs, etc.), and realistic outcomes [11].

Additional interventions before surgery are mostly dependent on the patient's needs and aim to improve their functional status. These do not substantially differ from the post-amputation interventions described earlier in this chapter.

Interventions Within the First Months After Nerve Transfer Surgery

Following surgery it takes about 3–6 months until the transferred nerves reach their target. Initial muscle contractions can be registered either by surface or needle EMG. The time directly after surgery is characterized by wound healing, standard post-surgical interventions, and re-fitting with a prosthesis (if requested). To ensure the highest levels of safety, the therapist should be informed about possible precautions and contraindications. Additionally, the therapist should request a surgery report to know where to expect which EMG signals. Apart from the general post-surgical interventions described before, the therapist may consider to facilitating the cortical representation of the lost limb to enhance selective muscle activation later. This can be supported by the use of Graded Motor Imagery [13, 15, 20] or mirror therapy [16].

Myo-training After Successful Re-innervation

In most cases, it takes about 3–6 months after surgery until the muscles show signs of re-innervation. This can be detected by observing slight muscle twitches or via surface or needle EMG [44] and marks the beginning of the myo-training phase. Still, re-innervation is a gradual process, and there might be a different time frame for transferred nerves. Therefore, it might be possible that one muscle can easily be controlled by the patient at a point in time, while another one is still not active [23]. If a direct control system (i.e., one electrode corresponds to one specific prosthetic movement) is used, evaluating the best electrode positions is indispensable. While defining the best hot spots for the electrodes and performing myo-training, one should follow the same principles as in conventional two-signal control systems. The additional number of myo-signals significantly increases the complexity of control. Thus, a longer period of time should be spent on myo-training and establishing good signal separation (in direct control prosthetic systems). Additionally, the use of surface EMG biofeedback systems for displaying the activity of all used muscles is highly recommended [43]. If pattern recognition systems are used for prosthetic control, defining and training distinct muscle activation pattern are needed. As soon as positions for electrodes are defined, the prosthetist can start with the production of test socket for training purposes and training with a table-top prosthesis (as shown in Fig. 20.13) can be started.

Prosthetic Training with the TMR Fitting

Approximately 9–18 months after surgery, the prosthetic fitting can be initialized. Here, the therapist assists the prosthetist in defining the best electrode positions in the socket. However, it is important to note that weight load has some impact on the patterns of muscle contraction [39]. Thus, it might be necessary to change signal amplification or other settings with the full fitting compared to the table-top prosthesis. When the patient is fit with a prosthesis, the prosthetic training starts. This part does not differ from prosthetic training with any other control method. However, if there are unintended prosthetic movements due to co-contraction of other muscles, it might be beneficial to go back to the myo-training with sEMG biofeedback.

In most TMR prostheses, like the Dynamic Arm Plus (Ottobock, Duderstadt, Germany), a mode of sequential and simultaneous control is possible. In sequential control, only one joint can be active at a certain time, while with simultaneous control all degrees of freedom (e.g., hand

Fig. 20.13 Table-top prosthesis training after TMR surgery. (Used with permission from Ottobock)

close, pronation, and elbow flexion) can be used at the same time. This allows a more natural and faster prosthetic control but is also more error-prone in cases of insufficient signal control.

After discharge from rehabilitation, the patient should be able to control the prosthesis and use it in daily life. Regular appointments, including functional tests, should be used to monitor medical status, prosthetic function, and device satisfaction. This allows the clinical team to intervene, if needed.

Outlook

In the recent years, research has led to many new solutions for individuals facing an amputation. This involves 3D-printed devices [46], more degrees of freedom for prosthetic hands [18], but also an improvement of the human-machine interface in regard to the mechanical attachment [22], as well as control [6, 34] and feedback [33, 37]. As described in the various chapters of this book, all these technological advances have the potential to enhance the possibilities of individuals with an amputation. Additionally, new patient groups as people with severe brachial plexus injuries or hand injuries beyond biological reconstruction may be fitted with prosthetic devices to restore their extremity function [1, 2, 19]. All of these developments pose new challenges in regard to user training that need to be addressed by the occupational and physical therapy of the future. With more advanced and complex technologies, it is expected that providing adequate therapy will become even more important [39, 44]. At the same time, the therapist will have the opportunity to use novel technologies, such as serious games [31], mobile applications, virtual reality, and telerehabilitation systems (see Chap. 21 for more details). While all of this will influence the techniques used by physical and occupational therapists, their goal will still remain to support individuals to regain autonomy, social role functioning, and participation in activities of their daily life.

Conclusion

The potential of the individual with upper limb loss is limitless, as often he/she is able to accomplish activities he/she never would have expected. The success of this individual is not only strategically intertwined with the occupational and physical therapist but also the prosthetist and a collaborative interaction with the surgeon, rehabilitation physician, and other key members of the team. The therapist can have a profound impact and influence on the acceptance or rejection of a prosthesis.

One of the most important "ingredients" of a successful prosthetic training program is the development of a protocol that continually focuses upon the individual with upper limb loss. Listening to their needs, goals, and aspirations, as well as cultivating their motivation, is valuable key to one's success. This is true for all types of prosthetic fittings, ranging from body-powered ones to modern myo-electric prosthesis, as well as these using advanced interfaces in the future. The impact of a therapist during this pivotal passage of recovery cannot be overstated as it will remain with this individual for life.

References

1. Aszmann OC, et al. Bionic reconstruction to restore hand function after brachial plexus injury: a case series of three patients. Lancet. 2015;385(9983):2183–9.
2. Aszmann OC, et al. Elective amputation and bionic substitution restore functional hand use after critical soft tissue injuries. Sci Rep. 2016;6:34960.
3. Atkins D. The upper extremity prosthetic prescription: conventional or electric components. Phys Disabil Spec Int OT Newslett. 1987;10(2).
4. Atkins DJ. Adult upper-limb prosthetic training. In: Atkins DJ, Meier RH III, editors. Comprehensive management of the upper-limb amputee. New York: Springer; 1989.
5. Batsford S, Ryan CG, Martin DJ. Non-pharmacological conservative therapy for phantom limb pain: a systematic review of randomized controlled trials. Physiother Theory Pract. 2017;33(3):173–83.
6. Bergmeister KD, et al. Broadband prosthetic interfaces: combining nerve transfers and implantable multichannel EMG technology to decode spinal motor neuron activity. Front Neurosci. 2017;11(421).
7. Biddiss EA, Chau TT. Multivariate prediction of upper limb prosthesis acceptance or rejection. Disabil Rehabil Assist Technol. 2008;3(4):181–92.
8. Bouwsema H, van der Sluis CK, Bongers RM. Changes in performance over time while learning to use a myoelectric prosthesis. J Neuroeng Rehabil. 2014;11(1):16.
9. Carey SL, et al. Compensatory movements of transradial prosthesis users during common tasks. Clin Biomech (Bristol, Avon). 2008;23(9):1128–35.
10. Carey SL, Lura DJ, Highsmith MJ. Differences in myoelectric and body-powered upper-limb prostheses: systematic literature review. J Rehabil Res Dev. 2015;52(3):247–62.
11. Carlsen BT, Prigge P, Peterson J. Upper extremity limb loss: functional restoration from prosthesis and targeted reinnervation to transplantation. J Hand Ther. 2014;27(2):106–13; quiz 114.
12. Cordella F, et al. Literature review on needs of upper limb prosthesis users. Front Neurosci. 2016;10:209.
13. Dickstein R, Deutsch JE. Motor imagery in physical therapist practice. Phys Ther. 2007;87(7):942–53.
14. Durance JP, O'Shea BJ. Upper limb amputees: a clinic profile. Int Disabil Stud. 1988;10(2):68–72.
15. Ehrsson H, Geyer S, Naito E. Imagery of voluntary movement of fingers, toes, and tongue activates corresponding body-part-specific motor representations. J Neurophysiol. 2003;90(5):3304–16.
16. Ezendam D, Bongers RM, Jannink MJA. Systematic review of the effectiveness of mirror therapy in upper extremity function. Disabil Rehabil. 2009;31(26):2135–49.
17. Ferraro F, et al. Diagnosis and treatment of pain in plexopathy, radiculopathy, peripheral neuropathy and phantom limb pain. Evidence and recommendations from the Italian Consensus Conference on Pain on Neurorehabilitation. Eur J Phys Rehabil Med. 2016;52:855–66.
18. Fu Q, Santello M. Improving fine control of grasping force during hand-object interactions for a soft synergy-inspired myoelectric prosthetic hand. Front Neurorobot. 2017;11:71.
19. Hruby LA, et al. Algorithm for bionic hand reconstruction in patients with global brachial plexopathies. J Neurosurg. 2017;127(5):1163–71.
20. Jackson PL, Lafleur MF, Malouin F, Richards C, Doyon J. Potential role of mental practice using motor imagery in neurologic rehabilitation. Arch Phys Med Rehabil. 2001;82(8):1133–41.
21. Jones LE, Davidson JH. Save that arm: a study of problems in the remaining arm of unilateral upper limb amputees. Prosthetics Orthot Int. 1999;23(1):55–8.
22. Jonsson S, Caine-Winterberger K, Branemark R. Osseointegration amputation prostheses on the upper limbs: methods, prosthetics and rehabilitation. Prosthetics Orthot Int. 2011;35(2):190–200.
23. Kuiken TA, et al. The use of targeted muscle reinnervation for improved myoelectric prosthesis control in a bilateral shoulder disarticulation amputee. Prosthetics Orthot Int. 2004;28(3):245–53.
24. Kuiken TA, et al. Targeted muscle reinnervation for the upper and lower extremity. Tech Orthop. 2017;32(2):109–16.
25. Malone JM, et al. Immediate, early, and late postsurgical management of upper-limb amputation. J Rehabil Res Dev. 1984;21(1):33–41.
26. Miguelez J, Conyers D, Keams N. Using meaningful data to optimize the quality of life for persons with upper limb amputation. In: American Academy of Orthotists & Prosthetists annual meeting; 2017.
27. Moseley GL. Graded motor imagery for pathologic pain: a randomized controlled trial. Neurology. 2006;67(12):2129–34.

28. Ortiz-Catalan M, et al. Phantom motor execution facilitated by machine learning and augmented reality as treatment for phantom limb pain: a single group, clinical trial in patients with chronic intractable phantom limb pain. Lancet. 2016;388(10062):2885–94.
29. Østlie K. Evidence-based guideline for rehabilitation after acquired upper-limb loss in Norway [Norwegian]. 2016. www.magicapp.org/public/guideline/Jn3zaL. Accessed 30 Mar 2016.
30. Ostlie K, et al. Assessing physical function in adult acquired major upper-limb amputees by combining the Disabilities of the Arm, Shoulder and Hand (DASH) Outcome Questionnaire and clinical examination. Arch Phys Med Rehabil. 2011;92(10):1636–45.
31. Prahm C, et al. PlayBionic: game-based interventions to encourage patient engagement and performance in prosthetic motor rehabilitation. PM&R. 2018;10(11):1252–60.
32. Ramachandran VS, Rogers-Ramachandran D. Synaesthesia in phantom limbs induced with mirrors. Proc Biol Sci. 1996;263(1369):377–86.
33. Rossi M, et al. HapPro: a wearable haptic device for proprioceptive feedback. IEEE Trans Biomed Eng. 2019;66(1):138–49.
34. Saikia A, et al. Recent advancements in prosthetic hand technology. J Med Eng Technol. 2016;40(5):255–64.
35. Salminger S, et al. [Prosthetic reconstruction of the upper extremity]. Unfallchirurg. 2016;119(5):408–13.
36. Salminger S, et al. [Prosthetic reconstruction in high amputations of the upper extremity]. Orthopade. 2015;44(6):413–8.
37. Schofield JS, et al. Applications of sensory feedback in motorized upper extremity prosthesis: a review. Expert Rev Med Devices. 2014;11(5):499–511.
38. Schweitzer W, Thali MJ, Egger D. Case-study of a user-driven prosthetic arm design: bionic hand versus customized body-powered technology in a highly demanding work environment. J Neuroeng Rehabil. 2018;15(1):1.
39. Simon AM, Lock BA, Stubblefield KA. Patient training for functional use of pattern recognition-controlled prostheses. J Prosthet Orthot. 2012;24(2):56–64.
40. Smith D, Michael J, Bowker J. Atlas of amputations and limb deficiencies. Rosemont, IL: American Academy of Orthopaedic Surgeons; 2004.
41. Smurr LM, et al. Managing the upper extremity amputee: a protocol for success. J Hand Ther. 2008;21(2):160–76.
42. Stubblefield KA, et al. Occupational therapy protocol for amputees with targeted muscle reinnervation. J Rehabil Res Dev. 2009;46(4):481–8.
43. Sturma A, et al. Recommendation for therapy after targeted muscle reinnervation. In: University of New Brunswick's myoelectric controls/powered prosthetics symposium. New Brunswick: Institute of Biomedical Engineering, University of New Brunswick; 2014.
44. Sturma A, et al. Rehabilitation of upper extremity nerve injuries using surface EMG biofeedback: protocols for clinical application. Front Neurosci. 2018;12:906.
45. Sturma A, Salminger S, Aszmann O. Proximale Amputationen des Armes: Technische, chirurgische und handtherapeutische Möglichkeiten. Zeitschrift für Handtherapie. 2018;21(1):18–25.
46. Vujaklija I, Farina D. 3D printed upper limb prosthetics. Expert Rev Med Devices. 2018;15(7):505–12.
47. Wakolbinger R, et al. Home based tactile discrimination training (TDT) reduces phantom limb pain. Pain Pract. 2017;18(6):709–15.
48. Weeks DLP, Anderson DIP, Wallace SAP. The role of variability in practice structure when learning to use an upper-extremity prosthesis. JPO J Prosthet Orthot. 2003;15(3):84–92.

Novel Technologies in Upper Extremity Rehabilitation

21

Cosima Prahm, Ivan Vujaklija, Fares Kayali, and Agnes Sturma

Structured and sufficient training is a key factor for successful fitting of an upper limb prosthesis. This is especially true for more advanced myoelectric control strategies, or for individuals with comorbidities that require additional treatment. With advances in technology, not only have the control strategies become more complex, but also possibilities for more tailored rehabilitation have increased. Novel rehabilitation technologies include virtual and augmented reality systems, as well as training systems relying on computers and smartphone apps. These technologies can be used within the clinical setting, enable telerehabilitation, and/or can support unsupervised home training. While most experts agree that novel rehabilitation technologies can be a good supplement for conventional therapy, one of the greatest challenges is to transfer the progress achieved in the technology-assisted realm into real-world situations and actual prosthetic function.

Background

Learning how to control a myoelectric prosthesis can be a discouraging experience for the amputee. Indeed, a lot of devices end up getting abandoned due to complications with their control and functionality [23]. After several months of recovery, the patient is fitted with a first prosthesis. By then, many amputees learn how to perform tasks unimanually and experience irritation of the residual limb when donning their prosthesis [5]. Moreover, the man-machine interface is little intuitive, and the cognitive demand for controlling a prosthesis is high, thus delaying the actual use of the device during day-to-day activities. This holds especially true for patients with higher amputation, i.e., above elbow, since the higher the amputation, the more degrees of freedom (DoF) need to be controlled. In fact, 50% of upper extremity amputees report problems with the prosthesis control and functionality. In order to overcome these possible stumbling blocks, sufficient training and therapy are essential for proficient handling of a myoelectric prosthesis. Thereby, training administered prior to final

C. Prahm (✉)
Department for Hand, Plastic, Reconstructive and Burns Surgery, BG Trauma Clinic Tuebingen, Eberhard Karls University, Tuebingen, Germany
e-mail: cprahm@bgu-tuebingen.de

I. Vujaklija
Department of Electrical Engineering and Automation, Aalto University, Espoo, Finland

F. Kayali
Center for Teacher Training, University of Vienna, Vienna, Austria

A. Sturma
Clinical Laboratory for Bionic Extremity Reconstruction, Medical University of Vienna, Vienna, Austria

Department of Bioengineering, Imperial College London, London, UK

O. C. Aszmann, D. Farina (eds.), *Bionic Limb Reconstruction*,
https://doi.org/10.1007/978-3-030-60746-3_21

fitting is a critical component influencing long-term success of prosthesis control [5, 73, 78].

Advances in technology continually influence the current medical practice and create new tools for therapy. Novel rehabilitation methods strive to take advantage of neuroplastic processes during the recovery period in the brain as well as the muscles while targeting motor deficiencies and aim at harnessing the learning potential of the brain [88]. Virtual reality and gaming systems were once ascribed to the domain of gamers, but it is one of those areas of advancement, where changes have spawned unique treatment approaches. Virtual rehabilitation has received increasing attention from researchers and clinicians alike, who recognize potential therapeutic benefits due to the immersive nature of the medium [27]. The primary focus of the application of virtual environments during upper limb amputee rehabilitation is the acquisition and sustenance of particular motor skills needed to control a prosthesis. It is best achieved by regular training of the specific muscles and improving their coordination and control. These skills of fine muscle coordination form the basis for prosthetic control, which is subsequently trained in a real-world environment with tasks of daily living [35]. However, as described in Chap. 20, other measures can be essential as well, depending on the individual patient's needs, preferences, and prosthetic fitting. Here, the therapist's task is to choose the right balance between conventional therapy and implementation of novel technologies, as well as guiding the training with these options.

In this context, novel technologies can be classified in three main categories depending on how they are used to support therapeutic efforts: systems to be used during clinical appointments, systems to be used within telerehabilitation settings/systems that enable remote therapy, and finally systems meant for (unsupervised) home exercises. In all of these settings, used technologies may support assessment of the patient's status, aid patient education, and/or actively guide rehabilitation exercises.

Clinically used systems include all technology the therapist/medical team have in their practice/hospital and can range from standard devices for electrical stimulation or EMG biofeedback to game-based virtual solutions and systems for body weight support or robotic assistance [7, 9, 79]. Furthermore, especially for advanced prosthetic control strategies, digital tools enable the clinician (and researcher) to assess the functionality of a control system and the patient's interaction with it already before prosthetic fitting [57, 80, 87] (for further details see Chap. 7). In contrast to the other categories, systems used in the clinical setting usually remain in one place and are operated by trained professionals. For this reason, they can be bulkier and more complex than systems meant for home use of a patient.

Telerehabilitation describes healthcare services that are delivered remotely via digital communication and devices. They are also referred to as digital practice, telehealth, or telemedicine. These systems have in common that they allow a therapist to deliver their services in (usually) one-to-one settings without the need to be located at the same place as the patient [92, 93]. Technologies for this range from simple video conferencing tools [22] to more elaborate sensor- or camera-based systems including gamification elements [71] with real-time feedback coming from the therapist. Depending on the patient's needs, telerehabilitation can have a focus on patient education or on exercise-based interventions. For both, preliminary evidence on general rehabilitation suggests a non-inferiority to face-to-face interaction and high acceptance levels from the patient and clinician side [2, 16, 68].

Finally, home training systems are meant to support (repetitive) home training exercises that the patient can do on their own without real-time feedback from the therapist. As a high number of repetitions are needed for motor learning, training outside the therapy sessions is considered essential. An important aspect to this end is the motivation of the patient, which tends to quickly decline during repetitive or frustrating rehabilitation tasks. To mitigate these effects, we can rely on technology which can offer an engaging experience to the patients. This way they can remain focused for prolonged periods of time while exercising in a non-traditional way. An additional advantage of virtual interventions is the opportunity to further quantify user performance: to be able to monitor the patient's progress and remotely provide this information back to the physiotherapist, who can in turn individualize the treatment, i.e., by adjusting the task difficulty or a number of mandatory repetitions. This usually happens in an asynchronous way, allowing patient and therapist to train/assess without time dependencies. Managing these parameters can prevent frustration, boredom, and fatigue when engaging the user in a repetitive exercise program [39]. While it is recommended to test these home training systems with the patients during their regular therapy sessions, it is an imperative that the solutions offered work in a simple plug-and-play manner. Otherwise, they would hinder patients in carrying out their practice sessions and become another source of frustration.

As described in more detail below, several different technologies can be used to enhance outcomes and motivation within prosthetic rehabilitation. They range from fully immersive 3D simulations, over showing a virtual arm on a computer screen, to remote communication and training options with a therapist, to EMG-controlled games on a computer or mobile phone.

Myoelectric Training Systems

Enabling amputees to volitionally generate separate muscle signals is imperative for skillfully controlling a prosthesis. Myoelectric training systems help to achieve this goal, ideally already in advance of the patients receiving their device. A number of solutions are already available on the market offering users and therapists a way to facilitate rehabilitation and training. They primarily rely on simple biofeedback which, if used in conjunction with structured rehabilitation protocols, can assist in managing even the more complex and less intuitive myocontrol paradigms [66]. However, optimization of motor learning can be further enhanced through use of more engaging technologies which can be divided into four major categories: (1) immersive 3D virtual reality systems using head-mounted displays, (2) camera-based augmented reality, (3) EMG-controlled video games on a computer or a console, and (4) mobile health applications. Enriched training environments can enhance the learning experience and assist in consolidating the practiced motor skills. By using treatment interventions created in virtual environments, several training parameters can be manipulated to explicitly engage motivation, performance, and effort [39, 54].

Commercial Products

Myocontrol-focused technology to support rehabilitation has been in great demand, and over the years, both major prosthetic manufacturers and some smaller players in this field have invested efforts in developing their own solutions. Some of the most relevant and representative commercial products used in prosthetic rehabilitation are discussed below.

MyoBoy

Ottobock's MyoBoy is a portable myoelectric visualization and training device displaying two LED bars that correspond to the activation of two sEMG electrodes, respectively. It is battery powered and connects via cable to two active Ottobock electrodes. The MyoBoy can be used as a device to find the correct site for electrode placement or train patients during early recovery in basic muscle strength and coordination [49]. The system is limited to only two electrodes and is a rather basic tool to display EMG activity.

Paula

Paula (Prosthetist's Assistant for Upper Limb Architecture) consists of a software package including settings for various control schemes, patient management, a virtual hand, and a computer game. During the game, the user's EMG graph is depicted as tracks of a car that need to be navigated through openings in walls, requiring the user to modulate their EMG amplitudes accordingly. The software is also able to store user parameters, including records of the patient's signal strength over time. It is available as part of the Ottobock Data Station for Windows systems [18, 50].

Myo Plus

The Myo Plus app by Ottobock is a mobile app with which users can operate and configure the Myo Plus prosthesis. The implemented pattern recognition control can be adapted to individual needs of the patient with a smartphone or a tablet. Movement patterns and the patient's EMG can be visualized through two different graphs as shown in Fig. 21.1. Additionally, prosthetic functions can be added or switched off [51].

biosim-i

The Touch Bionics biosim-i software is a desktop interface as well as a mobile app that enables prosthetic users to configure and adjust various features of their myoelectric hand, such as grip patterns and hand mode. It can check whether the hand is functioning properly and display the user's real-time EMG graph. The prosthesis connects to the interface via Bluetooth, and low battery can be indicated by an adjustable audio signal [83].

my i-limb

Similar to biosim-i software, the mobile app "my i-limb™" by Össur provides users with the flexibility to customize their prosthesis and to access pre-installed and customizable grips while on the go. It furthermore shows EMG signals in real time [84, 85].

Fig. 21.1 A screenshot of the Myo Plus training system: This application allows the visualization of the activation of all eight electrodes used in the pattern recognition system with a so-called spider plot (left). Furthermore, it guides through the calibration process (right). (Used with permission from Ottobock)

Desktop/Console

The high costs of VR systems, the necessity for substantial technical expertise, and the need for large vacant space have led researchers to pursue more accessible commercial technology, such as Xbox™, PlayStation™, Wii™, and computer games with hope to provide users with similar rehabilitation experience. There are several open-source games that have been altered to support myocontrol-relevant motor practice [54, 60]. Similar to AR, these rehabilitation systems rely on computer or TV screens to project the virtual scenes. However, they do not require a video stream, and the user can control the applications through EMG input.

Researchers have used modified versions of a number of gaming consoles to create suitable myocontrol therapeutic devices. The commercially available video games Guitar Hero [8] and Pong [67] had been interfaced using myoelectric signals. Nintendo Wiimote control has been adjusted in such a way as to substitute buttons with EMG-elicited commands [47]. While all these approaches surely offer a degree of engagement, the necessary control actions are limited and not strictly mapped to movements needed in everyday control activities. This further begs a question whether the developed skills are truly transferable to the handling of a prosthesis [12].

Virtual Reality (VR)

Virtual reality, as depicted in Fig. 21.2, is a technology used to simulate engaging 3D environments which users experience as being comparable to the real world [39]. By tracking their movements within a given space, amputees can interact with the simulated world and manipulate objects without needing the actual prosthesis. On the affected limb, it usually shows either a human arm or a robotic prosthesis. The user is wearing a head-mounted display (HMD), a wearable screen that projects the virtual scenes directly in front of the eyes independent of the viewing direction, thus creating the immersive ego perspective. Some users, however, report getting affected by cyber sickness, an induced nausea which can be attributed to delays in data transmission or incongruencies between movement of the user and what is seen on the display [86]. Also, the system itself is rather expensive and requires technical expertise. Nonetheless, a great benefit that comes with employing VR systems is the potential to flexibly implement embedded prosthesis software and to constantly adapt the simulated 3D world. VR has successfully been used as an extension to the mirror box in order to alleviate phantom limb pain [45]. Therapy can be facilitated by a controlled virtual environment, and the clinician can guide patients step by step through everyday activities, eliminating the risk of breaking objects and without the weight of the prosthesis [65, 91].

Ultimately, VR environments should support the rehabilitation process and help patients to orient themselves within their new reality and not be a substitute for using their actual prosthesis in the real world. After all, amputees should use their prosthesis as much as possible in order to benefit

Fig. 21.2 Overview of digital myoelectric training systems: personal computer, virtual reality, augmented reality, and mobile device. (© Aron Cserveny for Oskar Aszmann)

from all the associated advantages, such as the alleviation of phantom limb pain.

Augmented Reality (AR)

In contrast to virtual reality systems, the user is not wearing a HMD. Instead the scenes are displayed on a computer screen, where a computer-generated image superimposes on the user's view of the real world. One or multiple cameras provide a video stream of the patient and track the affected limb with the help of special markers. What can be seen on the screen now is the patient with a computer-rendered virtual arm or hand, which can be controlled and moved by the patient for as long as the camera is able to recognize the markers.

Similar to VR, AR has been applied to research of phantom limb pain relief as a replacement of the mirror box as described in Chap. 12. Although it is methodologically equivalent, distal movements of the phantom limb cannot be inferred by this system [48]. Other applications are training systems in which the patients can see themselves with a virtual arm attached to their residual limb and also have to perform a gaming task [5].

Mobile Devices

Research indicates that patients who underwent pre-prosthetic training before receiving their actual prosthesis achieve better functional results. However, the previously introduced training systems are usually stationary and bulky in comparison to a mobile phone. The mobile health sector is rapidly growing, and mobile phones are powerful tools that can be used for pre-prosthetic training outside of the clinical environment. Similar to desktop applications, a patient profile could be created and retrieved by the physiotherapist. In this way, the patient's progress can be monitored remotely while the therapist continues to provide feedback and support.

Advantages of mobile systems are familiarity with the device and its ease of use. In a recent study [53, 59], a rhythm-based game could be played by patients on a tablet. They wore a Myo Armband on their stump which is connected to the tablet via Bluetooth. The patients could subsequently control the game by using their myoelectric signals. Over the course of 4 weeks, during which the patients had used the system at home, a significant improvement was found in the parameters underlying successful myocontrol, which included muscle separation, co-contraction, and proportional control. In another study [90] that lasted 1 week with able-bodied participants, participants could choose between four different dinosaur-themed games to train varying aspects of their myocontrol. In both studies, both patients and able-bodied participants not only improved their game score, which is directly reflected in improvement of neuromuscular control, but also reported a high level of engagement and motivation to continue playing even after completion of the study.

Game-Based Rehabilitation

Rehabilitation concepts traditionally rely on the execution of repetitive movements [66]. Applying the concept of gamification, which describes assigning game design and game concepts to usually non-gaming training scenarios [20] can boost the user's engagement as well as entertainment. A set-up for such a game can be seen in Fig. 21.3. Video game-based therapies can be a stimulating way to increase patient motivation, effort, and performance during those otherwise monotonous neuromuscular exercises. Myoelectric signals needed to control an upper limb prosthesis are thereby intuitively trained, and the patient eventually benefits from increased prosthesis usability [56]. Even though established rehabilitation protocols address the necessary parameters needed for prosthesis control and offer direct functional benefit, some patients report a lack of motivation to keep up with the repetitive and lengthy process. While long training times for establishing good neuromuscular control are uncommon in otherwise healthy individuals, keeping up motivation over a long period of time might be more

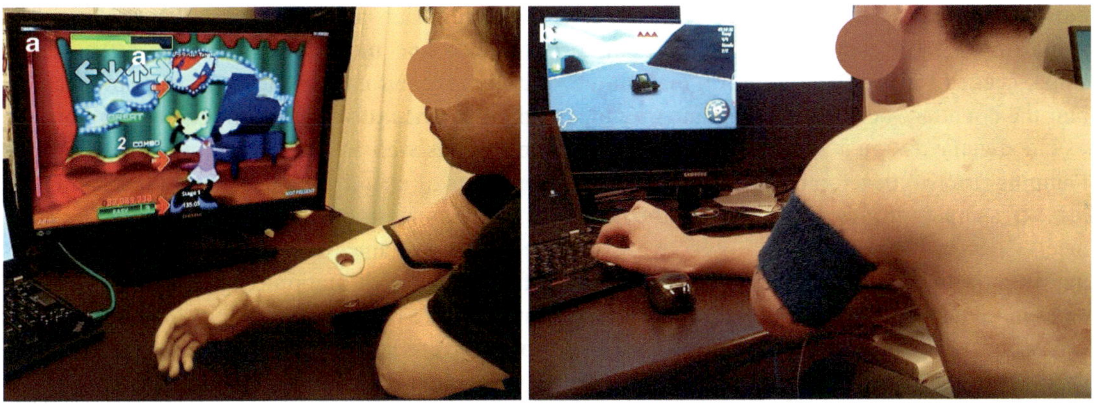

Fig. 21.3 Training in a virtual environment on a personal computer: (**a**) using the Michelangelo hand prosthesis (**b**) with Ottobock active electrodes attached to the skin with a wristband

relevant in people with additional comorbidities, complex control systems, or nerve transfers [77, 79]. Both extrinsic and intrinsic motivation play important roles in sustained engagement with game-based rehabilitation. Many examples of gamification in that space primarily focus on extrinsic motivators such as points, rewards, and progression in the game. The danger of extrinsic motivation is that, while necessary and a good motivator in itself, it potentially harms intrinsic motivation, as shown in a meta-study by Deci et al. [19]. To trigger intrinsic motivation, game-based rehabilitation can build on the concepts of competence, relatedness, and autonomy from self-determination theory [69] to make a game engaging in the long term. This further helps health-beneficial behaviors to persist beyond the context of a game for health used to train or sustain them. A systematic review of studies on active video games that promote physical activity in children and youth also describes self-initiation and choice as the most important factors for sustained engagement [10]. An approach related to self-determination theory, for example, is goal setting, which is also used in behavior change technology [44], where players are allowed to define their individual clinical outcomes and in-game goals. Additionally, patient education on the anticipated results of using the game and how they translate into patient-related outcomes is an important tool to further increase internal motivation.

Selecting Games for Rehabilitation

With the gaining popularity of eHealth games and apps, it is increasingly difficult to identify appropriate and feasible applications. The right game goes beyond being a fun and exciting diversion from regular activities during physical therapy. The game must accomplish a delicate balance: It should be demanding enough, adapted to the task at hand, but also take into consideration the patient's abilities. After nerve injuries or nerve transfers, months of intensive training are required to become proficient in controlling a myoelectric prosthesis, depending on factors such as available electrode sites. Users with insufficient training abandon the prosthesis before this mastery is achieved, because the training process has proven to be too difficult and tiresome leading to low user acceptance [23].

Delivering biofeedback using off the shelf video games is a viable low-cost alternative to virtual or augmented reality, as no further equipment is needed and training can be continued at home. Especially during research of neuromuscular disorders and stroke have commercially available video games been extensively used as a tool to incorporate repetitive movements during rehabilitation [14, 30, 37, 40]. They come with the advantage of professional game designers and marketing experts. However, should the game not be task-oriented or in line with the patient's individual background, needs, and goals, even the most elaborate incentive is futile. A custom-made game specifically aimed at improving motor skills, needed for prosthesis control, may further blur the boundary between playing and training.

Good game design is built on user experience. Planning prior to development and iterative testing on sample groups ensures that the game meets design and training objectives and improves with each cycle. One of the primary measurements of successful games is active engagement. Challenging gameplay that adjusts to the player's skill level and visibly linked actions and achievements encourage sustained engagement. Incorporating high scores and leader boards is a way to display achievements in a straightforward way which enables the players to chart their progress. Opportunities to further engagement that games can appropriate from behavior change technology are self-defined challenges and personal goal setting [44]. Occasional additions to pre-existing difficulty levels keep the content fresh and stimulating. Notifying and rewarding users encourages desirable behavior and further promotes the training effect. However, a delicate balance between extrinsic feedback or rewards and providing opportunities for intrinsic goals has to be maintained.

A study [54, 60], questioning a pool of patients on how should an ideal game for neuromuscular rehabilitation look like, concluded that it should be a version of a dexterity game with short reward periods, a global high-score system, engaging gameplay, and background music, which is playable with complete EMG control. All patients without exception had asked to be able to play the game at home in order to visit the clinic less often and also to share the playing experience together with their family [54].

Impact of Game-Based Rehabilitation on Clinical Outcomes

A particular obstacle to recovery, following a discharge from the rehabilitation facility, is for patients to continue exercising at home [27]. Therefore, engaging training is essential for ensuring that key concepts are assimilated and retained and motivation kept high [39]. Virtual training environments can aid this process, as they provide diverse incentives to support the patient as well as the therapist to achieve a high number of exercise repetition without losing the patient's investment and perseverance [1, 25, 42, 82]. Video games have been used by clinicians for motor rehabilitation especially when dealing with stroke [40, 64] and Parkinson's disease [30] as these are large patient populations that require long rehabilitation. However, in the last years, they also received some recognition as a method in upper limb amputee rehabilitation for myoelectric prosthesis control [3, 5, 6, 8, 13, 21, 47, 54, 56, 60, 67, 70, 81].

In recent studies [53, 56, 58, 60], clinically oriented, stationary programs and also mobile applications for a smartphone have been developed that train and evaluate the patient's neuromuscular capacity pre- and post-interventional for (1) maximum voluntary contraction, (2) proportional fine muscle control, and (3) isolated activation of different muscle groups. Participants were not only prompted to conduct repetitive flexor and extensor motions but also trained to continue muscle contraction over varying periods of time, perform precisely timed contractions, and execute simultaneous contractions of both muscle groups—similarly to how patients would control an actual prosthesis to interact with their everyday environment. Motivation and effort were assessed using the Intrinsic Motivation Inventory [32] and a user evaluation survey and subsequently compared to a standard myoelectric rehabilitation device, the MyoBoy (described above). All participants achieved a significant improvement in all three EMG assessment criteria: maximum voluntary contraction, proportional control, and muscle separation. Enjoyment, usefulness, and exerted effort were rated higher during the game-based rehabilitation training compared to the MyoBoy [55]. Surprisingly, the patient's strength, used as a baseline calibration, has also shown an increase instead of the expected decrease after playing the games, which could be attributed to warmth or sweat that would influence the electrode resistance. However, this is a solid indicator that the gaming session was not fatiguing for the patients. In fact, the patients became even more skillful during the assessments. The main advantage of game-based training compared to conventional therapy is clearly the motivational aspect. Though it is reasonable to assume that patients would also improve their EMG control by only being exposed to the basic assessments, the continued exposure would eventually lead to a loss of interest, which could be prevented by the engaging context of a video game [41]. Ideally, patients would also be able to continue the rehabilitative intervention at home, either in the form of a mobile device or as an easy stationary set-up, to achieve the maximum benefit for proficient prosthesis control [62].

The goal of game-based training systems is not only to improve the user's muscle strength, coordination, and separation but also to eventually transfer those skills to the handling of a myoelectric prosthesis. Several hand function tests have been mentioned in the previous chapters that are established measures of this transfer (e.g., SHAP, ARAT, or clothespin test). Transferability, task orientation, and personalized training arrangements contribute to effective and engaging motor training and thereby reduce the likelihood of prosthesis abandonment.

Digitally Assisted Diagnostics

Though virtual rehabilitation describes mainly the use of virtual environments for motor training purposes during rehabilitation, it can also be applied to the diagnostic process. Progress and impact of the physical therapy and medical interventions can be monitored, and qualitative data can be quantified. Using Internet queries instead of filling out questionnaires with pen and paper is already clinical practice, especially for long-term follow-ups where patients may not necessarily need to see the clinician face to face. Additionally, novel technologies allow assessments that were not possible in this form before but can be crucial in understanding patient's myocontrol performance as other traditional "offline" metrics tend to be misleading [87]. For this purpose tests such as the Box and Beans test [57], the virtual clothespin test [29], and target achievement control (TAC) test [72] can be used for evaluating the conventional myocontrol as well as more advanced machine learning algorithms. Finally, digital body image visualization tools can be utilized in order to understand and appropriately address a patient's own body representation.

Digital Testing

Especially during the evaluation of real-time advanced prosthetic control algorithms (see Chap. 13), digital environments can provide an accessible setting either for training a patient's neuromuscular performance or to assess novel control models.

As described in Chap. 7, the analogue clothespin test [31] involves the retrieval of three clothespins that are clamped on a horizontal bar in front of the patient and clamping them to a second, vertical bar. This task requires the use of elbow flexion/extension, wrist supination/pronation, and hand opening and closing for grasps and release. The virtual version [29] featured the same set-up, with a 3D model of a human arm. The control input was modeled after an electrode configuration suitable for targeted muscle reinnervation (TMR) patients. This digitized test was used to evaluate the performance of pattern recognition algorithms with regard to the system's usability [29].

During the target achievement control test, users were asked to move a digital prosthesis into a target posture, indicated by a human hand, and to hold that position for a period of time. If the user did not hit the target immediately, or overshot the target, the pose had to be corrected. The same is applied for any incorrect prosthetic movements, which had to be compensated again in order to succeed in this test [72]. A study with able-bodied participants evaluating the TAC test found that the visualization of more complex target acquisitions led to confusion and thus to poorer performance of the user, not necessarily the model. It also showed that the TAC test was rather difficult to learn for novices [26].

The Box and Beans test is the digital counterpart of the Box and Blocks test, which is a quick and commonly used tool to assess unilateral manual dexterity in neurological disorders [43]. During the analogue Box and Blocks test, patients are sitting/standing in front of the box and can grab the wooden squares to transport them from one compartment to the other using mainly the joints of the hand and elbow, which would correspond to two DoF. However, in front of a two-dimensional computer screen, one DoF is lost, the one associated with elbow flexion and extension. To make up for the lost dimension, in the digital Box and Beans test (see Fig. 21.4), the cubes were replaced by beans, which can only be grasped from a certain position of the virtual hand, therefore coercing the user to also use wrist rotation and adding a DoF [57].

The virtual grasper was controlled by using EMG signals taken from commercially available electrodes or the Myo Armband. The Myo Armband's built-in gyroscope allowed users to freely move their arm and transfer the beans to the opposite compartment. The quantitative performance score remained the same as in the Box and Blocks test: the number of moved beans within 1 min. Since 150 blocks would crowd the screen, 3 new beans would spawn when there was only 1 bean left in the start compartment. The new beans were oriented, placed, and colored randomly, but the size was always kept the same [57].

This test was developed to assess novel control algorithms for machine learning techniques in healthy participants as well as amputees; however, it can certainly be used to evaluate rehabilitation progress and the proficiency to control myoelectric prostheses. The digital Box and Beans test can be performed even without a prosthesis and also be used to train prospective myoelectric control.

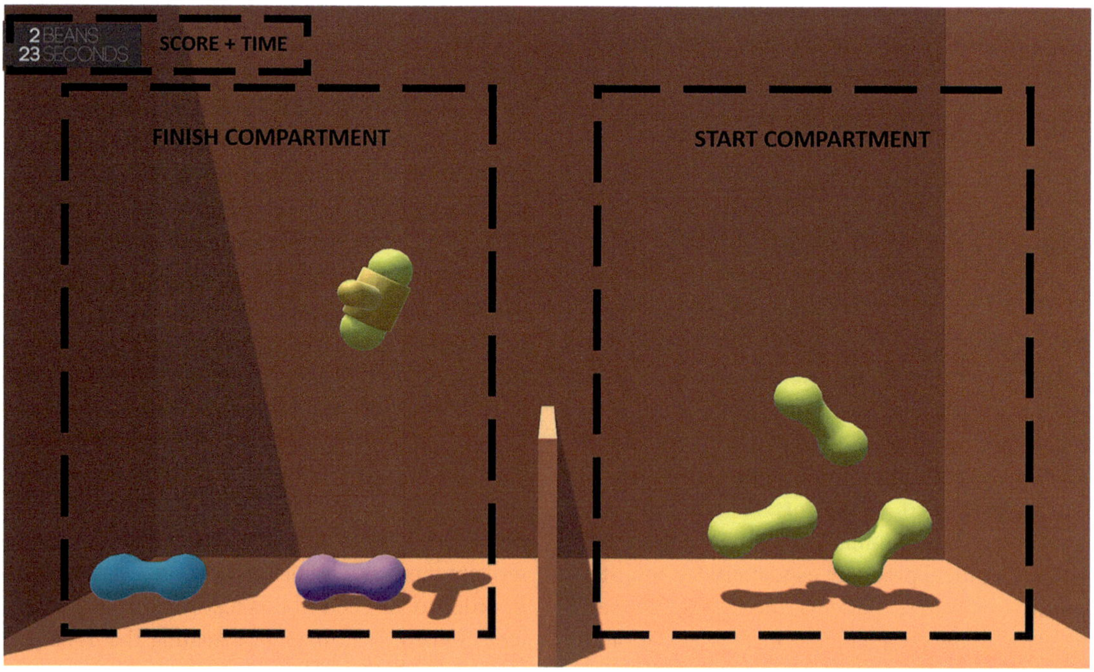

Fig. 21.4 Screenshot of the digital Box and Beans test [57]. Patients transport beans from the start compartment to the adjacent finish compartment during a set time limit of 60s. The transported beans directly translate into the score. (Used with permission from IEEE)

Body Image Visualization Tools

Upper limb amputation is often followed by the feeling that the affected limb is still present. This phantom limb representation is also reflected in the body image of the patient and can be associated with painful and non-painful sensations [33, 34]. Noble et al. reported as early as 1954 in their work "Psychiatric Disturbances following Amputation" that changes in body image occur after amputation [46]. It is common practice to assess the appearance and pain levels of the phantom limb using analogue methods such as questionnaires [15, 61, 89] or drawings on paper [15, 28]. However, while these tools are helpful to monitor pain, they are limiting patients in expressing the characteristics and sensations of their phantom limb. Even though questionnaires provide means of standardized comparison, their ability to adequately reflect a person's perception is constrained. Moreover, manual tasks that involve the participant to actively draw or build something are often subject to the researcher's interpretation.

Computer-aided instruments for visualizing body image have been previously popular with regard to eating disorders such as anorexia nervosa [4, 24, 38]. However, there are not many digital tools to evaluate the body image perception of amputees. In 2006 Brunel University published a pain identification software which displays pain areas of different quality on a 3D human avatar [75, 76]. The pain category could be adjusted, however, not the intensity. The commercial program "Navigate Pain" by Aglance Solutions allows to display and track the pain of different etiologies. In 2D and 3D views of the human body, both pain category and intensity can be displayed [11, 74]. The avatar itself remains static and cannot be re-positioned.

CALA (Computer Assisted Limb Assessment) is a digital, stand-alone program to visualize phantom limb appearance and its position, as well as pain and cramps. It can be used to diagnose and depict the state and location of a phantom limb, telescoping, and pain hot spots and to track changes during the rehabilitation process [52]. The CALA application adequately depicts the body image perception of patients with amputations or nerve injuries in a 3D environment and monitors the position and condition of the virtualized phantom limb over the course of physical therapy. As shown in Figs. 21.3d and 21.5 human avatar is created either according to the patient's self-perception or their actual body measurements. It assesses the body image through guided virtual distortion and manipulation of the upper extremities and part of the trunk. Specifically, the patients are able to model their affected arm, which can differ from their original one in being

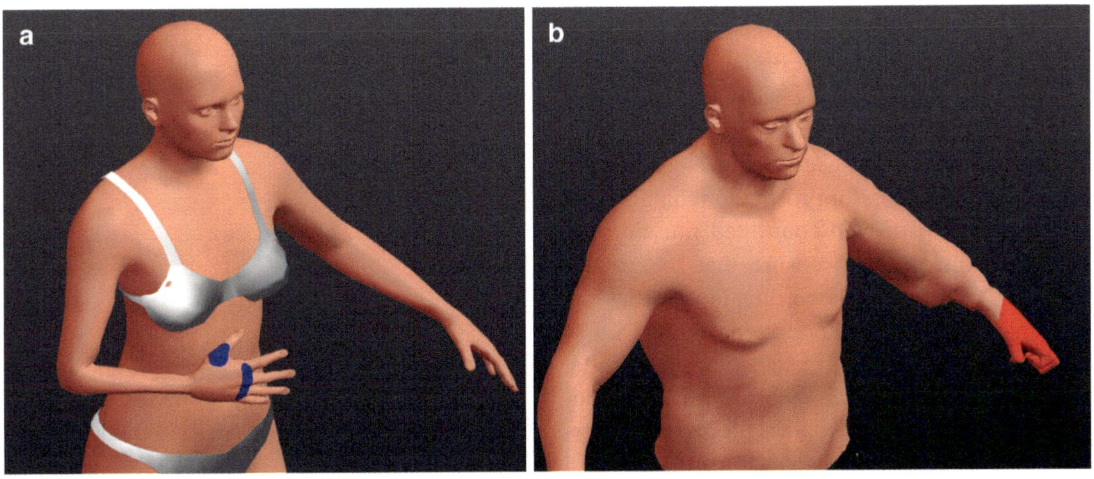

Fig. 21.5 Modeling and positioning the phantom limb on a 3D human avatar analogue to the body image of a patient with (**a**) a cramping fist and (**b**) a telescoping lower arm. (Used with permission from IEEE)

thinner or thicker, longer or shorter, or the patient's hand could be hinged directly at the elbow, depending on their sensation of the phantom limb. It is also possible to position the virtual arm in any natural or unnatural way, since phantom limbs do not adhere to anatomical limitations. Extremities can be maximally distorted and manipulated beyond physiologically possible degrees to take into account all forms of embodiment or phantom limbs. Since most patients also experience strong pain within their phantom limb, it is also possible to color code the affected area on the 3D model depending on the pain intensity [52].

Discussion

Most upper limb myoelectric prostheses are controlled via a direct two-electrode interface, up to five or six electrodes if the patient has received a TMR surgery (see Chap. 10). To effectively control a myoelectric prosthesis, it is important that the patient can intentionally and reliably elicit separable, proportional, fast, and sustained muscle contractions. Moreover, they should also be able to generate co-contractions of two muscles at will. For more advanced systems such as those based on machine learning approaches (which are discussed in detail in Chap. 13), training is still essential to establish robust and reliable muscular activation patterns [36]. In all control methods, for a sufficiently proficient prosthetic control, parameters such as timing of the respective muscle contractions for a series of connected tasks and estimation of needed muscle activation level for each step need to be considered. Furthermore, successful prosthetic rehabilitation does not end with well-controlled muscle contractions, but needs to ensure that the patient manages ergonomic coordination with the remaining joints and other extremities as well as dual-task situations as expected in daily life. Apart from that, every individual with an amputation presents with their own life story, comorbidities, preferences, and goals. Meeting all these requirements is not trivial and therefore requires structured and holistic training.

Technology for upper limb amputees or nerve damage patients should always be tested on the intended population, since there are several differences to healthy volunteers. What works for the able-bodied might not work for the amputee. A reduced number of electrode sites, early fatiguing, and possible cognitive deficits acquired through accidents could be limiting factors in comparing patient population.

Given the complex and sometimes repetitive nature of upper limb amputation rehabilitation, different digital tools can be used to support understanding, enhance motivation, and allow for additional training time. For pre-prosthesis training in the home setting, appropriate rehabilitation-focused games can offer adaptive challenges that can keep the patients engaged. Content-wise this means to support individualization by offering a variety of games. Game mechanics-wise this means to find the right balance between increasingly challenging patients while at the same time not discouraging them (see flow theory depicted in Fig. 21.6 [17, 63]) where a too high challenge leads to anxiety and too little challenge leads to boredom. The increase in challenge in pre-prosthesis training has to be carefully aligned with game progress on the one hand and rehabilitation process on the other. Furthermore, skills

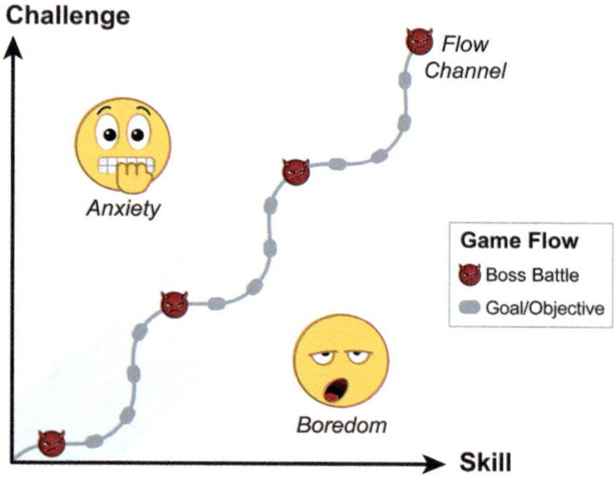

Fig. 21.6 The flow channel [63] in games illustrates how tension rises when the challenge increases (e.g., when facing a tough boss enemy at the end of a level) and how tension falls when players get more skilled, thus keeping them in this flow channel

trained within the game need to be relevant for the prosthetic control method that the patient is using after rehabilitation.

Sustaining this motivation long-term is one of the challenges yet to be fully addressed in game-based rehabilitation. A strong embedding of rehabilitation games in a personalized and individual context along with variety in game contents, rehabilitation goals, and supplementary training can maximize engagement and, not to forget, performance in handling the prosthesis. After all, the progress in-game needs to be transferred to real-world situations, and further therapeutic interventions still need to enable proficiency of prosthetic use in daily life. It is arguable that patients will improve in EMG control regardless of the feedback presentation, as long as there is some sort of biofeedback [12]. However, they will not keep at it for very long if the motivational aspect is neglected. One way of preventing a loss of interest could be by incorporating enriched virtual environments into the rehabilitation process.

References

1. Grigore C. Burdea, “Virtual reality technology for the clinician,” 2015 International Conference on Virtual Rehabilitation (ICVR). Valencia. 2015: p. 1–1. https://doi.org/10.1109/ICVR.2015.7358629.
2. Adamse C, Dekker-Van Weering MGH, van Etten-Jamaludin FS, Stuiver MM. The effectiveness of exercise-based telemedicine on pain, physical activity and quality of life in the treatment of chronic pain: a systematic review. J Telemed Telecare. 2018;24(8):511–26.
3. Al-Jumaily A, Olivares RA. Electromyogram (EMG) driven system based virtual reality for prosthetic and rehabilitation devices. In: Proceedings of the 11th international conference on information integration and web-based applications & services. New York: ACM; 2009. p. 582–6.
4. Alcañiz M, Perpiña C, Baños R, Lozano JA, Montesa J, Botella C, Palacios AG, Villa H, Alozano J. A new realistic 3D body representation in virtual environments for the treatment of disturbed body image in eating disorders. CyberPsychology Behav. 2000;3:433–439. https://doi.org/10.1089/10949310050078896.
5. Anderson F, Bischof WF. Augmented reality improves myoelectric prosthesis training. Int J Disabil Hum Dev. 2014;13:349–54. https://doi.org/10.1515/ijdhd-2014-0327.
6. Annett M, Anderson F, Bischof WF. Activities and evaluations for technology-based upper extremity rehabilitation. Virtual Real Enhanc Robot Syst Disabil Rehabil. 2016;307. https://doi.org/10.4018/978-1-4666-9740-9.ch015.
7. Aprile I, Cruciani A, Germanotta M, Gower V, Pecchioli C, Cattaneo D, Vannetti F, Padua L, Gramatica F. Upper limb robotics in rehabilitation: an approach to select the devices, based on rehabilitation aims, and their evaluation in a feasibility study. Appl Sci. 2019;9(18):3920. https://doi.org/10.3390/app9183920.
8. Armiger RS, Vogelstein RJ. Air-Guitar Hero: a real-time video game interface for training and evaluation of dexterous upper-extremity neuroprosthetic control algorithms. Biomed Circuits Syst Conf. 2008;121–4. https://doi.org/10.1109/biocas.2008.4696889.
9. Baker L. Neuro muscular electrical stimulation: a practical guide. 4th ed. Downey, CA: Los Amigos Research Institute; 2000.
10. Biddiss E, Irwin J. Active video games to promote physical activity in children and youth: a systematic review. Arch Pediatr Adolesc Med. 2010;164:664–72. https://doi.org/10.1001/archpediatrics.2010.104.
11. Boudreau SA, Badsberg S, Christensen SW, Egsgaard LL. Digital pain drawings: assessing touch-screen technology and 3D body schemas. Clin J Pain. 2016;32(2):139–45. https://doi.org/10.1097/AJP.0000000000000230.
12. Bouwsema H, van der Sluis CK, Bongers RM. The role of order of practice in learning to handle an upper-limb prosthesis. Arch Phys Med Rehabil. 2008;89:1759–64. https://doi.org/10.1016/j.apmr.2007.12.046.
13. Bouwsema H, van der Sluis CK, Bongers RM. Effect of feedback during virtual training of grip force control with a myoelectric prosthesis. PLoS One. 2014;9:e98301. https://doi.org/10.1371/journal.pone.0098301.
14. Burke JW, McNeill M, Charles D, Morrow P, Crosbie J, McDonough SM. Serious games for upper limb rehabilitation following stroke. In: Conference in games and virtual worlds for serious applications, 2009. VS-GAMES’09; 2009. p. 103–10.
15. Cleeland CS, Ryan KM. Pain assessment: global use of the Brief Pain Inventory. Ann Acad Med Singap. 1994;23(2):129–38.
16. Cottrell MA, Galea OA, O’Leary SP, Hill AJ, Russell TG. Real-time telerehabilitation for the treatment of musculoskeletal conditions is effective and comparable to standard practice: a systematic review and meta-analysis. Clin Rehabil. 2017;31(5):625–38.
17. Csikszentmihalyi M. Toward a psychology of optimal experience. In: Flow and the foundations of positive psychology: the collected works of Mihaly Csikszentmihalyi; 2014. p. 209–26.
18. Dawson MR, Carey JP, Fahimi F. Myoelectric training systems. Expert Rev Med Devices. 2011;8:581–9. https://doi.org/10.1586/erd.11.23.

19. Deci EL, Ryan RM, Koestner R. A meta-analytic review of experiments examining the effects of extrinsic rewards on intrinsic motivation. Psychol Bull. 1999;125:627–68. https://doi.org/10.1037/0033-2909.125.6.627.
20. Deterding S, Dixon D, Khaled R, Nacke L, Sicart M, O'Hara K. Gamification: using game design elements in non-game contexts. In: Proc 2011 Annu Conf Ext Abstr Hum Factors Comput Syst (CHI 2011); 2011. p. 2425–8. https://doi.org/10.1145/1979742.1979575.
21. Van Dijk L, Van Der Sluis CK, Van Dijk HW, Bongers RM. Task-oriented gaming for transfer to prosthesis use. IEEE Trans Neural Syst Rehabil Eng. 2015;24(12):1384–94. https://doi.org/10.1109/TNSRE.2015.2502424.
22. Donaghy E, Atherton H, Hammersley V, McNeilly H, Bikker A, Robbins L, Campbell J, McKinstry B. Acceptability, benefits, and challenges of video consulting: a qualitative study in primary care. Br J Gen Pract. 2019;69(686):e586–94. https://doi.org/10.3399/bjgp19X704141.
23. Dudkiewicz I, Gabrielov R, Seiv-Ner I, Zelig G, Heim M. Evaluation of prosthetic usage in upper limb amputees. Disabil Rehabil. 2004;26:60–3. https://doi.org/10.1080/09638280410001645094.
24. Ferrer-Garcia M, Gutiérrez-Maldonado J, Riva G. Virtual reality based treatments in eating disorders and obesity: a review. J Contemp Psychother. 2013;43(4):207–21. https://doi.org/10.1007/s10879-013-9240-1.
25. Flores E, Tobon G, Cavallaro E, Cavallaro FI, Perry JC, Keller T. Improving patient motivation in game development for motor deficit rehabilitation. In: Proceedings of the 2008 international conference on advances in computer entertainment technology; 2008. p. 381–4.
26. Gusman J, Mastinu E, Ortiz-Catalan M. Evaluation of computer-based target achievement tests for myoelectric control. IEEE J Transl Eng Heal Med. 2017;5:1–10. https://doi.org/10.1109/JTEHM.2017.2776925.
27. Halton J. Virtual rehabilitation with video games: a new frontier for occupational therapy. Occup Ther Now. 2008;10:12–4.
28. Handelzalts JE, Ben-Artzy-Cohen Y. The draw-a-person test and body image. Rorschachiana. 2014;35(1):3–22. https://doi.org/10.1027/1192-5604/a000042.
29. Hargrove L, Losier Y, Lock B, Englehart K, Hudgins B. A real-time pattern recognition based myoelectric control usability study implemented in a virtual environment. In: Annu Int Conf IEEE Eng Med Biol—Proc; 2007. p. 4842–5. https://doi.org/10.1109/IEMBS.2007.4353424.
30. Herz NB, Mehta SH, Sethi KD, Jackson P, Hall P, Morgan JC. Nintendo Wii rehabilitation ("Wii-hab") provides benefits in Parkinson's disease. Park Relat Disord. 2013;19:1039–42. https://doi.org/10.1016/j.parkreldis.2013.07.014.
31. Hussaini A, Kyberd P. Refined clothespin relocation test and assessment of motion. Prosthet Orthot Int. 2016;41(3):294–302. https://doi.org/10.1177/0309364616660250.
32. Intrinsic Motivation Inventory. Intrinsic Motivation Inventory (IMI). Intrinsic Motiv Invent Scale Descr. 1994;1–3. www.selfdeterminationtheory.org.
33. Jensen TS, Krebs B, Nielsen J, Rasmussen P. Phantom limb, phantom pain and stump pain in amputees during the first 6 months following limb amputation. Pain. 1983;17(3):243–56. https://doi.org/10.1007/BF01402796.
34. Jensen TS, Krebs B, Nielsen J, Rasmussen P. Non-painful phantom limb phenomena in amputees: incidence, clinical characteristics and temporal course. Acta Neurol Scand. 1984;70:407–14. https://doi.org/10.1111/j.1600-0404.1984.tb00845.x.
35. Johnson SS, Mansfield E. Prosthetic training: upper limb. Phys Med Rehabil Clin N Am. 2014;25(1):133–51.
36. Kuiken TA, Miller LA, Turner K, Hargrove LJ. A comparison of pattern recognition control and direct control of a multiple degree-of-freedom transradial prosthesis. IEEE J Transl Eng Heal Med. 2016;4:2100508. https://doi.org/10.1109/JTEHM.2016.2616123.
37. Lang CE, Macdonald JR, Reisman DS, Boyd L, Kimberley TJ, Schindler-ivens SM, Hornby TG, Ross SA, Scheets PL. Observation of amounts of movement practice provided during stroke rehabilitation. Arch Phys Med Rehabil. 2010;90:1692–8. https://doi.org/10.1016/j.apmr.2009.04.005.Observation.
38. Letosa-Porta A, Ferrer-Garcia M, Gutiérrez-Maldonado J. A program for assessing body image disturbance using adjustable partial image distortion. Behav Res Methods. 2005;37(4):638–43. https://doi.org/10.3758/BF03192734.
39. Levin MF, Weiss PL, Keshner EA. Emergence of virtual reality as a tool for upper limb rehabilitation: incorporation of motor control and motor learning principles. Phys Ther. 2015;95:415–25.
40. Lloréns R, Alcañiz M, Colomer C, Gil-Gomez J-A, Llorens R, Alcaniz M, Colomer C. Effectiveness of a Wii balance board-based system (eBaViR) for balance rehabilitation: a pilot randomized clinical trial in patients with acquired brain injury. J Neuroeng Rehabil. 2011;8:30. https://doi.org/10.1186/1743-0003-8-30.
41. Lohse K, Shirzad N, Verster A, Hodges N. Video games and rehabilitation: using design principles to enhance engagement in physical therapy. J Neurol Phys Ther. 2013;37(4):166–75. https://doi.org/10.1097/NPT.0000000000000017.
42. Lohse K, Shirzad N, Verster A, Hodges N, der Loos HFM V. Video Games and Rehabilitation. J Neurol Phys Ther. 2013;37:166–75. https://doi.org/10.1097/NPT.0000000000000017.
43. Mathiowetz V, Volland G, Kashman N, Weber K. Adult norms for the box and block test of manual dexterity. Am J Occup Ther. 1985;39:386–91. https://doi.org/10.5014/ajot.39.6.386.

44. Michie S, Ashford S, Sniehotta FF, Dombrowski SU, Bishop A, French DP. A refined taxonomy of behaviour change techniques to help people change their physical activity and healthy eating behaviours: the CALO-RE taxonomy. Psychol Health. 2011;26:1479–98. https://doi.org/10.1080/08870446.2010.540664.
45. Murray CD, Pettifer S, Howard T, Patchick EL, Caillette F, Kulkarni J, Bamford C. The treatment of phantom limb pain using immersive virtual reality: three case studies. Disabil Rehabil. 2007;29:1465–9. https://doi.org/10.1080/09638280601107385.
46. Noble D, Price DB, Gilder R. Psychiatric disturbances following amputation. Am J Psychiatry. 1954;110:609–13. https://doi.org/10.1176/ajp.110.8.609.
47. Oppenheim H, Armiger RS, Vogelstein RJ. WiiEMG: a real-time environment for control of the Wii with surface electromyography. In: Proceedings of 2010 IEEE International Symposium on Circuits and Systems (ISCAS); 2010. p. 957–60.
48. Ortiz-Catalan M, Gudmundsdottir RA, Kristoffersen MB, Zepeda-Echavarria A, Caine-Winterberger K, Kulbacka-Ortiz K, Widehammar C, Eriksson K, Stockselius A, Ragnö C, Pihlar Z, Burger H, Hermansson L. Phantom motor execution facilitated by machine learning and augmented reality as treatment for phantom limb pain: a single group, clinical trial in patients with chronic intractable phantom limb pain. Lancet. 2016;388:2885–94. https://doi.org/10.1016/S0140-6736(16)31598-7.
49. Ottobock. MyoBoy upgrade/exchange | Myo Software. 2016. https://professionals.ottobockus.com/Prosthetics/Upper-Limb-Prosthetics/Myo-Hands-and-Components/Myo-Software/MyoBoy-Upgrade-Exchange/p/757M11~5X-CHANGE. Accessed 1 Sept 2017.
50. Ottobock. PAULA 1.2 | Myo Software. 2016. https://professionals.ottobockus.com/Prosthetics/Upper-Limb-Prosthetics/Myo-Hands-and-Components/Myo-Software/PAULA-1-2/p/646C52~5V1~82. Accessed 1 Sept 2017.
51. Ottobock. Myo Plus App. 2020. https://www.ottobock.com/en/apps/myoplusapp/myo-plus-app-de.html. Accessed 20 May 2020.
52. Prahm C, Bauer K, Sturma A, Hruby L, Pittermann A, Aszmann O. 3D body image perception and pain visualization tool for upper limb amputees. In: IEEE proceedings on serious games and applications for health, Kyoto; 2019. p. 1–5.
53. Prahm C, Kayali F, Aszmann O. MyoBeatz: using music and rhythm to improve prosthetic control in a mobile game for health. In: IEEE proceedings on serious games and applications for health, Kyoto; 2019. p. 1–6.
54. Prahm C, Kayali F, Sturma A, Aszmann O. Recommendations for games to increase patient motivation during upper limb amputee rehabilitation. In: Converging clinical and engineering research on neurorehabilitation {II}. Cham: Springer; 2017. p. 1157–61.
55. Prahm C, Kayali F, Sturma A, Aszmann O. PlayBionic: game-based interventions to encourage patient engagement and performance in prosthetic motor rehabilitation. PM&R. 2018;10:1252–60. https://doi.org/10.1016/j.pmrj.2018.09.027.
56. Prahm C, Kayali F, Vujaklija I, Sturma A, Aszmann O. Increasing motivation, effort and performance through game-based rehabilitation for upper limb myoelectric prosthesis control. In: 2017 International conference on virtual rehabilitation (ICVR). Montreal, CA: IEEE; 2017. p. 1–6.
57. Prahm C, Schulz A, Paaben B, Schoisswohl J, Kaniusas E, Dorffner G, Hammer B, Aszmann O. Counteracting electrode shifts in upper-limb prosthesis control via transfer learning. IEEE Trans Neural Syst Rehabil Eng. 2019;27(5):956–62. https://doi.org/10.1109/TNSRE.2019.2907200.
58. Prahm C, Sturma A, Kayali F, Mörth E, Aszmann O. Smart Rehab: app-based rehabilitation training for upper extremity amputees—case report. Handchir Mikrochir Plast Chir. 2018;50:1–8. https://doi.org/10.1055/a-0747-6037.
59. Prahm C, Sturma A, Mörth E, Aszmann O. Interactive mobile training app after nerve transfer or amputation of the upper extremity. In: Journal of Hand Surgery, Federation of European Societies for Surgery of the Hand (FESSH), vol. 43. Copenhagen: Sage; 2018. p. 198–9.
60. Prahm C, Vujaklija I, Kayali F, Purgathofer P, Aszmann OC. Game-based rehabilitation for myoelectric prosthesis control. JMIR Ser Games. 2017;5:13. https://doi.org/10.2196/games.6026.
61. Price DD, McGrath PA, Rafii A, Buckingham B. The validation of visual analogue scales as ratio scale measures for chronic and experimental pain. Pain. 1983;17(1):45–56. https://doi.org/10.1016/0304-3959(83)90126-4.
62. Rand D, Yacoby A, Weiss R, Reif S, Malka R, Weingarden H, Zeilig G. Home-based self-training using video-games: preliminary data from a randomised controlled trial. In: Virtual Rehabil Proc (ICVR), 2015 Int Conf; 2015. p. 86–91. https://doi.org/10.1109/ICVR.2015.7358588.
63. Raymer R. Gamification: using game mechanics to enhance eLearning. 2011. https://elearnmag.acm.org/featured.cfm?aid=2031772. Accessed 30 Aug 2019.
64. Reinkensmeyer DJ, Housman SJ. "If I can't do it once, why do it a hundred times?": connecting volition to movement success in a virtual environment motivates people to exercise the arm after stroke. In: 2007 virtual rehabilitation. IEEE; 2007. p. 44–8.
65. Resnik L, Etter K, Klinger SL, Kambe C. Using virtual reality environment to facilitate training with advanced upper-limb prosthesis. J Rehabil Res Dev. 2011;48(6):707–18.
66. Roche AD, Vujaklija I, Amsüss S, Sturma A, Göbel P, Farina D, Aszmann OC. A structured rehabilita-

tion protocol for improved multifunctional prosthetic control: a case study. J Vis Exp. 2015;(105):e52968. https://doi.org/10.3791/52968.
67. la Rosa R, Alonso A, de la Rosa S, Abasolo D. Myo-Pong: a neuromuscular game for the UVa-Neuromuscular training system platform. In: 2008 virtual rehabilitation; 2008. p. 61.
68. Rush KL, Hatt L, Janke R, Burton L, Ferrier M, Tetrault M. The efficacy of telehealth delivered educational approaches for patients with chronic diseases: a systematic review. Patient Educ Couns. 2018;101(8):1310–21.
69. Ryan RM, Deci EL. Self-determination theory and the facilitation of intrinsic motivation, social development, and well-being. Am Psychol. 2000;55:68–78. https://doi.org/10.1037/0003-066X.55.1.68.
70. Seagal I, Morin E. A virtual training environment for prosthetic control. In: CMBES proceedings 39, Calgary, AB; 2016. p. 1–4.
71. Shani M, Feldman Y, Chared M. ReAbility online. http://www.reabilityonline.com/. Accessed 15 May 2020.
72. Simon AM, Hargrove LJ, Lock BA, Kuiken TA. Target achievement control test: evaluating real-time myoelectric pattern-recognition control of multifunctional upper-limb prostheses. J Rehabil Res Dev. 2011;48:619. https://doi.org/10.1682/JRRD.2010.08.0149.
73. Smurr LM, Gulick K, Yancosek K, Ganz O. Managing the upper extremity amputee: a protocol for success. J Hand Ther. 2008;21:160–75.; ; quiz 176. https://doi.org/10.1197/j.jht.2007.09.006.
74. Solutions Ag. Navigate pain. 2017. https://aglancesolutions.com/. Accessed 4 Apr 2017.
75. Spyridonis F, Ghinea G. 2D vs. 3D pain visualization: User preferences in a spinal cord injury cohort. In: Lecture notes in computer science (including subseries Lecture notes in artificial intelligence and lecture notes in bioinformatics); 2011
76. Spyridonis F, Ghinea G. 3-D pain drawings and seating pressure maps: relationships and challenges. IEEE Trans Inf Technol Biomed. 2011;15(3):409–15. https://doi.org/10.1109/TITB.2011.2107578.
77. Stubblefield KA, Miller LA, Lipschutz RD, Kuiken TA. Occupational therapy protocol for amputees with targeted muscle reinnervation. J Rehabil Res Dev. 2009;46:481. https://doi.org/10.1682/JRRD.2008.10.0138.
78. Sturma A, Goebel P, Herceg M, Gee N, Roche A, Fialka-Moser V, Aszmann O. Advanced rehabilitation for amputees after selective nerve transfers: EMG-guided training and testing. In: Jensen W, Andersen OK, Akay M, editors. Replace, repair, restore, relieve and bridging clinical and engineering solutions in neurorehabilitation. Cham: Springer; 2014. p. 169–77.
79. Sturma A, Hruby LA, Prahm C, Mayer JA, Aszmann OC. Rehabilitation of upper extremity nerve injuries using surface EMG biofeedback: protocols for clinical application. Front Neurosci. 2018;12:906. https://doi.org/10.3389/fnins.2018.00906.
80. Sturma A, Roche AD, Goebel P, Herceg M, Ge N, Fialka-Moser V, Aszmann O. A surface EMG test tool to measure proportional prosthetic control. Biomed Tech. 2015;60:207–13. https://doi.org/10.1515/bmt-2014-0022.
81. Tabor A, Bateman S, Scheme E, Flatla DR, Gerling K. Designing game-based myoelectric prosthesis training. In: CHI 2017, Denver, CO; 2017. p. 1–12.
82. Tatla SK, Shirzad N, Lohse KR, Virji-Babul N, Hoens AM, Holsti L, Li LC, Miller KJ, Lam MY, Van der Loos HFM. Therapists' perceptions of social media and video game technologies in upper limb rehabilitation. JMIR Ser Games. 2015;3:e2. https://doi.org/10.2196/games.3401.
83. Touch Bionics. Biosim-i. In: MA01178, Issue 2. 2014. http://www.touchbionics.com/sites/default/files/files/biosim-i_my i-limb datasheet June 2014.pdf. Accessed 1 Sept 2017.
84. Össur, i-limb Mobile Apps by Össur. http://www.ossur.com/en-gb/professionals/apps/i-limb-mobile-apps. Accessed Nov. 2020.
85. Touch Bionics Inc. my i-limb—Android Apps on Google Play. 2020. https://play.google.com/store/apps/details?id=com.touchbionics.myilimb.app. Accessed 22 May 2020.
86. Treleaven J, Battershill J, Cole D, Fadelli C, Freestone S, Lang K, Sarig-Bahat H. Simulator sickness incidence and susceptibility during neck motion-controlled virtual reality tasks. Virtual Real. 2015;19:267–75. https://doi.org/10.1007/s10055-015-0266-4.
87. Vujaklija I, Roche AD, Hasenoehrl T, Sturma A, Amsuess S, Farina D, Aszmann OC. Translating research on myoelectric control into clinics—are the performance assessment methods adequate? Front Neurorobot. 2017;11:1–7. https://doi.org/10.3389/fnbot.2017.00007.
88. Wheaton LA. Neurorehabilitation in upper limb amputation: understanding how neurophysiological changes can affect functional rehabilitation. J Neuroeng Rehabil. 2017;14(1):41.
89. Williamson A, Hoggart B. Pain: a review of three commonly used pain rating scales. J Clin Nurs. 2005;14(7):798–804.
90. Winslow BD, Ruble M, Huber Z. Mobile, game-based training for myoelectric prosthesis control. Front Bioeng Biotechnol. 2018;6:1–8. https://doi.org/10.3389/fbioe.2018.00094.
91. Woodward RB, Hargrove LJ. Adapting myoelectric control in real-time using a virtual environment. J Neuroeng Rehabil. 2019;16:11. https://doi.org/10.1186/s12984-019-0480-5.
92. American Occupational Therapy Association. Am J Occup Ther. 2018;72:7212410059p1. https://doi.org/10.5014/ajot.2018.72S219.
93. Alan L, Karen F, Lesley H, Diane M, Chris P. World confederation of physical therapy. In: Report, WCPT/INPTRA Digit. Pract. Final; 2019.

22 Conclusions and Future Outlook

Oskar C. Aszmann and Dario Farina

This book has summarized the significant advances that have taken place mainly in the past three decades in all aspects of bionic limb reconstruction. They relate to new surgical approaches for improving the biomechanical and neural interfacing of robotic limbs to the human body, precision mechanics and miniaturized electronics for robotic design, algorithms that interpret biosignals, implantation technology, as well as rehabilitation protocols and technology for promoting learning of new motor tasks. These advances have provided several new directions for research and clinical translation of bionic limbs. Nonetheless, despite the efforts and the achievements, the daily clinical reality indicates to us that the solutions offered to amputees, especially in case of complex limb deficiencies, are still far from satisfactory. The loss of a limb cannot yet be treated with assistive technologies that approach the functions of biological limbs. If we consider the challenges to be addressed, we may conclude that this goal is not even at the horizon.

As we have discussed in several of the chapters, the main challenge in providing robotic substitutes of biological limbs remains the same as during the early developments in this field: neural interfacing of the limb with the patient's central nervous system. The neural interface requires the establishment of a *rich* and *reliable* information transfer flow between the limb and the nervous system. This has proved to be a tremendous challenge. The problem of transferring enough information is associated to the large information flow in natural limb control. An example clarifies this concept. The human hand and arm is innervated by more than 350,000 individual nerve fibers, of which approximately 10% bring information to the extremity and the remaining 90% bring information back from the hand and arm to the spinal cord and the brain. The information transfer along the nerve fibers is biologically coded by the frequency of discharge of discrete events (action potentials), so that several action potentials per second travel along each nerve fiber. The amount of natural bidirectional information transfer per second is therefore extremely large. Re-establishing the same amount of information transfer is clearly far from our current technological potential. Besides the amount of information, the transfer should also be reliable and therefore be robust over time and with changes in

O. C. Aszmann (✉)
Clinical Laboratory for Bionic Extremity Reconstruction, Department of Surgery, Medical University of Vienna, Vienna, Austria

Division of Plastic and Reconstructive Surgery, Department of Surgery, Medical University of Vienna, Vienna, Austria
e-mail: oskar.aszmann@meduniwien.ac.at

D. Farina
Department of Bioengineering, Imperial College London, London, UK
e-mail: d.farina@imperial.ac.uk

O. C. Aszmann, D. Farina (eds.), *Bionic Limb Reconstruction*,
https://doi.org/10.1007/978-3-030-60746-3_22

the environment and the patients' body. This relates to the problem of training of the control algorithms and of the patient. Despite the exciting advances that we have described in this book, the quantity and quality of information transferred in current prostheses is clearly insufficient.

Despite the distance from the final goal, this book has provided our best current means to address the open problems, focusing on a unique aspect of the modern developments in prosthetic technologies: the close integration of the disciplines of surgery, neural interfacing, and robotics. Modern approaches to solve the conundrum of information transfer in artificial limbs have relied on combining surgical interventions to facilitate neural interfacing and robotic design to facilitate control. As an example, the developments in nerve transfers in the past 20 years following the pioneering work of Todd Kuiken have determined improvements in neural interfacing far more important than specific advances in machine learning algorithms per se. Moreover, these surgical interventions have redirected the research efforts in implanted technology and artificial intelligence and have stimulated new research ideas in bionic reconstruction. For example, it is theoretically possible to decode single nerve efferent fiber activity from muscle recordings following nerve transfers, which provides the basis for new neural interfaces. This integration of disciplines will undoubtedly further increase in the future, so that breakthrough advances will be mostly achieved by research centers that will include co-located expertise in all these disciplines.

The book has also underlined a slow trend in prosthetic research toward implanted technologies, which have characterized many of the recent advances in neural interfacing. Despite adding another level of clinical complexity, there is no doubt that implanted technologies have the potential to mitigate the challenges associated to the quantity and quality of transferred information. Nerve and muscle implants are becoming feasible in relatively large clinical populations due to miniaturized electronics and advances in material sciences. Difficulties still remain in terms of wireless transmission of information and power to and from the body, but they can be addressed by modern electronics and communication engineering. The transfer of the achievements reached in the past with noninvasive systems to implanted devices will boost the quality of control and sensation in artificial limbs at a large clinical scale in the very near future.

Much of the activities compiled in this book have been triggered by individual patients and their tremendous need after loss of an extremity or its function. Any novel reconstructive method, however, has to be evaluated against the more traditional options available and determine the specific risk-benefit profile. Bionic replacement of obviously irreversibly damaged body parts offer tremendous possibilities that we have tried to present in an objective manner. Even though there are still many obstacles to overcome to provide a fluent prosthetic replacement, it is our conviction that man-machine interfacing as is presented in this book provides the foundation to further expand the indications of bionic reconstruction. We have started with classic limb loss and moved on to our first series of elective amputations in patients after severe brachial plexus lesions and shortly after reported restoration of hand use after massive soft tissue damage. Loss of neural connectivity after brain stroke with loss of hand function due to spasticity has very recently triggered a larger research effort to explore the possibilities and limits of cognitive nerve transfers and their potential value in bionic reconstruction. Each subset of patients have their own set of challenges, however, if managed correctly have achieved an outstanding level of functionality, and the many reports that have been published and are in part presented in this book reveal the enormous potential that bionic extremity reconstruction may have in the treatment of this challenging patient group. Finally, this book has been a journey through an active medical and research area that includes multiple disciplines and therefore requires an integrated collaborative approach among professionals with different backgrounds. The chapters have been authored by physiotherapists, neuroscientists, surgeons, engineers, roboticists, rehabilitation doctors, psychologists, and computer scientists, just to mention some of the professional profiles. It has been conceptualized from collaborative work over many years during which we have learned to listen to each other, even if from our very different perspectives, and to enrich our differences by combining our ideas. For us, this book is a reminder of this important achievement. We hope the reader will be inspired by this approach and will find this overview as useful as it has been for us writing it.

GPSR Compliance

The European Union's (EU) General Product Safety Regulation (GPSR) is a set of rules that requires consumer products to be safe and our obligations to ensure this.

If you have any concerns about our products, you can contact us on ProductSafety@springernature.com

In case Publisher is established outside the EU, the EU authorized representative is:

Springer Nature Customer Service Center GmbH
Europaplatz 3
69115 Heidelberg, Germany

Batch number: 10731385

Printed by Printforce, the Netherlands